Advanced Structured Materials

D1827086

Volume 159

Series Editors

Andreas Öchsner, Faculty of Mechanical Engineering, Esslingen University of Applied Sciences, Esslingen, Germany

Lucas F. M. da Silva, Department of Mechanical Engineering, Faculty of Engineering, University of Porto, Porto, Portugal

Holm Altenbach ⓘ, Faculty of Mechanical Engineering, Otto von Guericke University Magdeburg, Magdeburg, Sachsen-Anhalt, Germany

Common engineering materials reach in many applications their limits and new developments are required to fulfil increasing demands on engineering materials. The performance of materials can be increased by combining different materials to achieve better properties than a single constituent or by shaping the material or constituents in a specific structure. The interaction between material and structure may arise on different length scales, such as micro-, meso- or macroscale, and offers possible applications in quite diverse fields.

This book series addresses the fundamental relationship between materials and their structure on the overall properties (e.g. mechanical, thermal, chemical or magnetic etc.) and applications.

The topics of *Advanced Structured Materials* include but are not limited to

- classical fibre-reinforced composites (e.g. glass, carbon or Aramid reinforced plastics)
- metal matrix composites (MMCs)
- micro porous composites
- micro channel materials
- multilayered materials
- cellular materials (e.g., metallic or polymer foams, sponges, hollow sphere structures)
- porous materials
- truss structures
- nanocomposite materials
- biomaterials
- nanoporous metals
- concrete
- coated materials
- smart materials

Advanced Structured Materials is indexed in Google Scholar and Scopus.

More information about this series at http://www.springer.com/series/8611

Aleksander N. Guz

Eight Non-Classical Problems of Fracture Mechanics

 Springer

Aleksander N. Guz
S. P. Timoshenko Institute of Mechanics
National Academy of Sciences of Ukraine
Kyiv, Ukraine

ISSN 1869-8433   ISSN 1869-8441   (electronic)
Advanced Structured Materials
ISBN 978-3-030-77503-2   ISBN 978-3-030-77501-8   (eBook)
https://doi.org/10.1007/978-3-030-77501-8

This Springer imprint is published by the registered company Springer Nature Switzerland AG
The registered company address is: Gewerbestrasse 11, 6330 Cham, Switzerland

Foreword

In this monograph, the analysis of the main results on studying some non-classical problems of fracture mechanics is considered in a sufficiently brief form (in the form of review). These results were obtained by the author of monograph as well as his pupils and followers for the last 50 years in the Department of Dynamics and Stability of Continua of the S. P. Timoshenko Institute of Mechanics of the National Academy of Sciences of Ukraine (NASU).

At that, the non-classical problems of the fracture mechanics are defined as those in which the approaches and fracture criteria of the classical fracture mechanics are not applicable (not working).

The specificity of the results of the author and his pupils is the application of the **three-dimensional theories** of stability, dynamics, and statics of the mechanics of deformed bodies to the study of non-classical problems of fracture mechanics. The overwhelming majority of other authors apply in the study of non-classical problems of fracture mechanics the various approximate theories of shells, plates, and rods, and other approximate approaches.

In this monograph, the main scientific results on the eight non-classical problems of the fracture mechanics are presented. They are obtained under the aforementioned approach (the three-dimensional theories of the mechanics of the deformable body), as noted above, in a very brief form. The main attention is focused on the statement of the problems with analyzing the corresponding experiments, developing the solutions within the approach under consideration, and discussing the turn-key results. Because of the above, the mathematical aspects of the methods of solving and their computer implementation are not discussed in this monograph. The information on this issue is presented in the form of annotation in a short form.

The following eight non-classical problems of the fracture mechanics (the results of the author and his pupils) are discussed in this monograph:

The first problem is the fracture in composite materials under compression along the reinforcing elements.
The second problem is the model of short fibers in the theory of stability and the fracture mechanics of composite materials under compression.

The third problem is the end-crush fracture of the composite materials under compression.

The fourth problem is the brittle fracture of materials with cracks, taking into account the action of the initial (residual) stresses along the cracks.

The fifth problem is the brittle fracture in the form of separation into the slender parts of the composite materials under tension or compression along the reinforcing elements.

The sixth problem is the fracture under compression along the parallel cracks.

The seventh problem is the brittle fracture of materials with cracks under the action of dynamic loads (taking into account the contact interaction of the crack sides).

The eighth problem is the fracture of the thin-walled bodies with cracks under tension in the case of a preliminary loss of stability.

About 525 monographs and main scientific articles published by the author and his pupils on the above **eight** non-classical problems of the fracture mechanics are included in the references of this monograph.

As it seems to the author of this monograph, the results *on the eight non-classical problems of the fracture mechanics*, obtained at the S. P. Timoshenko Institute of Mechanics of the NASU for the last 50 years, **have by the solved problems any analog in the modern world science:** *on the generality and stringency of the statement of the problems, on the exactness and commonality of the obtained concrete results, as well as on the thoroughness and validity of the formulated conclusions related to the study of the analyzed mechanical phenomena.*

The present monograph consists of three parts, which include the Foreword, Introduction, Chaps. 1–10, and the References, which includes the names of 599 primary sources.

The Part I has the subtitle "General Problems" and consists of the Foreword, Introduction, Chap. 1 "Division into Classical and Non-Classical Problems of Fracture Mechanics" and Chap. 2 "Brief Statement of Foundations of Three-Dimensional Linearized Theory of the Deformed Bodies Stability (TLTDBS)".

The Part I is published in Russian in the journal "Prikladnaya mekhanika" (2019, vol.55, N2, pp. 8–72) and in English in the journal "International Applied Mechanics" (2019, vol. 55, N2, pp. 129–174).

The Part II has the subtitle "Fracture in Composite Materials Under Compression" and consists of Chap. 3 "Problem 1. Fracture in Composite Materials Under Compression Along Reinforcing Elements", Chap. 4 "Problem 2. Short Fibers Model in Theory of Stability and Fracture Mechanics of Composite Materials Under Compression", and Chap. 5 "Problem 3. End-Crush Fracture of Composite Materials Under Compression".

The Part II is published in Russian in the journal "Prikladnaya mekhanika" (2019, vol. 55, N3, pp. 5–91) and in English in the journal "International Applied Mechanics" (2019, vol. 55, N3, pp. 239–296).

The Part III has a subtitle "Other Non-Classical Problems of Fracture Mechanics" and consists of Chap. 6 "Problem 4. Brittle Fracture of Materials with Cracks Taking into Account ACTION of Initial (Residual) Stresses Along Cracks", Chap. 7

"Problem 5. Separation into Slender Parts of Composite Materials Under Tension or Compression Along Reinforcing Elements", Chap. 8 "Problem 6. Fracture Under Compression Along Parallel Cracks", Chap. 9 "Problem 7. Brittle Fracture of Materials with Cracks Under Action of Dynamic Loads (With Allowance for Contact Interaction of Sides of Cracks)", and Chap. 10 "Problem 8. Fracture of Thin-Walled Bodies with Cracks Under Tension with Pre-Buckling". The third part is published in Russian in the journal "Prikladnaya mekhanika" (55, No. 4, 2019, pp. 3–100) and in English in the journal International Applied Mechanics (55, No. 4, 2019 343–415).

In the entire monograph (Chaps. 1–10) for all formulas, figures, notes, and tables, the double numbering (within each chapter) is adopted. At that, the number before the dot corresponds to the chapter number and the second number (after the dot) corresponds to the object number within this chapter. Thus, it is possible to consider the results of each chapter practically independently of the results of the other chapters, focusing on the References that is common for all chapters and is presented at the end of the monograph.

This monograph was published in Russian by the Publishing House Akademperiodika in 2020 (Guz A.N. Vosiem nieclassicheskikh problem mekhaniki razrushenija. Kiev, Akademperiodika (Ukraine) and in 2020, 400 pp. ISBN 978-966-360-403).

Preface

In this monograph, the description and corresponding analysis of the main results on the non-classical problems of fracture mechanics are proposed in a sufficiently brief form (in the form of review). These results were obtained by the author of monograph and his pupils for the last 50 years in the S. P. Timoshenko Institute of Mechanics of the National Academy of Sciences of Ukraine.

The non-classical problems of the fracture mechanics are defined as those in which the approaches and fracture criteria of the classical fracture mechanics are not applicable (not working).

The presented in this monograph results have any analog in the modern world scientific literature on the fracture mechanics.

Kyiv, Ukraine   Aleksander N. Guz
2020

Introduction

It is now overall accepted that the fundamental work of A. A. Griffith [1] of 1920 marked the beginning of a new scientific direction in natural science and technology—the fracture mechanics, which is one of the most actively developed areas of fundamental and applied character in mechanics in the second half of twentieth and early twenty-first centuries. This direction in our time can be comparable on topicality, fundamentality, and applicability to engineering *in mechanics*, apparently, only with the mechanics of composites.

As some confirmation of the above consideration is the fact that in the middle of the second half of twentieth century, two eight-volume collective monographs of an encyclopedic nature were created: The monograph [2] is devoted to the problems of composite materials, including the mechanics of composites; the monograph [3] is devoted to the problems of fracture of materials, including the fracture mechanics.

In the nearly 100-year development of the fracture mechanics in the numerous scientific centers of different countries around the world, starting with 1920, the various problems of fracture mechanics were explored and nowadays are actively considered. By the results of researches, hundreds of monographs and tens of thousands of articles are published.

It should be noted that these problems of fracture mechanics can be divided into the classical and non-classical ones.

At that, the classical problems of fracture mechanics are at present already relatively in detail researched and developed, which has led some scientists to think about the impending crisis in the scientific direction under discussion. The non-classical problems of fracture mechanics have remained relatively under-researched and developed, which is likely to continue in the coming years due to the multiplicity of the problems noted.

In a brief consideration of the non-classical problems of the fracture mechanics can be identified as the problems to which analysis and research the approaches and criteria of the classic problems of the fracture mechanics are not applicable.

In more detail and consistently the question of dividing the problems of fracture mechanics into the classical and non-classical problems is discussed in Chap. 1 of this monograph.

In the Department of Dynamics and Stability of Continua of the S. P. Timoshenko Institute of Mechanics of the National Academy of Sciences of Ukraine (NASU) as early as more than 50 years, the author of this monograph together with pupils are engaged in researching many non-classical problems of fracture mechanics.

The first publications in this scientific direction were the articles of the author's [4, 5] of 1969, related to one of the non-classical problems of fracture mechanics, although the results of the article [5] were actually published in the article [6] of 1967 concerning the orthotropic elastic body, by which in the continuum approach the composite material is modeled.

Over the next 50 years, the Department of Dynamics and Stability of Continua Environments actively researched many non-classical problems in fracture mechanics. The main results were obtained for *eight* non-classical problems of fracture mechanics, which are summarized in subsection 1.2 of this monograph.

In connection with the above, the offered to readers monograph is devoted to a sufficiently brief analysis of the obtained results. This monograph is prepared with the allowance for the thoughts on the presentation style of the material, which is outlined below in this Introduction.

It is worth noting that the References of this monograph includes about 523 monographs and main publications in the scientific journals and the proceedings of international conferences published by the author of the monograph and his pupils. By the obtained results, **14** dissertations for the Doctor of Sciences in Physics, Mathematics, or Engineering are defended.

Thus, it can be considered that the relatively short style of presentation and discussion of the results obtained in the Department of Dynamics and Stability of Continua is partly predetermined by the above number of publications and defensed Doctor of Sciences dissertations.

Also, as it seems to the author, the results on non-classical problems of the fracture mechanics presented in the scientific articles and monographs, for the traditional reasons about the style of presentation of scientific results of a fundamental nature in mechanics, are presented in a form that is well perceived by the representatives of the certain scientific directions and is not always sufficiently informative for a wide range of specialists on the fracture problems.

It should be noted that the fracture problems are dealt with by the representatives of various scientific directions, including mechanics, physics, materials science, etc., as well as the representatives of numerous scientific and technical directions involved in the development of various aspects of engineering. In this regard, it seems that *the presentation of results on non-classical problems of fracture mechanics in a form that would be informative enough for a wide range of specialists interested in the various fundamental and applied aspects of fracture problem is looking very actual.*

At the same time, the above representation should include the main aspects of the problems under consideration (a non-classical nature of the mechanisms of fracture under consideration, the rigor of description, the analysis of the main approaches and obtained results, the description of new mechanical effects, etc.). It is quite obvious that when trying to implement the above-mentioned style of presentation of results, it is necessary to abandon the detailed information on the purely mathematical aspects of the methods of solving and the excessive richness of the specific results in the form of numerous graphs and tables.

Thus, *the purpose* of this monograph can be defined as follows—*a brief description and corresponding analysis of the main results on non-classical problems of fracture mechanics, obtained by the author of the monograph and his pupils over the past 50 years in the Department of Dynamics and Stability of Continua of the S. P. Timoshenko Institute of Mechanics of the NASU, with the involvement of the above-mentioned style of writing review articles.*

In this case, the focus is on the statement of the problems with the analysis of corresponding experiments, the development of a method of solution in the approach under consideration, and discussion of the final results. Therefore, the mathematical aspects of the methods of solving and their computer implementation are not discussed in this monograph, and information on these issues is provided in a brief form in the form of annotations.

The streaming for preparing the review articles on non-classical problems of fracture mechanics, based on the results of the author and his pupils and adhering to the above style of the review articles, is formed in the preparation of an invited lecture [7] for the IUTAM Symposium (Cambridge, UK, 1995), in the preparation of an article [8] for the Encyclopedia on Fracture [9] (USA, 1998), lectures at Yildiz Technical University (Istanbul, March 1998) and at the Institute of Mechanics (Hanoi) December 1998) and a report at a seminar at the University of Technology (Vienna, December 1999). The above approach was partially implemented in the lectures [10], which can be considered to be the beginning of the work on the formation of review articles on the non-classical problems of fracture mechanics (mainly on the publications of the author and his pupils) within the framework of the above-mentioned style of preparation of the review articles.

In the following years, with allowance for the accepted *style* of preparation, the review articles were prepared and published on the non-classical problems of the fracture mechanics (mainly by the works of the author and his pupils), such as [11] of 2000, [12] of 2009, and [13] of 2014, as well as several others. Moreover, the article [12] in the form of its reduced version was also published in the journal "ANNALS. THE EUROPEAN ACADEMY OF SCIENCES" owing to awarding the author by the 2007 BLAISE PASCAL MEDAL in Materials Sciences of the EAS. Thus, the article [14] can be considered as the extended PASCAL MEDAL LECTURE on the ceremony of awarding. The commonly accepted Pascal Medals Lecture (written presentation) is published in the journal "Prikladnaya Mekhanika" [15].

It should be noted, however, that the above review articles [11–14] on the non-classical problems of fracture mechanics relate to the partial non-classical problems or a shorter period, although they are prepared within the framework of the review style under discussion.

In this regard, the publication of this monograph (mainly based on the results of the author and his pupils over the past 50 years), prepared following the style under discussion, can be considered quite appropriate. Certainly, in the preparation of this monograph the review articles [11–14] and several others are also used.

Note

In this monograph, only Chap. 2 does not correspond to the style of results discussed. So, the mathematical apparatus of the three-dimensional linearized theory of the stability of deformed bodies in a very brief form is set out in Chap. 2, taking into account the basic relations. This exception is related to that five of the problems (1, 2, 3, 4, and 6) of the eight problems are used a three-dimensional linearized theory of the stability of deformed bodies. At that, this theory of stability is less well known and less widely used than other parts of the mechanics of deformed bodies.

Note some specificities of terminology, which is used in mechanics concerning the study of the fracture phenomena of materials and structural elements.

"The fracture" refers to the destruction, which is determined by the propagation of one or more cracks. At that, "the fracture mechanics" is engaged in the study of the destruction of materials or structural elements, which is also determined by the propagation of one or more cracks.

"The failure" refers to the destruction, which is determined by the exhaustion of the loading capability of a material or structural element and, in general, manifests itself not only in the propagation of the single cracks. "The failure mechanics" is also involved in the destruction of materials and structures, which is also determined by the exhaustion of the loading capability of materials and structural elements and manifests itself mainly not only in the propagation of the single cracks.

"The damage" refers to the destruction, which manifests itself in the accumulation of damage in the form of diffusely located developing or incipient cracks or other damages. At that, "the damage mechanics" is engaged in the study of regularities (kinetics) of the accumulation of damage, mainly within the framework of the various continuum representations involving the selected in certain way "damage indicator".

Of course, the above classification is quite conditional and at the same time quite useful and informative, from the author's point of view of this monograph, when analyzing the various results in the fracture mechanics in the broad sense of this term. So, for example, the results on the non-classical problems of the fracture mechanics of materials, which are analyzed in this monograph, refer to the fracture mechanics and failure mechanics, which in the text of the monograph will not be distinguished.

It is reasonable to emphasize that the scientific results on the non-classical problems of fracture mechanics, obtained over the last 50 years in the Department of Dynamics and Stability of Continua of the S. P. Timoshenko Institute of Mechanics of the National Academy of Sciences of Ukraine, a brief analysis of which in the chosen style of presentation is given in this monograph, *are* apparently sufficiently presented to the world scientific community. The above consideration is confirmed by the information that can be obtained by analyzing the References in this monograph. As noted in this Introduction, the References includes 523 monographs and main publications in the scientific journals and the proceedings of the international conferences written by the author of this monograph and his pupils.

Three conclusions follow from the analysis of these publications.

1. The 38 of the 523 publications were published in the journal "DAN USSR", which was translated into English and was one of the most authoritative scientific journals.
2. The 320 of the 523 publications are published in English-language scientific journals.
3. The 53 of the 523 publications are published in the proceedings of the international congresses, conferences, and symposiums that correspond to reports in these international forums.

The information provided in this Introduction relates to all the eight non-classical problems of fracture mechanics, which makes it possible to avoid a certain repetition concerning each problem.

References

1. Griffith, A.A.: The phenomena of rupture and flow in solids. Phil. Trans. Roy. Soc., Ser. A. **211**(2), 163–198 (1920)
2. R.H. Krock, L.J. Broutman (eds.), Kompozitnye materialy (Composite materials), [Russian translation], (Vol. 1, 2, 5, 6 – Mir, Vol. 3, 4, 7, 8 – Mashinostroenie, Moscow, 1978–1979), T. 1. Poverkhnosti razdela v metallicheskikh kompozitakh (V.1. Interfaces in metal matrix composites) (1978), T. 2. Mekhanika kompozitnykh materialov (V.2. Mechanics of composite materials) (1978), T. 3. Primenenie kompozitnykh materialov v tekhnike (V.3. Engineering applications of composites) (1979), T. 4. Kompozitnye materialy s metallicheskoi matritsei (V.4. Metal matrix composites) (1978), T.5. Razrushenie i ustalost (V.5. Fracture and fatigue) (1978), T. 6. Poverkhnosti razdela v polimernykh kompozitakh (V. 6. Interfaces in polymer matrix composites) (1978), T. 7. Analiz i proektirovanie konstruktsii. Ch. 1 (T. 7. Structural design and analysis. Part 1) (1979), T. 8. Analiz i proektirovanie konstruktsii. Ch. 2 (T. 8. Structural design and analysis. Part 2) (1979)
3. G. Libowitz (ed.), Razrushenie, v 7 tomakh (Fracture, in 7 volumes) [Russian translation] (Vol. 1, 2, 3, 7 - Mir, Vol. 4, 5 - Mashinostroenie, Vol.6 - Metallurgia, Moscow, 1973–1977), T. 1. Mikroskopicheskie i makroskopicheskie osnovy mekhaniki razrusheniia (Vol.1. Microscopic and macroscopic foundations of fracture mechanics) (1973), T. 2. Matematicheskie osnovy mekhaniki razrusheniia (Vol.2. Mathemastical foundations of fracture mechanics) (1975), T. 3. Inzhenernye osnovy i vozdeistvie vneshnei sredy (Vol.3. Engineering foundations and the impact of the external environmen) (1976), T. 4. Issledovanie razrusheniia dlia inzhenernykh raschetov (Vol.4. Investigation of fracture for engineering calculations) (1977), T. 5. Raschet konstruktsii na khrupkuiu prochnost (Vol.5. Calculation of structures for brittle strength) (1977), T. 6. Razrushenie metallov (Vol.6. Fracture of metals) (1976), T.7, ch. 1. Razrushenie nemetallov i kompozitnykh materialov. Neorganicheskie materialy (stekla, gornye porody, kompozity, keramiki, led) (Vol.7, Part 1. Fracture of non-metals and composite materials. Inorganic materials (glass, rocks, composites, ceramics, ice)) (1976), T.7, ch. 2. Razrushenie nemetallov i kompozitnykh materialov. Organicheskie materialy (stekloobraznye polimery, elastomery, kost) (Vol.7, Part 2. Fracture of non-metals and composite materials. Organic materials (glassy polymers, elastomers, bone)) (1976)
4. Guz, A.N.: O postroenii teorii ustoichivosti odnonapravlennykh voloknistykh materialov (On stability theory construction for unidirectional fibrous materials). Prikladnaya Mekhanika **5**(2), 62–70 (1969)

5. Guz, O.M.: Pro vyznachennia teoretychnoi hranytsi mitsnosti na stysk armovanykh mate-
    rialiv (Determining the theoretical compressive strength of reinforced materials). Dopovidi
    Akademii Nauk URSR, Ser. A. **3**, 236–238 (1969)
6. Guz, A.N.: The stability of orthotropic bodies. Sov. Appl. Mech. **3**(5), 17–22 (1967)
7. A.N. Guz, The study and analysis of non-classical problems of fracture and failure mechanics.
    Abstracts of IUTAM Symposium of nonlinear analysis of fracture (Cambridge, 3–7 Sept.,
    1995), p.19
8. A.N. Guz, Some modern problems of physical mechanics of fracture. In ed. by G.P.
    Cherepanov. FRACTURE. A Topical Encyclopedia of Current Knowledge (Krieger Publ.
    Company, Malabar, Florida, 1998), pp. 709–720
9. G.P. Cherepanov (ed.), FRACTURE. A Topical Encyclopedia of Current Knowledge (Krieger
    Publishing Company, Malabar, Florida, 1998)
10. A.N. Guz, Study and Analysis of Non-classical Problems of Fracture and Failure Mechanics
    and Corresponding Mechanisms. Lecture presented at Institute of Mechanics. (HANOI,
    1998)
11. Guz, A.N.: Description and study of some nonclassical problems of fracture mechanics and
    related mechanisms. Int. Appl. Mech. **36**(12), 1537–1564 (2000)
12. Guz, A.N.: On study of nonclassical problems of fracture and failure mechanics and related
    mechanisms. Int. Appl. Mech. **45**(1), 1–31 (2009)
13. Guz, A.N.: O postroenii osnov mekhaniki razrusheniia materialov pri szhatii vdol treshchin
    (obzor) (On the construction of the foundations of the fracture mechanics of materials in
    compression along cracks (review)). Prikladnaya Mekhanika **50**(1), 5–89 (2014)
14. Guz, A.N.: On Study of Nonclassical Problems of Fracture and Failure Mechanics and
    Related Mechanisms, pp. 35–68. ANNALS of the European Academy of Sciences, Liège,
    Belgium (2006–2007)
15. A.N. Guz, Pascal Medals Lecture (written presentation). Int. Appl. Mech. **44**(1), 6–11 (2008)

New Mechanical Effects Detected by Prof. Aleksander N. Guz and His Collaborators for the First Time

1. **Field: Stress concentration near holes**

Effect Non-monotonic increasing of the stress concentration under approach of the holes as applied to static problems of shells and dynamic problems of plates.

2. **Field: Elastic waves diffraction**

Effect 1 The existence of "conditional resonances" in multiconnected elastic bodies.

Effect 2 The existence of the "Wood-type" resonances in elastic bodies with periodical structures.

3. **Field: Three-dimensional theory of deformable bodies stability**

Effect 1 Asymptotical accuracy of the two-dimensional applied theory of shells and plates stability based on the Kirchhoff–Love hypothesis.

Effect 2 Stability or instability of the elastic simply-connected isotropic body with elastic potential of arbitrary form under omni-directional compression in case of different boundary conditions on separate parts of surfaces.

4. **Field: Elastic waves in bodies with initial stresses**

Effect 1 The regularities of wave propagation may be described only within the framework of the elastic potential depending on the first, second, and third invariants.

Effect 2 The existence of some frequencies for each layer and each cylinder, when the values of velocity of wave propagation don't depend on values of initial stresses.

5. **Field: Mechanics of composite materials**

Effect 1 In continual approximation, the composite materials with initial stresses cannot be described within the framework of classical linear elasticity.

Effect 2   In case of the unidirectional fibrous composites, the form of stability loss along a helical path is not realized.

6.   **Field: Contact problems for elastic bodies with initial stresses**

Effect   The existence of the "resonance-type features" when initial stresses are approaching to the values corresponding to the surface instability.

7.   **Field: The problems of hydroelasticity for some shells in liquid**

Effect   Non-monotonic change of dynamic processes under approaching the shells.

8.   **Field: Non-destructive ultrasonic methods of stresses determination**

Effect   Basic regularities in case of two-axial and three-axial stresses.

9.   **Field: Mechanics of brittle fracture of materials with initial stresses**

Effect 1   In case of the planar and anti-planar static problems for bodies with elastic potentials of arbitrary form and in cases of the spatial static problems and the planar dynamic problems for bodies with elastic potential of particular form, the order of singularity in the tip of crack coincides with results of classical linear mechanics of brittle fracture.

Effect 2   An existence of the "resonance-type features" when initial stresses are approaching to the value corresponding to the surface instability.

Effect 3   An existence of the "resonance-type features" for moving cracks in the brittle materials with initial (residual) stresses where the value of velocity of moving cracks is approaching to the value of velocity of Rayleigh wave in the above-mentioned materials.

10.   **Field: Mechanics of failure in compression along cracks in one plane**

Effect   The beginning of failure coincides with appearing the surface instability of this material.

11.   **Field: Dynamics of rigid bodies in compression**

Effect   Radiation force in compressible viscous fluid resulting in from interaction of rigid body with acoustical wave may be by several orders of magnitude greater than radiation force in compressible ideal fluid.

12.   **Field: Mechanics of failure in compression along interface cracks in composite materials**

Effect   The beginning of failure coincides with appearing the surface instability of the first or the second materials (components).

Contents

Part I   General Problems

**1   Division into Classical and Non-classical Problems of Fracture
    Mechanics**   3
  1.1   Classical Problems of Fracture Mechanics   3
  1.2   Non-classical Problems of Fracture Mechanics   5
  1.3   Eight Non-classical Problems of Fracture Mechanics   6
  1.4   Additional Discussion of Non-classical Problems
      of Fracture Mechanics 10
    1.4.1   Brief Discussion of Models and Approaches
         in Non-classical Problems of Fracture Mechanics
         (Problems 1–8, Sect. 1.3) 11
    1.4.2   On Consideration of Non-classical Problems
         of Fracture Mechanics from the Point of View
         of Classical Problems of Fracture Mechanics 16
    1.4.3   On Some Other Publications 17
  References 19

**2   Brief Statement of Foundations of Three-Dimensional
    Linearized Theory of the Deformable Bodies Stability
    (TLTDBS)** 21
  2.1   On the Formation of TLTDBS 22
  2.2   Classification of Approaches (Variants of Theory)
      in the TLTDBS 26
    2.2.1   Theory of Large (Finite) Subcritical Deformations 27
    2.2.2   The First Variant of Theory of Small Subcritical
         Deformations 29
    2.2.3   The Second Variant of Theory of Small Subcritical
         Deformations 31
    2.2.4   On the Linearized Theory of Stability at Small
         Deformations and Small Averaged Angles
         of Rotation 33

| | | 2.2.5 | On the Theory of Incremental Deformations .... | 34 |

2.2.5   On the Theory of Incremental Deformations 34
2.2.6   Approximate Approach in the ILIDBS 35
2.2.7   Notes 38
2.3   On Stability Criteria in the TLTDBS 41
2.3.1   Elastic Bodies 42
2.3.2   Plastic Bodies 43
2.3.3   Bodies with Rheological Properties 47
2.4   General Questions of the TLTDBS 51
2.4.1   General Formulation of the TLTDBS Problems
        for Various Models of Deformable Bodies 51
2.4.2   Sufficient Conditions for Applicability
        of the Euler's Method (Statical Method) 54
2.4.3   Sufficient Conditions of Stability 56
2.5   On Variational Principles of the TLTDBS for Elastic
      and Plastic Bodies 58
2.5.1   Variational Principle of Hu–Vashizu Type
        in the TLTDBS for Incompressible Bodies
        with "Dead" External Loads. Unified General Form
        for Theories 2.1, 2.2, and 2.3 60
2.5.2   Variational Principle of the TLTDBS
        for Compressible Bodies Under "Following" Load.
        Results for Theory 2.3 63
2.6   General Solutions of the TLTDBS Under Homogeneous
      Subcritical States 65
2.6.1   General Solutions of the TLTDBS for Compressible
        Bodies 67
2.6.2   The General Solutions of the TLTDBS
        for Incompressible Bodies 69
2.6.3   The Complex Potentials in Plane Problems
        of the TLTDBS. Preliminary Discussion 72
2.6.4   Basic Relations and General Solutions
        of the TLTDBS in Coordinates of Initial State 74
2.6.5   Complex Potentials in Plane Linearized Problems
        in Coordinates of Initial State 78
2.6.6   Complex Potentials in Dynamical Plane Linearized
        Problems in Coordinates of Initial State for Moving
        Cracks and Loads 83
References 89

Part II   Fracture in Composite Materials Under Compression

**3   Problem 1. Fracture in Composite Materials Under
      Compression Along the Reinforcing Elements** 97
3.1   General Concept and Key Directions of Research 97
3.1.1   General Concept 98

3.1.2 The First Direction (Very Approximate Approaches) 102
3.1.3 The Second Direction (Strict Sequential
Approaches Based on the TLTDBS) 104
3.2 Analysis of Experimental Results on Compression
of Composites 108
3.2.1 Experimental Results on the Loss of Stability
in the Internal Structure of Composites Under
Compression 109
3.2.2 Experimental Results on Fracture of Composites
Under Compression Along the Reinforcing
Elements 112
3.2.3 About the Study of the Phenomenon of "Kinking" 120
3.3 Main Results of the Second Direction (Strict Sequential
Approaches Based on the TLTDBS) 122
3.3.1 Introductory Information 123
3.3.2 Continuum Theory of Fracture 125
3.3.3 Layered Composites: Model of a Piece-Wise
Homogeneous Medium 133
3.3.4 Fibrous Unidirectional Composites: Model
of Piece-Wise Homogeneous Medium 148
3.4 Conclusion 159
References 160

4   **Problem 2. Model of Short Fibers in Theory of Stability
and Fracture Mechanics of Composite Materials Under
Compression** 171
4.1 Experimental Results on Loss of Stability in the Internal
Structure of Composites Under Compression: Case
of Short Fibers 172
4.2 Statement of Problems 173
4.3 Classification of Design Schemes. About Analogies 181
4.3.1 Model of Infinitely Long Fibers and Layers
in the First Direction of Research 181
4.3.2 Model of Infinitely Long Fibers and Layers
in the Second Direction of Research 182
4.3.3 Model of Short Fibers and Layers in the Framework
of the Second Direction of Research 184
4.4 Statement of Plane Problems of Brittle Fracture Mechanics
of Composites with Short Reinforcing Elements Under
Compression 186
4.4.1 On Statement of Problems 186
4.4.2 On the Method of Numerical Study of Problems
of Sect. 4.4 187

4.5   Results of Studies of Plane Problems of Brittle Fracture
       Mechanics of Composites with Short Fibers Under
       Compression 189
       4.5.1 Asymptotic Transition to the Model of "Infinitely
              Long Fibers" 190
       4.5.2 Results for Single Fiber Under Compression Along
              Fibers 193
       4.5.3 Results for Two Sequentially Located Fibers Under
              Compression Along the Fibers 196
       4.5.4 Results for Two Parallel Fibers Under Compression
              Along the Fibers 199
       4.5.5 Results for One Periodic Row of Sequentially
              Located Fibers Under Compression Along
              the Fibers 201
       4.5.6 Results for One Periodic Row of Parallel Fibers
              Under Compression Along the Fibers 205
       4.5.7 Results for Single Fiber Located Close to Surface
              Under Compression Along the Fiber (Analysis
              of Near-the-Surface Instability) 207
4.6   Conclusion 212
References 213

5   Problem 3. End-Crush Fracture of Composite Materials
    Under Compression 215
    5.1   Introduction 215
    5.2   Experimental Researches 216
    5.3   Theoretical Researches 218
           5.3.1 General Concept 218
           5.3.2 On Studies Within the Model of Piecewise
                  Homogeneous Medium 220
           5.3.3 On Studies Within the Continuum Medium Model
                  (Continuum Approximation) 225
    References 229

Part III   Other Non-Classical Problems of Fracture Mechanics

6   Problem 4. Brittle Fracture of Materials with Cracks Taking
    into Account the Action of Initial (Residual) Stresses Along
    Cracks 235
    6.1   Introduction 235
    6.2   Preliminary Discussion. Statement of Problems 237
    6.3   Plane and Anti-plane Statical Problems. Criteria of Fracture 242
           6.3.1 Order of Singularity 242
           6.3.2 Effects of Resonant Character 243
           6.3.3 Criteria of Fracture 245
    6.4   Spatial Statical Problems 247

6.4.1   On Statement of Spatial Statical Problems
in Mechanics of Brittle Fracture of Materials
with Initial (Residual) Stresses Acting Along Cracks   247
6.4.2   To the Method of Research of Spatial Statical
Problems   248
6.4.3   Concrete Results (Using Both Exact Solutions
and Computer Methods) Obtained for the Following
11 Design Schemes   248
6.4.4   On Phenomena of Resonant Character for Spatial
Statical Problems of Non-classical Problem 4
of Fracture Mechanics   249
6.5   On Dynamical Plane and Anti-plane Problems
in Mechanics of Brittle Fracture of Materials with Initial
(Residual) Stresses Along Cracks   250
6.6   Repetition of Results   252
6.7   On Increasing the Objectivity of Citation   256
References   258

7   Problem 5. Brittle Fracture in the Form of Separation
into the Slender Parts of Composite Materials Under Tension
or Compression Along Reinforcing Elements   263
7.1   Introduction   263
7.2   Experimental Researches   265
7.3   Explanation of Mechanism of Fracture in the Form
of "Separation into the Slender Parts"   267
7.4   On Development of Fundamentals of Mechanics
of Composites with Curved Structures   272
7.4.1   Introduction   273
7.4.2   Continuum Theories and Results Based on Them   274
7.4.3   Model of a Piece-Wise Homogeneous Medium
and Results Based on It   277
References   285

8   Problem 6. Fracture Under Compression Along Parallel
Cracks   289
8.1   Introduction   289
8.2   General Statement of Problems: General Concept
and Basic Approaches   291
8.2.1   General Statement of Problems   291
8.2.2   General Concept   294
8.2.3   General Approaches   296
8.3   Results for Homogeneous Materials with Cracks Under
Brittle and Plastic Fracture: The Second General Approach   302

8.3.1   Results for Brittle and Plastic Fracture
of Homogeneous Materials with Cracks Located
in the Same Plane. The Second General Approach.
Exact Solutions   303
8.3.2   Results for Brittle and Plastic Fracture
of Homogeneous Materials with Cracks Located
in Parallel Planes: The Second General Approach   304
8.3.3   Results for Brittle and Plastic Fracture
of Homogeneous Materials with Cracks Located
in Parallel Planes. Second General Approach: The
Combined Approach for Problems 4 and 6   306
8.4   Results for Layered Composites with Cracks at Interface
Under Brittle and Plastic Fracture. The Second General
Approach   310
8.4.1   Introduction   310
8.4.2   Results for Brittle and Plastic Fracture of Layered
Composites with Microcracks at Interfaces: The
Second General Approach   311
8.4.3   Results for the Brittle Fracture of Layered
Composites with Macrocracks at the Interfaces:
The Second General Approach   314
8.5   Results for Brittle Fracture of Homogeneous Materials
with Cracks Located in the Close Arranged Parallel Planes.
Passage to the Limit. The Second General Approach   318
8.5.1   Short Description of Developed Research Method   318
8.5.2   Near-the-Surface Crack   321
8.6   On Results for Viscoelastic Fracture   324
References   327

9   Problem 7. Brittle Fracture of Materials with Cracks Under
Action of Dynamic Loads (with Allowance for Contact
Interaction of the Crack Edges)   335
9.1   Introduction   335
9.2   Substantiation of Statement of Problems. Method of Solving ...   336
9.2.1   Substantiation of the Discussed Problem Statement   337
9.2.2   On Research Method   339
9.3   Concrete Results   340
9.3.1   Two-Dimensional Problems   341
9.3.2   Three-Dimensional (Spatial) Problems   342
References   346

**10   Problem 8. Fracture of Thin-Wall Bodies with Cracks Under
      Tension in the Case of Preliminary Loss of Stability** 351
      10.1   Introduction 351
      10.2   Statement of Problems 352
      10.3   Research Methods and Results Obtained 355
      References 359

General Conclusion to the Monograph (Parts I, II, III) 363

Bibliography 365

About the Author

Aleksander N. Guz was born January 29, 1939 in Ichnia of Chernigov region of Ukraine and graduated from the Mechanics and Mathematics Department of Kiev State University in 1961. He is working at the Institute of Mechanics of the NASU (National Academy of Sciences of Ukraine) since 1960, receiving an appointment as Head of the Department of Dynamics and Stability of Continuum Media (1967) and serving as Director of the Institute of Mechanics of the NASU (from 1976 until present).

He received the Candidate of Sciences degree (1962) and the Doctor of Sciences degree (1965). He was named Professor (1969).

He is Academician of the NASU (1978), Member of the Academia Europaea (1992), Fellow of the New York Academy of Sciences (1997), Fellow of the World Innovation Foundation (2001), Member of the European Academy of Sciences (2002).

His principal scientific results have been obtained in mechanics of deformable solids and related problems of continuum mechanics: the three-dimensional theory of stability of deformable bodies, the theory of propagation and diffraction of elastic waves in multiconnected bodies and bodies with initial stresses, stress concentration around holes in shells, mechanics of composites materials and structural members utilizing them, aerohydroelasticity, non-classical problems of fracture mechanics, rock mechanics, dynamics of viscous compressible liquid, contact problems, mechanics of nanocomposites, and non-destructive methods of stress determination.

He is the author or coauthor of 73 books (including 21 solo) and about 1000 scientific papers (including more than 400 solo).

He has trained 36 doctors and about 100 candidates of sciences.

Awards (Prizes): **Medal BLAISE PASCAL of the European Academy of Sciences** (2007), **Medal LIFETIME ACHIEVEMENT of the ICCEF (International Conference on Computational and Experimental Engineering and Sciences)** (2012), **DIPLOMA di MERITO con MEDALIA D'O RO of the European Scientific—Industrial Chamber** (2013), **Albert Nelson Marquis Lifetime Achievement Award** (2017), **Order "HONOR" of the Kremlin Fund (Russia)** (2007), **State Prize of USSR** (1985), **State Prizes of Ukraine** (1979, 1988), **V. I. Vernadsky GOLD MEDAL of the National Academy of Sciences of Ukraine** (2014), **Prize of the National Academy of Sciences of Ukraine** (1979, 1983, 2000, 2014, 2016, 2017), **Lenin Komsomol Prize** for young scientists (1967), **Lenin Komsomol of Ukraine Prize** for young scientists (1973).

He serves in the editorial boards of several international scientific journals and is the editor-in-chief of the international scientific journal **Prikladnaya Mekhanika** (since 1976). He is the Chairman of the National Committee of Ukraine of Theoretical and Applied Mechanics (since 1993).

Multivolume collective fundamental books: *Methods of Shells Theory* in five volumes (1980–1982), *Mechanics of Composites Materials and Structural Members* in three volumes (1982–1983), *Three-Dimensional Problems of the Theory of Elasticity and Plasticity* in six volumes (1984–1986), *Mechanics of Coupled Fields in Constructive Members* in five volumes (1987–1989), *Non-Classical Problems of Fracture Mechanics* in four volumes (1990–1994), *Mechanics of Composites* in twelve volumes (1993–2003), *Advances of Mechanics* in six volumes (2005–2012), *Modern problems of mechanics* in three volumes (2016–2018) were supervised and coauthored by Prof. Aleksander N. Guz.

Some main scientific results of A.N.GUZ were presented (in English) in book "SERIES "Classics of World Science", vol.11, **Olexander M GUZ"**

(TIMPANI, Ukraine, 2006, 521p.). This book is Vol. 11 of series "Classics of World Science" publishing by Austria, Czech Republic, Slovakia, and Ukraine.

List of Monographs
of Prof. Aleksander N. Guz

1. Stress Around Curvilinear Holes in Shells (transl. Suryanaranan). Bangalor Aeronautical Laboratory, 1967, 144p. Co-author G. N. Savin.
2. Spherical bottoms weakened by holes. Naukova Dumka, Kiev, 1970, 324p. Co-authors I. S. Chernyshenko, K. I. Shnerenko
3. Stability of three-dimensional deformed bodies. Naukova Dumka, Kiev, 1971, 276p.
4. Diffraction of elastic waves in multi-connected bodies. Naukova Dumka, Kiev, 1972, 256p. Co-author V. T.Golovchan.
5. Stability of elastic bodies under finite deformations. Naukova Dumka, Kiev, 1973, 272p.
6. Diffraction of Elastic Waves in Multiply Connected Bodies. FOREIGN TECHNOLOGY DIV WRIGHT—PATTERSON AFB, OHIO, 1973, 280p. Co-author V. T. Golovchan.
7. Cylindrical shells weakened by holes. Naukova Dumka, Kiev,1974, 272p. Co-authors I. S. Chernyshenko, Val. N. Chekhov, Vic. N. Chekhov, K. I. Shnerenko
8. Foundations of ultrasonic non-destructive method of stresses determination in solid bodies. Naukova Dumka, Kiev, 1974, 108p. Co-authors O. I. Guscha, F. G. Makhort, V. K.Lebedev
9. Conical shells weakened by holes. Naukova Dumka, Kiev, 1976, 164p. Co-authors P. Z. Lugovoy, N. A. Shulga
10. Waves in layer with initial stresses. Naukova Dumka, Kiev, 1976, 104p. Co-authors A. P. Zhuk, F. G. Makhort
11. Foundations of stability theory of mine workings. Naukova Dumka, Kiev, 1977, 204p.
12. Introduction in acoustoelasticity. Naukova Dumka, Kiev, 1977, 152p. Co-authors F. G. Makhort, O. I. Guscha

13. Diffraction of elastic waves. Naukova Dumka, Kiev, 1978, 308p. Co-authors V. D. Kubenko, M. A. Cherevko

14. Stability of elastic bodies under omni-directional compression. Naukova Dumka, Kiev, 1979, 144p.

15. Strength of rocket engine construction with solid-state propellant. Mashinostroenie, Moscow, 1980, 248p. Co-authors A. G. Makarenkov, I.S. Chernyshenko

16. Three-dimensional stability theory of rods, plates and shells. Vyshcha Shkola, Kiev, 1980, 168p. Co-author I. Yu. Babich

17. Methods of shells design. In 5 vol. V.1. Theory of thin shells weakened by holes. Naukova Dumka, Kiev, 1980, 636p. Co-authors I. S. Chernyshenko, Val. N. Chekhov, Vic. N. Chekhov, K. I. Shnerenko

18. Non-destructive testing of materials and constructions. Naukova Dumka, Kiev, 1981, 276p. Co-authors M. E. Garf, S. V. Malashenko, A. O. Rasskazov and others

19. Failure and stability of thin bodies with cracks. Naukova Dumka, Kiev, 1981, 184p. Co-authors M. Sh. Dyshel', G. G. Kuliev, O. B. Milovanova

20. Perturbation method in spatial problems of elasticity theory. Vyshcha Shkola, Kiev, 1982, 350p. Co-author Yu. N. Nemish

21. Methods of shells design. In 5 vol. V.5. Theory of non-stationary hydroelasticity of shells. Naukova Dumka, Kiev, 1982, 400p. Co-author V. D. Kubenko

22. Mechanics of composite materials and constructions. In 3 vol. V.1. Mechanics of materials. Naukova Dumka, Kiev, 1982, 368p. Co-authors L. P. Khoroshun, G. A. Vanin and others

23. Mechanics of composite materials and constructions. In 3 vol. V.2. Mechanics of constructions. Naukova Dumka, Kiev, 1983, 464p. Co-authors Ya. M. Grigorenko, I. Yu. Babich and others

24. Mechanics of composite materials and constructions. In 3 vol. V.3. Applied investigations. Naukova Dumka, Kiev, 1983, 264p. Co-authors I. V. Ignatov, A. G. Girchenko and others

25. Mechanics of brittle fracture of materials with initial stresses. Naukova Dumka, Kiev, 1983, 296p.
26. Experimental investigations of thin-walled constructions. Naukova Dumka, Kiev, 1984, 240p. Co-authors V. A. Zarutsky, I. Ya. Amiro and others
27. Hydroelasticity of shells systems. Vyshcha Shkola, Kiev, 1984, 208p. Co-authors V. D. Kubenko, A. E. Babaev
28. Spatial problems of elasticity and plasticity theory. In 6 vol. V.2. Static of elastic bodies with non-canonical form. Naukova Dumka, Kiev, 1984, 280p. Co-author Yu. N. Nemish
29. Spatial problems of elasticity and plasticity theory. In 6 vol. V.4. Three-dimensional stability theory of deformed bodies. Naukova Dumka, Kiev, 1985, 280p. Co-author I. Yu. Babich
30. Spatial problems of elasticity and plasticity theory. In 6 vol. V.5. Dynamics of elastic bodies. Naukova Dumka, Kiev, 1986, 286p. Co-authors V. T. Golovchan, V. D. Kubenko, N. A. Shulga and others
31. Foundations of three-dimensional stability theory of deformed bodies. Vyshcha Shkola, Kiev, 1986, 512p.
32. Elastic waves in bodies with initial stresses. In 2 vol. V.1. General problems. Naukova Dumka, Kiev, 1986, 374p.
33. Elastic waves in bodies with initial stresses. In 2 vol. V.2. Propagation regularities. Naukova Dumka, Kiev, 1986, 536p.
34. Mechanics of coupled fields in constructions. In 5 vol. V.3. Acoustoelectromagnetoelas ticity. Naukova Dumka, Kiev, 1988, 288p. Co-author F. G. Makhort
35. Technological stresses and strains in composite materials. Vyshcha Shkola, Kiev, 1988, 270p. Co-authors V. T. Tomashevsky, N. A. Shulga, V. S. Yakovlev
36. Method of perturbation of surface form in mechanics of continuum. Vyshcha Shkola, Kiev, 1989, 352p. Co-author Yu.N.Nemish
37. Failure mechanics of composite materials under compression. Naukova Dumka, Kiev, 1990, 630p.

38. Dynamics of bodies connected with media.
     Naukova Dumka, Kiev, 1991, 392p. Co-authors
     S. Markus, L. Pust, V. D. Kubenko and others

39. Non-classical problems of fracture mechanics. In
     4 vol. V.2. Brittle fracture of materials with initial
     stresses. Naukova Dumka, Kiev, 1991, 288p.

40. Dynamics and stability of laminated composite
     materials. Naukova Dumka, Kiev, 1992, 368p. Co-
     authors S. Markus, I. Kabelka, I. Yu. Babich and
     others

41. Non-classical problems of fracture mechanics.
     In 4 vol. V.4, Book 1. Fracture and stability of
     materials with cracks. Naukova Dumka, Kiev,
     1992, 454p. Co-authors M. Sh. Dyshel', V. M.
     Nazarenko

42. Mechanics of composites. In 12 vol. V.1. Statics
     of materials. Naukova Dumka, Kiev, 1993, 454p.
     Co-authors V. T. Golovchan, Yu. V. Kokhanenko,
     V. N. Kusch

43. Mechanics of composites. In 12 vol. V.2.
     Dynamics and stability of materials. Naukova
     Dumka, Kiev, 1993, 430p. Co-authors I. Yu.
     Babich, N. A. Shulga, A. S. Kosmodamiansky

44. Non-classical problems of fracture mechanics. In
     4 vol. V.4, Book 2. Brittle fracture of materials
     under dynamical loading. Naukova Dumka, Kiev,
     1994, 240p. Co-author V. V. Zozulya

45. Mechanics of composites. In 12 vol. V.4.
     Mechanics of materials with curved structures.
     Naukova Dumka, Kiev, 1995, 320p. Co-authors S.
     D. Akbarov, E. A. Movsumov, S. M. Mustafaev

46. Contact problems for elastic bodies with initial
     stresses. Vyshcha Shkola, Kiev, 1995, 304p. Co-
     authors S. Yu. Babich, V. B. Rudnitsky

47. Mechanics of composites. In 12 vol. V.5. Frac-
     ture mechanics. "A.C.K.", Kiev, 1996, 340p. Co-
     authors A. A. Kaminsky, V. M. Nazarenko, I. A.
     Guz and others

48. Mechanics of composites. In 12 vol. V.6. Techno-
     logical stresses and strains in materials. "A.C.K.",
     Kiev, 1997, 396p. Co-authors N. A. Shulga, V. T.
     Tomashevskii and others

49. Mechanics of composites. In 12 vol. V.7. Stress
     concentration. "A.C.K.", Kiev, 1998, 387p. Co-
     authors A. S. Kosmodamiansky, V. P. Shevchenko

50. Dynamics of compressible viscous fluid. "A.C.K.", Kiev, 1998, 350p.
51. Fundamentals of the Three-Dimensional Theory of Stability of Deformable Bodies. Springer-Verlag, 1999, 556p.
52. Mechanics of Curved Composites. Kluwer Academic Publishers, 2000, 464p. Co-author S. D. Akbarov
53. Mechanics of composites. In 12 vol. V.10. Stability of members of structures. "A.C.K.", Kiev, 2001, 376p. Co-authors I. Ju. Babich, D. V. Babich, I. A. Guz and others
54. Mechanics of composites. In 12 vol. V.12. Applied investigations. "A.C.K.", Kiev, 2004, 400p. Co-authors L. P. Khoroshun, M. I. Mikhailova, D. V. Babich, and others
55. Contact problems for elastic bodies with initial (residual) stresses. "Melnik", Khmelni tsky, 2004, 692p. Co-author V. B. Rudnitsky
56. Elastic waves in bodies with initial (residual) stresses. "A.C.K.", Kiev, 2004, 672p.
57. Fundamentals of the contact interaction of elastic bodies with initial (residual) stresses. "Melnik", Khmelnitsky, 2006, 710p. Co-author V. B. Rudnitsky
58. Statics and dynamics of the elastic grounds with initial (residual) stresses. Kremenchug, "Press-line", 2007, 796p. Co-authors S. Yu. Babich, Yu. P. Glukhov
59. Fundamentals of the compressive fracture mechanics of composites. In 2 vols. V.1. Fracture in structure of materials. "Litera", 2008, 592p.
60. Fundamentals of the compressive fracture mechanics of composites. In 2 vols. Related mechanisms of fracture. "Litera", 2008, 736p.
61. Dynamics of compressible viscous fluid. Cambridge Scientific Publishers, 2009, 428p.
62. Introduction to mechanics of nanocomposites. "Academperiodika", 2010, 398p. Co-author J. J. Rushchitsky, I. A. Guz
63. Analysis of estimate systems of scientific publications. Kiev. Institute of mechanics, 2013, 274p. Co-author J. J. Rushchitsky

64. Short introduction to mechanics of nanocomposites. Scientific & Academic Publishing, USA, 2013, 281p. Co-author J. J. Rushchitsky

65. Model of short fibers in the theory of stability of composites. LAMBERT Academic Publishing, Germany, 2015, 315p. Co-author V. A. Dekret

66. Mixed problems for elastic ground with initial stresses. LAMBERT Academic Publishing, Germany, 2015, 468p. Co-author S. Yu. Babich, Yu. P. Glukhov

67. Elastic waves in bodies with initial (residual) stresses. In 2 parts. Part 1. LAMBERT Academic Publishing, Germany, 2016, 501p.

68. Elastic waves in bodies with initial (residual) stresses. In 2 parts. Part 2. LAMBERT Academic Publishing, Germany, 2016, 505p.

69. United approach in nonclassical problems of fracture mechanics. LAMBERT Academic Publishing, Germany, 2017, 528p. Co-author V. L. Bogdanov, V. M. Nazarenko

70. Introduction in compressible viscous liquid dynamics. LAMBERT Academic Publishing, Germany, 2017, 240p.

71. To 100-th Anniversary of the S. P. Timoshenko Institute of Mechanics of the NASU (National academy of sciences of Ukraine). "Litera LTD", Kiev, 2018, 160p.

72. Fracture of Materials Under Compression Along Cracks, Springer, 2020, 505p., Co-author V. L. Bogdanov, V. M. Nazarenko

73. Eight Non-classical Problems of Fracture Mechanics, Kyiv, Akademperiodyka, 2020, 400p.

Part I
General Problems

Chapter 1
Division into Classical and Non-classical Problems of Fracture Mechanics

In this chapter, the division of the problems of fracture mechanics into the classical and non-classical problems is presented in a relatively consistent and clear form, *eight* non-classical problems of fracture mechanics (the subject of research in the Department of Dynamics and Stability of Continua of the S. P. Timoshenko Institute of Mechanics of the National Academy of Sciences of Ukraine) are formulated rather briefly, and the examples of investigating situations are given that cannot be classified as the non-classical problems of fracture mechanics.

It should be noted that the study of non-classical problems of fracture mechanics requires the development of not only new approaches and fracture criteria but also the development of new research methods. In the case of the application of research methods that are characteristic of classical problems of mechanics, the physically incorrect results can be obtained, which is discussed in the article [1].

It is also advisable to note that the classification of problems of fracture mechanics considered in this chapter with a clear separation of the classical and non-classical problems, apparently, was first published in 1990 and presented in the Introduction to a 4-volume (in 5 books) monograph [2], which is devoted to the presentation of the results on non-classical problems of fracture mechanics obtained at that time at the S. P. Timoshenko Institute of Mechanics. The noted Introduction refers to the entire collective monograph and is placed in volume 1 [2].

1.1 Classical Problems of Fracture Mechanics

At present, apparently, it can be considered that in fracture mechanics (in a broad sense) the basic concepts and approaches to the formulation of fracture criteria are formed. *The main concepts and approaches in fracture mechanics (in a broad sense) include the following results.*

A. N. Guz, *Eight Non-Classical Problems of Fracture Mechanics*,
Advanced Structured Materials 159,
https://doi.org/10.1007/978-3-030-77501-8_1

1.   the Griffiths' fundamental theory of brittle fracture [3];
2.   the concept of quasi-brittle fracture (Irwin, Orowan, and others);
3.   the Griffiths' energy criterion of fracture [3] or the equivalent (but more easily realized) Irwin's force criterion [4, 5];
4.   the concept of an integral independent of the contour of integration (J-integral, Γ-integral, Eshelby [6], Cherepanov [7, 8], Rice [9]);
5.   criterion of the critical crack opening.

The above concepts and approaches assume that certain conditions are met or implemented under certain conditions, which include *the following conditions.*

Condition 1.1 Tension or shear occurs in the vicinity of cracks. At that, **the action of compression is excluded.**

Condition 1.2 In the process of deformation of a body with cracks, the abrupt changes in the configuration of the body do not occur (e.g., the phenomenon of loss of stability does not precede the phenomenon of fracture of a body with cracks).

Condition 1.3 In the process of deformation of a body with cracks, the abrupt changes in the character of deformation before fracture do not occur (e.g., there is no change in the boundary conditions during deformation).

It should be noted that the **above** Condition 1.1 **is fundamental since, in the case of compression along the cracks, all the above concepts and approaches do not work.**
In the situation corresponding to Condition 1.2, all the above concepts and approaches can work, but it is first necessary to investigate the stress–strain state, taking into account a sharp change in the configuration of the body during deformation. At present, in the overwhelming majority of studies on fracture mechanics (in a broad sense), the above analysis is not carried out.
In the situation corresponding to Condition 1.3, all the above concepts and approaches can work, but it is first necessary to investigate the stress–strain state of the body, taking into account a sharp change in the nature of deformation before fracture (e.g., taking into account the changes in the boundary conditions during deformation). Currently, in the vast majority of studies on mechanics, the above analysis is not carried out.
Taking into account the above considerations, the mentioned results and problems corresponding to the noted five concepts or approaches and obtained when Conditions 1.1–1.3 are fulfilled can be considered as *the classical problems of fracture mechanics,* to which at present the following studies can be attributed.

1.   The determination of stress intensity factors for the bodies of complex shape containing cracks under various power, thermal, and electromagnetic actions. At that, to obtain these results, the analytical, numerical (with the involvement of computers), experimental, and experimental–theoretical methods are used. The results of these studies (stress intensity factors), together with the noted fracture criteria, provide the necessary information on the fracture of materials and structural elements in cases where these fracture criteria are applicable;

2.   The experimental study of complex cases of fracture of materials and structural
     elements, ending in most cases with descriptive results without proper anal-
     ysis and attempts to formulate new criteria for fracture corresponding to the
     phenomena under consideration.

It should be noted that at present the overwhelming number of publications refers
to the classical problems of fracture mechanics in the above sense. Apparently, in
connection with the noted situation, many scientists came to the conclusion about
the existence of an ideological crisis in fracture mechanics at the present stage of its
development.

It should also be noted that the second direction in the classical problems of
mechanics can serve as the first stage in studies of non-classical problems of fracture
mechanics.

1.2   Non-classical Problems of Fracture Mechanics

The following studies can be conditionally referred to as *the non-classical problems
of fracture mechanics.*

1.   a study of new mechanisms of fracture that are not described within the frame-
     work of the main five above-mentioned concepts and approaches (taking into
     account the fulfillment of Conditions 1.1–1.3), with the proper analysis and
     an attempt to formulate new criteria for the fracture corresponding to the
     phenomena under consideration;
2.   a study of certain classes of problems for materials and structural elements
     concerning the studied new mechanisms of fracture and with the involvement
     of the corresponding specially formulated criteria of fracture.

As already noted, the above classification (division into the classical and non-
classical problems) is rather conditionally and not always one-valued. Nevertheless,
this classification relatively clearly defines the direction of research and the form of
their novelty, which seems to be very significant in the analysis of research results.

It should also be noted that the number of fracture mechanisms significantly
expands when the microstructure of materials is taken into account at various levels of
its description. This feature, first of all, refers to the fracture mechanics of composites,
which are characterized by taking into account the microstructure at different levels.

For the scientists involved in the study of non-classical problems and mechanisms
of fracture, it is typical to use very approximate design schemes and models. In the
case of composite materials, approximate design schemes and models are used to
analyze fracture in the microstructure of composites. The use of the approximate
design schemes and models leads to significant quantitative errors and, in many cases,
to qualitative differences. Therefore, it is very difficult to perform a reliable analysis of
the non-classical problems and fracture mechanisms using the approximate schemes
and models. The noted situation determines the essential importance of the results on

the study of non-classical problems and fracture mechanisms, which were obtained with sufficiently rigorous design schemes and models.

As already noted in the introductory part of this chapter, the above division (Sects. 1.1 and 1.2) of the problems of fracture mechanics into the classical and non-classical was set out in the Introduction to the 4-volume (in 5 books) collective monograph [2], which is placed in vol. 1 of this monograph. In subsequent years, this separation of the classical and non-classical problems of fracture mechanics was included without significant change in the review articles [10–13] and several others.

1.3 Eight Non-classical Problems of Fracture Mechanics

In the last 50 years, the staff of the Department of Dynamics and Stability of Continua of the S. P. Timoshenko Institute of Mechanics, along with other studies, is researching eight non-classical problems of fracture mechanics.

Below (in this section) the discussed problems with an indication of the phenomena corresponding to these problems are briefly formulated. In the subsequent chapters (Chaps. 3–10) of this monograph, each of the eight non-classical problems will be considered in a separate chapter with a clear indication of the results obtained by the staff of the Department of Dynamics and Stability of Continuous Media.

Problem 1 *Fracture in composite materials under compression along reinforcing elements*: The experimental studies in many scientific centers around the world have shown that in the situation under discussion, the fracture in composite materials (the beginning or start of fracture) can be determined by the loss of stability of the equilibrium state in the structure of the composite, which is a general concept in the problem under consideration. With the general concept, *two approaches* are developed to study the stability and, consequently, fracture under compression of the composite materials.

The first approach is based on the use of various approximate theories of rods, plates, and shells, as well as other approximate design schemes. Due to the complexity of the analyzed phenomenon, such an approach, apparently, cannot lead to reliable results.

The second approach is based on the application of a three-dimensional linearized theory of stability of deformable bodies, constructed with the accuracy usually accepted in mechanics. The results within the framework of the second approach belong mainly to the author of the monograph and his pupils.

Problem 2 *Model of short fibers in the stability theory and fracture mechanics of composite materials under compression*: The composite materials are created both with relatively long fibers (the subject of Problem 1) and with rather short fibers (reinforcing elements). The experimental studies in the case of the composite materials with short fibers under compression along the fibers have revealed the phenomenon of loss of stability in the structure of the composite material with the modes of loss of stability that are not periodic along the fiber axis and which are characteristic of the short fibers in the matrix. The theory of stability in this case of compression and the corresponding mechanics of brittle fracture was developed using the three-dimensional linearized theory of stability of elastic bodies, constructed with the accuracy usually accepted in mechanics. The main results were obtained by the author of the monograph and his pupils, including the results in the framework of the plane problem for composites of various structures.

Problem 3 *End-crush fracture under compression of composite materials*: In the mechanics of structural elements, the phenomenon of crumpling of the ends of a structural element under compression along the reinforcing elements is known, when the fracture occurs only near the ends. Using the three-dimensional linearized theory of stability of deformable bodies, the fundamentals of fracture mechanics are constructed in the form of crumpling of ends under compression along reinforcing elements. This fracture mechanism is described by the phenomenon of near-surface instability of the loaded ends of the composite material when the modes of loss of stability attenuate with distance from the ends. The main results (concerning the plane and spatial problems) are obtained by the author of the monograph and his pupils.

Problem 4 *Brittle fracture of materials with cracks taking into account the action of the initial (residual) stresses along the cracks.* The discussed mechanics of brittle fracture refers to the isotropic and orthotropic materials, while the composite materials are modeled in the continuum approximation by the orthotropic materials. In the case of orthotropic materials, it is assumed that the cracks are located in the planes of symmetry of the material properties. In this situation, the brittle fracture mechanics of such materials cannot be constructed using the basic relations of the classical linear theory of elasticity, since the singular part of the well-known Inglis–Muskhelishvili solution for a material with a crack does not include the initial (residual) stresses acting along the cracks. In this regard, to construct the mechanics of brittle fracture of materials with cracks, taking into account the action of the initial (residual) stresses along the cracks, the three-dimensional linearized theory of elasticity is used for the finite (large) and small initial strains. With this approach, the fundamentals of the mechanics of brittle fracture of materials with cracks are constructed, taking into account the action of the initial (residual) stresses along the cracks, including the formulation of the basic relations and fracture criterion, as well as the development of methods for solving and research of the main problems.

These results are mainly obtained by the author of the monograph and his pupils.

Problem 5 *Brittle fracture in the form of separation of composite into slender parts under tension or compression of composite materials along the reinforcing elements.* In this case, "separation" means a division into the separate elongated (along the reinforcing elements) parts. In articles [11, 13], this problem is called "Shredding fracture of composites stretched or compressed along the reinforcing elements."

Since in this type of fracture, in fact, the separation (along the reinforcing elements and along the line of tension or compression) of the composite into separate sufficiently elongated (comparable to the length of the sample) parts occurs, the Foreword of this monograph uses a more accurate name of the present problem "separation into slender parts of composites under tension or compression along reinforcing elements." In the studies, a model of the cause of the discussed phenomenon of fracture of the composite under tension or compression is proposed—*the presence of periodic curvatures in the internal structure of the composite*, since the presence of these periodic curvatures leads to the occurrence of breaking normal or tangential stresses at the interface, *both under tension and under compression by the equal external loads*. In connection with the above, the statics of composites with internal curvatures studies is developed. The main results belong apparently to the author of the monograph and his pupils and followers.

Problem 6 *Fracture under compression along parallel fractures*: This problem refers to the metals, alloys, composites, and other materials that are modeled in the continuum approximation by the isotropic or orthotropic bodies. In the case of orthotropic materials, it is assumed that the plane cracks are located in the planes of symmetry of material properties. In the situation under the discussion of compression along a system of plane cracks lying in parallel planes, *all three stress intensity factors are equal to zero* (due to the symmetry of material properties and loading conditions) for any linear and nonlinear models of deformable bodies (taking into account elastic, plastic, and viscous deformations).

In connection with the above, in the situation under discussion, *all five basic approaches of classical fracture mechanics* described in Sect. 1.1 *do not work*. This situation (the presence of cracks in the parallel planes) is quite realistic for the engineering materials, in particular for the composite materials, since the composites are characterized by the presence of parallel surfaces of the interface of the properties of the filler and binder, in which the presence of various types of adhesion violation is observed, including the presence of the cracks.

Considering the above, *a general concept* is adopted in Problem 6—the start of fracture is determined by the reaching by compressive loads the values corresponding to the local loss of stability of the equilibrium state near the cracks. In Problem 6, as in Problem 1, with taking into account the above general concept, *two approaches* are developed to study the stability and therefore fracture under compression of materials with a system of parallel cracks.

The first approach is based on the application of various approximate theories of rods, plates, and shells, as well as other approximate schemes. Due to the complexity of the analyzed phenomenon, such an approach cannot lead, apparently, to reliable results.

The second approach is based on the application of the three-dimensional linearized theory of stability of deformable bodies, constructed with the accuracy usually accepted in mechanics. Therefore, the second approach allows for obtaining reliable results even in cases when the first approach is not applicable. The results in the framework of the second approach belong mainly to the author of the monograph and his pupils and followers.

Problem 7 *Brittle fracture of materials with cracks under the action of dynamic loads (with taking into account the contact interaction of the crack ends)*: In this problem, a model of the linear elastic isotropic body is used for the material and the cracks, as in other problems of fracture mechanics, are modeled by the mathematical cuts that have no thickness. In this situation, the distribution of stresses and displacements near the cracks is formed by the incident and reflected elastic waves that arise under the action of external dynamic loads.

Under the action of loads determined by the propagating elastic waves on the crack sides, the displacements of the crack sides arise, which change the sign in time. The noted phenomenon occurs, for example, in the case of the harmonic waves during each period, so the extension and compression phases appear in the longitudinal waves, and the phases with transverse displacements of different signs appear in the transverse waves.

So, in the simplest example (the normal incidence of a plane longitudinal harmonic wave on a plane crack), in the first half-period, the extension phase occurs (the crack is opened and the crack sides do not interact), and in the second half-period, a compression phase occurs (the crack is closed, and the contact interaction of the crack sides occurs). Note that the numbering of half-periods is conditional and the noted phenomenon *always exists regardless of the intensity of the external load*.

Thus, it can be considered that the change in time of the signs of displacements of the crack sides is apparently a characteristic feature of the physics of phenomena arising under the action of dynamic loads. This phenomenon *must be taken into account* when constructing the dynamic fracture mechanics. In the classical mechanics of dynamic fracture, the discussed phenomenon is not taken into account. As an example, one can point to the monographs [14, 15] and numerous other monographs, as well as the overwhelming number of publications in the scientific journals and conference proceedings.

It should be noted that this phenomenon (a change in the boundary conditions on the crack sides—a smooth transition from free cracks to the dynamically contacting crack sides for each period) can be taken into account only within the framework of the nonlinear dynamic fracture mechanics, which takes into account the change in boundary conditions during deformation (failure to fulfill the Condition 1.3 of application of the classical fracture mechanics).

In the works of the author and his pupils, the simplest variant of the nonlinear dynamic mechanics of brittle fracture is developed, which consists of the application of the *linear* dynamic equations of a linear elastic isotropic body, describing the propagation of elastic waves in the material, and *nonlinear* boundary conditions, describing the change in contact interaction during deformation. In these works, a numerical method for solving the formulated nonlinear dynamic problems was also developed and implemented. As a result of solving and analyzing a number of the plane, spatial and anti-plane problems, the main regularities that characterize the qualitative and quantitative difference between the classical and non-classical problems of dynamic fracture mechanics are identified and formulated.

Problem 8 *Fracture of thin-walled bodies with cracks under tension in case of preliminary loss of stability*: The non-classical nature of the considered problem of fracture mechanics is that in this case Condition 1.2 of applicability of the classical fracture mechanics (loss of stability does not precede the fracture), which is formulated in Sect. 1.1, is not satisfied.

Really, it is tacitly assumed in the classical fracture mechanics of materials with cracks under tension and shear that the fracture begins from the configuration of the body that it had in the undeformed state. Consequently, it is assumed that in the process of deformation before the start of fracture, there is no sharp change in the configuration of the body; that is, loss of stability does not precede fracture. In reality, even under tension in the case of thin-walled bodies, the local loss of stability of the equilibrium state near cracks may precede the fracture.

The reference problems in this non-classical problem of fracture mechanics are the problems of stretching the thin-walled plates and shells perpendicular to the crack. In the case of cylindrical shells, as a rule, the situation is investigated when the crack is located along the guiding line of the shell and the shell is stretched along the axis. In this situation, as a result of the stress concentration near the crack, the local zones of compressive stresses arise, which can lead to a local loss of stability near the cracks before the start of the fracture process.

Various authors in the study of the above situation use the various approximate design schemes when analyzing the local zones of loss of stability near cracks and holes. In the works of the author and his pupils, the strict equations of the mechanics of thin-walled systems are used to analyze this problem with the following use of the variational and numerical methods. At that, considerable attention is paid to the experimental studies and the use of their results in the analysis of these problems.

The information of this section in a brief formulation of the eight non-classical problems of fracture mechanics, indicating the phenomena corresponding to these problems, seems to be sufficient.

As already noted in the introductory part of this section, in the subsequent chapters of this monograph (Chaps. 3–10), each of the eight discussed problems will be considered in a separate section with a clear and concise indication of the results obtained by the author and his pupils.

1.4   Additional Discussion of Non-classical Problems of Fracture Mechanics

This subsection (Sect. 1.4) briefly quotes:

> an additional discussion of the models and approaches used in the study of the eight above-stated non-classical problems of fracture mechanics in the works of the author and his pupils;

discussion of one of the approximate approaches aimed at reducing the non-classical Problem 6 (Sect. 1.3) to the classical problems of fracture mechanics; discussion of one of the problems that at first glance is related to the non-classical problems of fracture mechanics.

1.4.1 Brief Discussion of Models and Approaches in Non-classical Problems of Fracture Mechanics (Problems 1–8, Sect. 1.3)

Concerning the considered non-classical Problems 1–8, the numerous approximate approaches, models and calculation schemes have been proposed in the publications of various authors. The approaches and models will be only considered below which were proposed and used in the publications of the author and his pupils.

Concerning Problem 1 (Fracture of composite materials under compression along reinforcing elements), the three-dimensional linearized theory of stability of deformable bodies at the finite and small subcritical deformations was used to construct the fracture mechanics.

The studies were carried out for the models of deformable bodies taking into account the elastic, plastic, and viscous deformations. The results are obtained for the composite materials for the models of a piecewise homogeneous body, which is the most rigorous model in the framework of the mechanics of a deformable body, and for the continuum approximation (the model of the homogeneous body with averaged values of parameters).

Concerning Problem 2 (The model of short fibers in the theory of stability and fracture mechanics of composite materials under compression), the three-dimensional linearized theory of stability at the small subcritical deformations was applied. The studies were carried out only taking into account the elastic deformations (brittle fracture) for a piecewise homogeneous body model.

Concerning Problem 3 (End-crush fracture under compression of composite materials), the three-dimensional linearized theory of stability of deformable bodies at the small subcritical deformations was applied. The studies were carried out for the models of deformable bodies taking into account the elastic and plastic deformations. The results were obtained for the continuum approximation when analyzing the phenomenon of the near-the-surface instability of the loaded ends of the composite material.

Concerning Problem 4 (Brittle fracture of materials with cracks, taking into account the action of initial (residual) stresses along the cracks), the three-dimensional linearized theory of elasticity was applied at the finite and small initial (residual) deformations for the hyperelastic isotropic and orthotropic materials with an arbitrary structure of the elastic potential.

It is advisable to note that the stated above linearized theory is equivalent to the three-dimensional linearized theory of stability of deformable bodies. The results were obtained for a model of a homogeneous material (continuum approximation) with various cracks, along which the initial (residual) stresses act.

Concerning Problem 5 (Brittle fracture in the form of *"separation into slender parts"* under tension or compression of composite materials along reinforcing elements), the three-dimensional linear theories of elasticity and viscoelasticity are applied for the isotropic and orthotropic materials under static loads. The studies were carried out for a model of a piecewise homogeneous material and a model of a homogeneous material (in the continuum approximation).

At that, for the model of a homogeneous material (continuum approximation), the special approaches (models and theories) are developed that make it possible *to determine stresses on areas whose dimensions are less than the periods of curvature in the structure of the composite material.*

Thus, the above continuum theories made it possible to determine the distribution of stresses within the period of curvature. Note that the continuum theories proposed by other authors make it possible to determine stresses only on areas whose dimensions are much larger than the periods of curvature in the structure of composite materials.

The results were obtained for the layered and fibrous materials.

Concerning Problem 6 (Fracture under compression along parallel cracks), the three-dimensional linearized theory of stability of deformable bodies under finite (large) and small subcritical deformations is applied. The studies were carried out for the homogeneous materials with cracks, which are located in the parallel planes, for the various models of deformable bodies, taking into account the elastic, plastic, and viscous deformations.

It should be noted that the results for models taking into account only the elastic and plastic deformations were obtained with a greater degree of completeness and generality compared to the results taking into account the viscous deformations.

This consideration is discussed in more detail in Chap. 2 in the presentation of the main relations and general problems of the three-dimensional linearized theory of stability of deformable bodies for the models taking into account the elastic, plastic, and viscous deformations.

In the case of the layered composite materials, the studies were also carried out for a model of piecewise homogeneous materials, when the cracks are located at the interfaces of the components. Both the exact solutions were obtained using analytical methods and the solutions using numerical methods and computers.

Concerning Problem 7 (*Brittle fracture of materials with cracks under the action of dynamic loads (taking into account the contact interaction of the crack ends)*, the equations of the dynamics of a linear elastic isotropic body and new nonlinear boundary conditions on the crack sides corresponding to the changing contact interaction during deformation are applied. To solve the formulated nonlinear dynamic problems, the method of successive approximations was developed, when in each of the approximations the problems are reduced to the problems with a fixed zone of contact interaction, the dimensions of which are determined from the previous approximations. In fact, as if the problem of classical dynamic fracture mechanics in each of the approximations is formulated.

Concerning Problem 8 (Fracture of thin-walled bodies with cracks under tension in the case of preliminary loss of stability), the theoretical and experimental studies were carried out for the thin-walled elements (plates and cylindrical shells).

In the theoretical studies, the brittle fracture was analyzed (the linear theory of elasticity was applied), and in the experimental studies, the fracture under elastic and plastic deformations was considered. In the study of the brittle fracture of thin-walled plates with cracks under tension (taking into account the possibility of local loss of stability near cracks), the linearized theory of the stability of thin-walled plates was used, which was constructed using the Kirchhoff–Love hypothesis.

The above information in this section with a very brief discussion of models and approaches in eight non-classical problems (Problems 1–8, Sect. 1.3), which are considered in the publications of the author and his pupils, seems to be sufficient.

It follows from this information that from of the eight formulated problems the three-dimensional linearized theory of stability of deformable bodies was applied in *five* problems (Problems 1, 2, 3, 4, and 6), the three-dimensional linear theory of elasticity or viscoelasticity under static loading was applied in Problem 5, the three-dimensional linear theory of elasticity under dynamic loading and nonlinear boundary conditions on the crack sides were used in Problem 7, and, the two-dimensional linearized theory of the stability of thin-walled plates, built using the Kirchhoff–Love hypothesis was used in Problem 8.

Thus, in most of the eight considered non-classical problems of fracture mechanics (5 of 8), the three-dimensional linearized theory of stability of deformable bodies was used.

At that, it should be noted that the above theory of stability, apparently, is less well-known and less widely used in comparison with other branches of mechanics of deformable bodies.

In connection with the above-mentioned situation, Chap. 2 is included in this monograph, in which a very brief survey of the formation of the three-dimensional linearized theory of stability of deformable bodies is given, following mainly the monographs [16, 17] and other monographs of the author, as well as review articles [18–21].

Also, in Chap. 2, the mathematical apparatus of the three-dimensional linearized theory of stability of deformable bodies under finite (large) and small subcritical deformations is presented in a very brief form with a statement of the basic relations, following [16–21], which is an exception and does not correspond to the style of this monograph. The noted situation is highlighted in the **Note** set out in the **Introduction**.

In the final part of this section, several general issues that arise in the mechanics of composites, including the mechanics of nanocomposites will be considered. Thus, when researching the discussed non-classical problems of fracture mechanics concerning the composite materials, two well-known approaches or models were applied (were used).

The first approach uses a piecewise homogeneous medium model. In this case, to describe the motion (equilibrium) of each of the reinforcing elements (filler) and matrix (binder), the three-dimensional relations of the mechanics of deformable bodies are generally used under certain (taking into account defects) conditions at the interface. The first approach is the most rigorous and accurate within the framework of the mechanics of deformable bodies and with its involvement, the stress state, dynamics, and stability in the structural elements, i.e., in the microstructure of the composite material, can be studied.

It is obvious that *the application of the first approach is necessary* in the study of stress and strain fields, the phenomena of wave propagation and loss of stability, when *the characteristic parameters* of the phenomena under consideration (the distances at which stresses and strains change significantly in the static problems; the wavelengths in the wave dynamics; the wavelengths of forms loss of stability) *are the quantities of the same order or significantly less than the geometric dimensions* (minimal dimensions) of *the structural elements of composite materials.*

In the second approach, the composite material is modeled as a homogeneous anisotropic material with the averaged constants. In this case, the microstructure of the composite material is taken into account by determining the averaged constants, which depend on the physical and mechanical properties and the geometric shape of the filler and binder, as well as on their volume fraction. It is obvious that *the application of the second approach is substantiated when the characteristic parameters* of the phenomena under consideration (they are shown in the first approach) *are much larger than the geometric dimensions (maximal dimensions) of the structural elements of composite materials.*

It should be emphasized that the methods of determining the averaged constants, which are also called the reduced constants, are currently sufficiently developed and presented in well-known numerous publications, but they will be not considered here, since this monograph is devoted to other issues. It is nevertheless advisable to emphasize that when considering the problem of determining the averaged constants of composite materials, the experimental methods for determining the indicated constants (as for the homogeneous anisotropic material) cannot be excluded from consideration, although in the domestic and foreign scientific literature much more attention is paid to the theoretical methods for determining the values of averaged constants.

However, only in the experimental determination of the values of the averaged constants (as for the homogeneous anisotropic material), it is possible to take into account the influence of various imperfections and defects in the internal structure of the composite material, which arise practically in any technology of their production.

It is worth noting that *the first approach* was described above, which was used in the study of eight considered non-classical problems of fracture mechanics as applied to the composite materials and which is the most exact and rigorous in the framework of the piecewise homogeneous medium model. Nevertheless, within the framework of the model of a piecewise homogeneous medium, the approximate approaches also exist, when the approximate design schemes are used for a filler or a binder (or both for a filler and a binder).

In general, *when passing from a model of a piecewise homogeneous material to a model of a homogeneous material with averaged constants, the application of the homogenization procedure (principle, process, or concept) is implemented,* and, as a result, a homogeneous (homogeneous) material is obtained.

It should be noted that when developing the foundations of the mechanics of nanocomposites with a polymer matrix (e.g., monographs [22, 23], article [24], and other publications indicated in the references to [22–24]), in addition to a *procedure (principle, process, or concept) homogenization the procedure (principle, process, or concept) of continualization* is also applied.

The point is that in nanocomposites with a polymeric matrix, the fillers are usually single-walled and multi-walled carbon nanotubes (CNT). By their internal structure, the CNT is formed of individual atoms in each atomic layer. In this case, naturally, the individual atoms are located at certain distances from each other and are held at these distances due to the forces of interatomic interaction, which determines the ordered (zigzag, armchair, or chiral) structures. Thus, each CNT, as a filler in a nanocomposite with a polymer matrix, is a discrete system that is deformed together with the polymer matrix. At that, the deformation of the latter is usually described within the framework of ordinary continuum representations (as a continuum). In this regard, to describe the common deformation of nanotubes and the polymer matrix (common motion), it is advisable to use a uniform description of the motion (deformation) of the nanotubes and matrix, which is achieved by applying the principle of continualization for nanotubes.

The procedure (principle, process, or concept) of continualization consists of replacing (modeling) the discrete system with the continuous system (medium) with the determination of averaged constants within the continuous system (medium).

It is advisable to note that the overwhelming majority of researchers involved in the experimental determination of the properties of CNT provide information on the properties of nanotubes in values usually adopted for the continuous representations (essentially after the application of the principle of continualizations) (e.g., in monograph [25], vol. 1, pp. 80–83). Thus, it can be considered that one of the most popular approaches in the study of composites is the application of the homogenization procedure for the entire composite. In the case of nanocomposites, the appropriate approach is the initial application of the continualization procedure to the nanotube and then application of the homogenization principle to the entire nanocomposite.

Additional information on the general issues briefly discussed above in the final part of this section is given in the Preface to monograph [26], in the Preface and Introduction to the 2-volumes monograph [25] and monographs [22, 23] (as applied to the construction of the foundations of the mechanics of nanocomposites).

1.4.2   On Consideration of Non-classical Problems
of Fracture Mechanics from the Point of View
of Classical Problems of Fracture Mechanics

Apparently, it can be considered that at present the attempts exist to consider or investigate certain non-classical problems of fracture mechanics from the point of view of the classical problems of fracture mechanics (models and approaches) with the introduction of the corresponding approximate design schemes.

In this subsection, one of the above proposals to investigate Problem 6 (*Fracture under compression along parallel cracks*) (Sect. 1.3) by introducing the approximate design schemes corresponding to the classical fracture mechanics is presented following several publications by other authors.

Consider the proposed proposal as applied to the right tip ($x_1 = + a$, Fig. 1.1) for a material with a crack length $2a$ under compression along the axis $0x_1$ (Fig. 1.1). Restrict the analysis to the framework of plane deformation in the plane $x_1 0x_2$.

The essence of the proposal is that when analyzing the crack propagation, one should take into account the microstructure of the material. Thus, the crack will not propagate along a straight line $x_2 = 0$, but along some broken line, which is determined by the influence of the microstructure and is close to the line $x_2 = 0$ (Fig. 1.1).

It should be noted that Fig. 1.1 corresponds to a polycrystalline material when the microstructure of the material near the right tip of the crack is taken into account (the presence of the monocrystals, which are marked with the thicker shading), and the crack propagates along a broken line, bypassing the monocrystals. In Fig. 1.1, the designation is also indicated: β is the angle between the vertical axis and the normal to one of the segments of the considered broken line, along which following the discussed proposal the fracture propagation is predicted. Under the action of external loads in the form shown in Fig. 1.1, *the shear stresses*, determined by the angle β in Fig. 1.1, will occur. In this regard, taking into account *the shear stresses* introduced in this way, it is already possible to apply the fracture criterion of the classical fracture mechanics (the Griffiths–Irwin approach, Sect. 1.1).

When analyzing the prospects for the development of the considered approach for the *polycrystalline materials* or the *composite materials* that have an internal structure close to the granular one, it is advisable to take into account the following three considerations.

Fig. 1.1

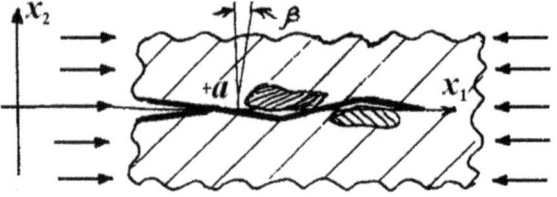

1.   Taking into account the influence of the microstructure of the above materials at the tip of a crack concerning the fracture mechanics for the case presented in Fig. 1.1 seems to correspond to the next stage in the cognitive process as applied to the fracture mechanics, which usually develops within the framework of continuum concepts for the various materials.
2.   When trying to implement the discussed approach for the above materials, it is necessary to carry out the most complex studies to identify the phenomena occurring at the tip of a crack at the microstructural level and determining the broken line—the crack propagation line, with the model concepts used in the fracture mechanics.
3.   With a slight difference in Fig. 1.1 of the broken line from the straight line (at small angles β in Fig. 1.1), the introduced (with the considered approach) shear stresses will be significantly lower than the compressive stresses σ_{11} in Fig. 1.1.

The above considerations (especially the second one), at least in the opinion of the author of this monograph, do not make it possible to expect to obtain (in the near future) the results for the materials under discussion (polycrystalline materials and composite materials with the internal structure close to the granular one) in their finished form when applied the discussed approach.

In the case of *the fibrous and layered composite materials* under compression along the reinforcing elements, the propagation of cracks in the interface is characteristic. In this regard, when studying the propagation of such cracks, the influence of the structure or microstructure of the composite at the crack tip is already taken into account in advance. Therefore, it seems that it makes no sense to additionally take into account the effect of the material structure at the crack tip.

Thus, the proposal discussed in this subsection can be considered as directly related to the non-classical problems of fracture mechanics, which are analyzed in this monograph. Nevertheless, as noted above, this proposal cannot be considered promising from the point of view of obtaining new concrete results when analyzing the consideration of the discussed eight non-classical problems of the fracture mechanics using the classical problems of fracture mechanics.

1.4.3   On Some Other Publications

Currently, several publications in the scientific literature exist in which the specific problems are investigated, which at first glance seem to be related to the non-classical problems of fracture mechanics discussed in this monograph. In reality, however, these results and the problems under study do not in any way relate to the eight non-classical problems of fracture mechanics discussed in this monograph but have independent significance.

The above considerations will be demonstrated using the example of a specific publication [27], the title of which indicates that the study is being carried out "... when the plate is compressed along the crack line." This publication does indeed

Fig. 1.2

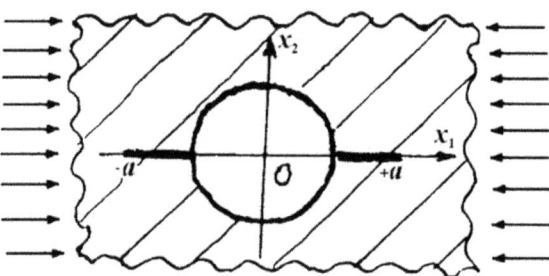

show the results of the experimental studies under compression along two cracks that emerge from the contour of a circular hole and are located on the continuation of the same hole diameter (located on the same line).

Figure 1.2 shows the design scheme, corresponding to research [27]. It should be noted that Fig. 1.2 fully corresponds to Fig. 1.1a [27], where information related to the description of the samples used in the experimental studies is omitted.

In the case shown in Fig. 1.2, the compression along cracks that lie in the same plane is indeed investigated. For the external loads, the situation in Fig. 1.2 seems to correspond to Problem 6 (*Compression fracture along parallel cracks*) following the terminology of Sect. 1.3. However, in the case shown in Fig. 1.2, *the cracks are in a complex stress field caused by the stress concentration around the hole*. So, in the case [27], the tensile stresses σ_{22} arise near the crack tips (when $x_1 = \pm a$, Fig. 1.2) corresponds to the design scheme), caused by the stress concentration near the hole, which makes it possible to apply the fracture criteria of the classical fracture mechanics (Griffiths–Irwin approach).

The above is also confirmed by the title of the publication [27] "Experimental determination of the value κ_I...", where is the well-known stress intensity factor.

It follows from the above information and the considerations of subsection 1.4.3 that it is necessary to determine clearly whether a particular publication relates to the non-classical problems of fracture mechanics discussed in this monograph.

In the conclusion of the considered chapter, we note that it contains in rather brief form information related to the division of the problems of fracture mechanics into the classical and non-classical problems, as well as to the formulation and preliminary discussion of *eight* non-classical problems of fracture mechanics, which have been studied for 50 years in the Department of Dynamics and Stability of Continua at the S. P. Timoshenko Institute of Mechanics of NASU.

As already noted in the introductory part of Sect. 1.3, in the subsequent Chaps. 3–10 of this monograph, each of eight non-classical problems of fracture mechanics will be considered in a separate chapter with a clear indication of the results obtained in the Department of Dynamics and Stability of Continua.

References

1. Guz, A.N.: On physical incorrect results in fracture mechanics. Int. Appl. Mech. **45**(10), 1041–1051 (2009)
2. Guz, A.N., (ed.): Neklassicheskie problemy mekhaniki razrusheniya, v 4 tomah, 5 knigah (Non-Classical Problems of Fracture Mechanics, in 4 volumes, 5 books). Naukova Dumka, Kyiv (1990−1993), Kaminsky, A.A., (ed.): T.1. Razrushenie viazkouprugikh tel s treshchinami (Vol. 1. Fracture of Viscoelastic Bodies with Cracks) (1990), Guz, A.N., (ed.): T.2. Khrupkoe razrushenie materialov s nachalnymi napriazheniiami, (T.2. Brittle Fracture of Materials with Initial Stresses) (1991), Kaminsky, A.A., Gavrilov, D.N., (eds.): T.3. Dlitelnoe razrushenie polimernykh i kompozitnykh materialov s treshchinami (T.3. Long-Term Fracture of Polymer and Composite Materials with Cracks) (1992), Guz, A.N., Dyshel, M.Sh., Nazarenko, V.M., (eds.), T.4, kniga 1. Razrushenie i ustoichivost materialov s treshchinami (V.4, Book 1. Fracture and Stability of Materials with Cracks) (1992), Guz, A.N., Zozulya, V.V., (eds.), T.4, kniga 2. Khrupkoe razrushenie materialov pri dinamicheskikh nagruzkakh (V.4, Book 2. Brittle Fracture of Materials under Dynamic Loads) (1993)
3. Griffith, A.A.: The phenomena of rupture and flow in solids. Phil. Trans. Roy. Soc., Ser. A. **211**(2), 163–198 (1920)
4. Irwin, G.R.: Analysis of stresses and strains near end a crack traversing a plate. J. Appl. Mech. **24**(3), 361–364 (1957)
5. Irwin, G.R.: Fracture. In: Flügge, S. (ed.) Handbuch der Physik, Bd. 6, pp. 551–590. Springer, Berlin (1958)
6. Eshelby, I.D.: The force on the elastic singularity. Phil. Trans. Roy. Soc. Ser. A. **244**, 87 (1951)
7. Cherepanov, G.P.: O rasprostranenii treschin v sploshnoy srede (On propagation of cracks in continuous medium). Prikladnaya Matematika Mekhanika **31**(3), 476–488 (1967)
8. Cherepanov, G.P.: Mekhanika khrupkogo razrusheniya (Brittle Fracture Mechanics). Nauka, Moscow (1974)
9. Rice, I.R.: Path independent integral and the approximate analysis of strain concentration by notches and cracks. J. Appl. Mech. **35**(4), 340–350 (1968)
10. Guz, A.N.: Description and study of some nonclassical problems of fracture mechanics and related mechanisms. Int. Appl. Mech. **36**(12), 1537–1564 (2000)
11. Guz, A.N.: On study of nonclassical problems of fracture and failure mechanics and related mechanisms. Int. Appl. Mech. **45**(1), 1–31 (2009)
12. Guz, A.N.: O postroenii osnov mekhaniki razrusheniia materialov pri szhatii vdol treshchin (obzor) (On the construction of the foundations of the fracture mechanics of materials in compression along cracks (review)). Prikladnaya Mekhanika **50**(1), 5–89 (2014)
13. Guz, A.N.: On study of Nonclassical Problems of Fracture and Failure Mechanics and Related Mechanisms, pp. 35–68. ANNALS of the European Academy of Sciences, Liège, Belgium (2006–2007)
14. Parton, V.Z., Borisovsky, V.G.: Dinamicheskaya mekhanika razrusheniya (Dynamic Fracture Mechanics). Mashinostroenie, Moscow (1985)
15. Parton, V.Z., Borisovsky, V.G.: Dinamika khrupkogo razrusheniya (Brittle Fracture Dynamics). Mashinostroenie, Moscow (1988)
16. Guz, A.N.: Osnovy trekhmernoi teorii ustoichivosti deformiruemykh tel (Fundamentals of the Three-Dimensional Theory of Stability of Deformable Bodies). Vyshcha Shkola, Kyiv (1986)
17. Guz, A.N.: Fundamentals of the Three-Dimensional Theory of Stability of Deformable Bodies. Springer, Berlin Hiedelberg New York (1999)
18. Guz, A.N.: Construction of the three-dimensional theory of stability of deformable bodies. Int. Appl. Mech. **37**(1), 1–37 (2001)
19. Guz, A.N.: Elastic waves in bodies with initial (residual) stresses. Int. Appl. Mech. **38**(1), 23–59 (2002)
20. Guz, A.N.: Establishing the fundamentals of the theory of stability of mine working. Int. Appl. Mech. **39**(1), 20–48 (2003)

21. Guz, A.N.: Stability of elastic bodies under omnidirectional compression (review). Int. Appl. Mech. **48**(3), 241–293 (2012)
22. Guz, A.N., Rushchitskyi, Ya.Ya., Guz, I.A.: Vvedenie v mekhaniku nanokompozitov (Introduction to Mechanics of Nanocomposites). S.P. Timoshenko Institute of Mechanics, Kyiv (2010)
23. Guz, A.N., Rushchitskii, J.J.: Short Introduction to Mechanics of Nanocomposites. Scientific & Academic Publishing Co. LTD, USA (2013)
24. Guz, A.N., Rushchitskii, J.J., Guz, I.A.: Establishing fundamentals of the mechanics of nanocomposites. Int. Appl. Mech. **43**(3), 247–271 (2007)
25. Guz, A.N.: Osnovy mekhaniki razrusheniia kompozitov pri szhatii: V 2-kh tomakh (Fundamentals of the Fracture Mechanics of Composites under Compression: In 2 volumes). Litera, Kyiv (2008)
26. Guz, A.N.: Mekhanika razrusheniia kompozitnykh materialov pri szhatii (Fracture Mechanics of Composite Materials under Compression). Naukova Dumka, Kyiv (1990)
27. Pisarenko, G.S., Naumenko, V.P., Mitchenko, O.V., Volkov, G.S.: Eksperimentalnoe opredelenie velichiny K1 pri szhatii plastiny vdol linii treschiny (Experimental determination of the value of K1 under compression of the plate along the crack line). Probl. Prochn. **11**, 3–9 (1984)

Chapter 2
Brief Statement of Foundations of Three-Dimensional Linearized Theory of the Deformable Bodies Stability (TLTDBS)

This chapter provides, in a very brief form, information on the foundations of the three-dimensional linearized theory of stability of deformable bodies, including the basic relations and information about the mathematical apparatus of this theory. The expediency of including this material into this monograph follows from the Notes in the Introduction and the information set out in subsection 1.4.1. The main consideration can be formulated as follows—in the *five* problems (Problems 1, 2, 3, 4, and 6 of the *eight* non-classical problems of fracture mechanics), a three-dimensional linearized theory of stability of deformable bodies is applied, which is less well-known and less widely used in comparison with other branches of mechanics of deformable bodies.

Note also that the inclusion in this chapter of the mathematical aspects of the three-dimensional linearized theory of stability of deformable bodies is an exception to the style of writing the review on non-classical problems of fracture mechanics, which was also noted in the Introduction. In the subsequent chapters (Chaps. 3–10) of this monograph, the main results on the eight discussed non-classical problems of fracture mechanics obtained in the Department of Dynamics and Stability of Continua will be stated briefly in the accepted style of presentation noted in the Introduction to this monograph.

When constructing this chapter, we will follow the monographs [1, 2], where the three-dimensional linearized theory of stability of deformable bodies is presented in a unified general form for the theory of finite (large) and small subcritical deformations for the various models of deformable bodies with constitutive equations of a sufficiently general structure, taking into account the elastic, plastic, and viscous deformations.

Additionally, the information will be used which is presented in monographs [3–7], as well as in the publications that are included in the bibliography for the monographs [2–7]. By the style of presentation of the results, this chapter corresponds to the style of presentation of survey papers [8, 9], but the results of this chapter are presented in a much short form. Therefore, more detailed information can also

A. N. Guz, *Eight Non-Classical Problems of Fracture Mechanics*,
Advanced Structured Materials 159,
https://doi.org/10.1007/978-3-030-77501-8_2

be obtained from modern reviews on the partial problems of the discussed theory [10–13].

2.1 On the Formation of TLTDBS

Starting with the famous work of Euler of 1744, traditionally in the mechanics of deformable bodies, the studies of the phenomenon of loss of stability were carried out (exclusively) and are carried out (in most publications for several traditional scientific directions) concerning the thin-walled structural members (rods, plates, and shells). At that, the approximate applied (one-dimensional for the rods, two-dimensional for the plates and shells) theories, constructed using the hypotheses of plane sections and Kirchhoff–Love, as well as other approximate design schemes, are used to carry out the research.

Only at the beginning of the twentieth century, the three-dimensional theory of stability of deformable bodies began to study to develop a general approach to the study of the phenomenon of loss of stability in the mechanics of deformable bodies and from the middle of the twentieth century to ensure the study of the phenomenon of loss of stability in the mechanics of a deformable body concerning the unconventional new scientific areas (e.g., geotectonics, mechanics of composite materials, the theory of stability of the local state of equilibrium near the mine workings, and several others).

At present, apparently, it can be considered that the first publication on the construction of a three-dimensional theory of stability of deformable bodies was the article by Southwell [14], published in 1913.

In the subsequent years, the results were published on the construction of a three-dimensional theory of stability of deformable bodies, obtained by several authors. At that, two approaches were established for the construction of this theory.

In the *first* approach, the basic equations and boundary conditions of the three-dimensional theory of stability of deformable bodies are formed based on the corresponding nonlinear theory, by applying a rigorous mathematical procedure—by linearizing the basic relations of the corresponding nonlinear theory.

In connection with the foregoing, the basic relations of the three-dimensional theory of stability of deformable bodies obtained in the *first* approach can be considered sufficiently rigorous and exact, as well as obtained consistently. The discussed results obtained in the *first* approach can be called the three-dimensional linearized theory of stability of deformable bodies.

In the *second* approach, the basic equations of motion and the corresponding boundary conditions of the three-dimensional theory of stability of deformable bodies are formed starting with certain considerations of a physical nature. At that, the results obtained (the equations of motion and boundary conditions) may differ slightly from each other, since the corresponding considerations of the physical nature of various authors may differ somewhat.

Let us illustrate the above *first* method of obtaining the basic relations of the three-dimensional linearized theory of stability of deformable bodies using the simplest example. Let, within the framework of the *considered version of nonlinear mechanics of a deformable body*, the following relation hold

$$y = f(x), \tag{2.1}$$

the counting which is carried out from the first reference state (in the case of elastic bodies—from the natural, undeformed, state). Consider the relation (2.1) as applied to the second (initial, unperturbed) and the third (perturbed) states. Taking into account the above notation (2.1) as applied to the second and third states, we can write

$$y_0 = f(x_0), \; y_0 + y = f(x_0 + x), \tag{2.2}$$

where y and x are the perturbations of the corresponding quantities. Taking into account the smallness of the perturbations, the following condition is introduced

$$|x| << |x_0|. \tag{2.3}$$

Within the framework of the first approach (an application of the linearization procedure), the following expression can be obtained from (2.2), taking into account (2.3)

$$y = \left[\left(\frac{df}{dx} \right) \bigg|_{x=x_0} \right] x. \tag{2.4}$$

Thus, the basic relations of the three-dimensional linearized theory of stability of deformable bodies are usually understood as the relations of the type (2.4), i.e., the relationships among the disturbances. Moreover, the equal sign is taken approximately in the relations of the type (2.4).

It is advisable to note that the above three-dimensional linearized theory of stability of deformable bodies corresponds to *the considered variant of nonlinear mechanics of a deformable body*.

In the middle of the twentieth century, many scientists in quite numerous publications proposed the various versions of the three-dimensional theory of stability of deformable bodies, which differ from each other in their approaches to obtaining them:

By invoking considerations of a physical nature;
By applying the principle of linearization;
By involving the variational principles;
By application of the theory of large (finite) subcritical deformations and various variants of the theory of small subcritical deformations;
By the formulation of the relations in Cartesian coordinates;

By the formulation of the relations in an arbitrary curvilinear coordinate system using the apparatus of tensor analysis;

By the use of stress tensors, in which the components refer to the dimensions of the areas in the reference configuration (the first state) or the actual configuration (the third state) and several other approaches.

Apparently, the article [15], published in 1952, in which the basic relations of the three-dimensional linearized theory of stability of deformable bodies under finite (large) subcritical deformations, obtained with the involvement of tensor analysis, and special cases are considered considered as the finishing publications in the above-mentioned historical process. In this article, for the linearized three-dimensional theory, the name "theory of small deformations superposed on finite deformations" was also introduced for the first time, which is often used in subsequent publications by other authors, especially by English-speaking authors.

Note 2.1 It should be noted that the basic relations of the three-dimensional linearized theory of the deformable bodies stability (TLTDBS) in the historical aspect were formed based on the basic relations of the three-dimensional linearized theory of elastic stability (TLTES). The fact is that the basic relations of the TLTES (equations of motion and boundary conditions written in stresses) before the introduction of linearized elasticity relations in them are common for both the TLTES and the TLTDBS. With the introduction of the linearized elasticity relations into the above basic relations, the basic relations of the TLTES in displacements are obtained. With the introduction of the linearized constitutive relations for any other model of deformable bodies into the above basic relations, the basic TLTDBS relations for the model under consideration are obtained.

Note 2.2 The discussed basic relations of TLTDBS include the stresses of the subcritical state, usually marked with the index "zero," which are unknown and are determined when solving each specific problem of this theory. If to consider the stresses marked by the index "zero" in the discussed basic relations of TLTDBS as given values, then in this case the basic relations of TLTDBS are the basic relations of three-dimensional linearized mechanics of deformable bodies (TLMDB) with the initial or residual stresses. Then the latter are the stresses marked with the index "zero." In the above sense, the three-dimensional linearized theory of elasticity at the finite (large) and small initial deformations, which is involved in **Problem 4** (*brittle fracture of materials with cracks, taking into account the action of the initial (residual) stresses along the cracks*) (in the terminology of Sect. 1.3 of this monograph), corresponds to TLTES.

The historical sketch of the formation and development of the three-dimensional theory of stability of deformable bodies in 1913–1985, including the TLTDBS, with a list of the main publications, is presented in the monograph [1] in Russian; in English, this sketch is presented in the monograph [2], published in 1999.

A historical sketch of the development and formation of the three-dimensional theory of stability of deformable bodies in 1913–2000, including the TLTDBS, with a list of the main publications, is presented in the article [8], originally published in 2001 in Russian in the journal "Prikladnaya Mekhanika" and subsequently in 2002 published in English in the journal "International Applied Mechanics" by the Springer, which is currently translating and publishing in English the journal "Prikladnaya Mekhanika" as the journal "International Applied Mechanics."

Additional information can be obtained from the monograph of 2002 [9] devoted to a related problem—the three-dimensional (in the spatial variables) theory of propagation of elastic waves in materials with initial (residual) stresses, as well as from the review articles [11] of 2003 and [13] of 2004, devoted to other related problems of three-dimensional linearized mechanics of deformable bodies (TLMDB).

It is worth noting that the review articles [2, 8, 9, 13] of the author of this monograph were published during the 2000–2009 period when the journal "Prikladnaya Mekhanika" was carrying out the action dedicated to **the Beginning of the III Millennium**.

During this action, 174 review articles prepared by scientists from 26 countries around the world were published in the journal "Prikladnaya Mekhanika" in Russian. These articles in 2001–2010 were published in English by "International Applied Mechanics." The above-mentioned action of the journal "Prikladnaya Mekhanika" (174 review articles over 10 years, the authors are scientists from 26 countries of the world) *has no analog in the world scientific literature in periodicals on mechanics.*

The S.P. Timoshenko Institute of Mechanics of the NASU published, in 2005–2011, the multivolume collective monograph "**Advances in Mechanics**" (in six volumes, in seven books), which included 174 generalizing review articles published in 2000–2009 in the journal "Prikladnaya Mekhanika" in Russian and in 2001–2010 in the "International Applied Mechanics" in English during the campaign dedicated to **the Beginning of the III Millennium**. This edition also includes the general review articles [8, 9] related to the subject of this monograph. The multivolume collective monograph "Advances in Mechanics" (174 review articles over the past 10 years, the authors are scientists from 26 countries around the world) has no analog in the world monographic scientific literature on mechanics.

The S.P. Timoshenko Institute of Mechanics of the NASU published, in 2016–2018, the multivolume collective monograph "**Modern Problems of Mechanics**" (in three volumes), dedicated to the 100th anniversary (1918–2018) of the National Academy of Sciences of Ukraine and the S.P. Timoshenko Institute of Mechanics of the NASU. This edition includes the generalizing review articles of the leading scientists of the Institute of Mechanics in the scientific areas that are developing at the Institute of Mechanics in the last few decades.

These generalizing review articles of the leading scientists of the Institute of Mechanics were published in 2010–2017 in the journal "Prikladnaya Mekhanika" in Russian and in the "International Applied Mechanics" in 2011–2018 in English. This edition also includes the articles of the author of this monograph ([16] of 2011, [12] of 2012), which are the modern reviews on certain areas of three-dimensional linearized mechanics of deformable bodies (TLMDB).

Thus, the information on the formation and development of the three-dimensional theory of stability of deformable bodies in 1913–2011, including the TLTDBS, with the lists of the main publications can be obtained from the generalizing review articles [2, 8, 9, 12, 13, 16] of the author of this monograph, which were published in 2001–2012 in Russian in the journal " Prikladnaya Mekhanika " and in English in the "International Applied Mechanics."

These review articles were also included in the multivolume collective monographs "Advances in Mechanics" (in six volumes, in seven books) of 2005–2011 and "Modern Problems of Mechanics" (in three volumes) of 2016–2018.

Undoubtedly, the most informative is the generalizing review article [8], which briefly presents the historical aspect of the formation and development of the three-dimensional theory of stability of deformable bodies, including the three-dimensional linearized theories of stability of deformable bodies, with the list of the main publications as applied in 1913–2000.

The results of the author of this monograph on the development of a three-dimensional linearized theory of stability of deformable bodies are presented in a series of monographs [1–6], the first of which was published in 1971, and in articles partly included in the list of references to this monographs [17–48] as well as several others.

It should be noted that the more detailed list of the discussed articles is presented in monographs [1–6] and reviews [8, 9, 11, 12, 13, 16]. Moreover, the first publication of the author of this monograph on the three-dimensional linearized theory of stability of deformable bodies was the paper [27], published in 1967.

2.2 Classification of Approaches (Variants of Theory) in the TLTDBS

This classification is most clearly stated in the review articles [8, 9], although the main positions of such a classification were proposed by the author in 1972 in the article, which is indicated at No. 35 in the list of references of the review [8]. In the subsequent years, this classification was used in the monographs [1, 2], which, apparently, can be considered the final in a series of the author's monographs on the three-dimensional linearized theory of deformable bodies stability.

In presenting the considered classification and further in this chapter, we will apply the Lagrangian method of describing the motion of a continuum and use the method of the comoving coordinate system. All relations will be given using the tensor analysis built based on the basis vectors and the metric tensor, which are introduced in the reference (the first) state. Also, the stresses will be used that act in the actual configuration (in the disturbed, the third, state), but are related to the sizes of the corresponding areas in the reference (the first) state. When using other stress tensors, the arising situation will be special.

The above method for describing the basic relationships is used in most publications in the nonlinear mechanics of deformable bodies with finite (large) and small deformations.

Taking into account the above introductory information, below, in a very brief form, the classification of approaches (variants of the theory) in the three-dimensional linearized theory of deformable bodies stability (TLTDBS) is presented. At that, the attention will be paid not only to the most rigorous and consistent approaches (theories) but also to the insufficiently rigorous and insufficiently consistent approaches (theories), as well as to one very approximate approach, which has nothing to do with TLTDBS (does not follow when applying the linearization principle) and gives even to errors of a qualitative character.

2.2.1 Theory of Large (Finite) Subcritical Deformations

The approach used here (a version of the theory) is the most rigorous and consistent. As already noted in Sect. 2.1, the discussed version of the theory in its most general form was formed in the publication [15] of 1952 with the use of the tensor analysis. Within the framework of this theory, the main relations are as follows:

The equations of motion

$$\nabla_i \left[\left(g_n^j + \nabla_n u_0^j \right) S^{in} + S_0^{in} \nabla_n u^j \right] - \rho \ddot{u}^j + \rho F^j = 0; \tag{2.5}$$

the relations between the contravariant components of the non-symmetric Kirchhoff stress tensor t and the contravariant components of the symmetric stress tensor S

$$t^{ij} = \left(g_n^j + \nabla_n u_0^j \right) S^{in} + S_0^{in} \nabla_n u^j; \tag{2.6}$$

the incompressibility condition

$$q^{nj} \nabla_n u_j = 0, \tag{2.7}$$

where the notation is introduced

$$q^{nj} = g_{*0}^{nm} \left(g_m^j + \nabla_m u_0^j \right); \tag{2.8}$$

the covariant components of the Green strain tensor

$$2\varepsilon_{nm} = \left[\left(g_m^j + \nabla_m u_0^j \right) \nabla_n + \left(g_n^j + \nabla_n u_0^j \right) \nabla_m \right] u_j; \tag{2.9}$$

the boundary conditions in stresses on a part of the surface S_1 (taking into account the notation (2.6))

$$Q^j \big|_{S_1} = P^j, \, Q^j \equiv N_i t^{ij}; \tag{2.10}$$

the boundary conditions in displacements on a part of the surface S_2

$$u^j\big|_{S_2} = f^j;\tag{2.11}$$

the boundary conditions for the dynamical boundary problems

$$u^j\big|_{\tau=\tau_1} = f_1^j,\ u^j\big|_{\tau=\tau_2} = f_2^j;\tag{2.12}$$

the initial conditions for the dynamical boundary problems with the initial conditions

$$u^j\big|_{\tau=0} = h_1^j,\ \dot{u}^j\big|_{\tau=0} = h_2^j;\tag{2.13}$$

the expressions for determining the contravariant components of the vectors of the right-hand sides of the boundary conditions in stresses (2.10) (quantities P^j) under action on S_1 of an intensity "following" load $\tilde{P}$, which is directed along the normal to S_1

$$
\begin{aligned}
P^j = \tilde{P}N_k\sqrt{g_0^* g^{-1}}\Big[&\left(g_n^j + \nabla_n u_0^j\right)\left(g_m^\alpha + \nabla_m u_0^\alpha\right)\left(g_{*0}^{kn} g_{*0}^{m\beta}\right.\\
&\left.- g_{*0}^{nm} g_{*0}^{\beta k} - g_{*0}^{n\beta} g_{*0}^{mk}\right) + g_{*0}^{k\beta} g^{j\alpha}\Big]\nabla_\beta u_\alpha;
\end{aligned}\tag{2.14}
$$

Note first of all that:

the index "zero" throughout this chapter marks all quantities related to the subcritical state (to the second - unperturbed, initial state according to the terminology applied to the linearization process). The notation S concerning the symmetric stress tensor is introduced following the notations of article [15].

In the studies of other authors, other notations are also used for the mentioned symmetric stress tensor. All other notations in (2.5)–(2.14) correspond to the monographs [1, 2] and survey articles [8, 9] of the author of this monograph.

Note also that in (2.14) and further, the intensity of the surface load calculated per unit area in the subcritical state (in the second state) is denoted through $\tilde{P}$. Naturally, in this case, the indicated value is the same for the subcritical and perturbed (in the actual configuration) states (due to the smallness of the perturbations, condition (2.3)), but it can differ significantly from the analogous quantity calculated per unit area in the reference (the first) state, due to the application of the theory of finite subcritical deformations.

Also, in the considered theory 2.1 and below, the following designations are introduced:

u^j—the contravariant components of the displacement vector disturbance;
F^j and P^j—the contravariant components of perturbations of vectors of external mass and surface forces;
$f^j, f_1^j, f_2^j, h_1^j,$ and h_2^j—contravariant components of the perturbations of the vectors of the right-hand sides in the boundary and initial conditions;

N^j—the contravariant components of the unit vector of the outer normal to the body surface in the reference configuration (in the first state);

g^{nm}—the contravariant components of the metric tensor of the comoving coordinate system in the reference configuration (in the first state);

g_*^{nm}—the contravariant components of the metric tensor of the comoving coordinate system in the actual configuration (in the third state).

We restrict ourselves to the above information in a very brief discussion of theory 2.1 (*the theory of finite subcritical deformations*). Additional information can be obtained from the reviews [8, 9] and the references cited therein.

2.2.2 The First Variant of Theory of Small Subcritical Deformations

First of all, consider the Fundamental Statement (definition, simplification) of the theory of small deformations and the Consequences that follow (are proving) from the Fundamental Statement.

The Fundamental Statement (definition, simplification) of the theory of small deformations is that the relative elongations and shears are small values in comparison with unity and they can be neglected in comparison with unity.

The Consequences of the Fundamental Statement are as follows:

1. The components of the Green strain tensor are small values in comparison with unity, and they can be neglected in comparison with unity.
2. The change in the elongations, areas of the oriented area elements, and volumes can be ignored.
3. The change in the components of the metric tensor of the comoving coordinate system during deformation can be ignored.

The more detailed information about the nonlinear theory of small deformations is presented in the monographs [1, 2] and in a more reduced form in the reviews [8, 9]. The above transition is obtained starting from the nonlinear theory of deformation of a continuum at the finite deformations.

For the first time, these simplifications and a consistent transition from the theory of finite deformations to the theory of small deformations were proposed in the articles [49, 50] ([50] p. 344, below) of 1939.

In the monograph [51] of 1948, the discussed approach was presented consistently and in full and was also considered in the monographs [1–3].

Note that the above information refers to the construction of nonlinear mechanics of small deformations of a continuum.

When constructing theory 2.2 (*the first variant of the theory of small subcritical deformations*), which is a linearized theory, **two methods** (two approaches) are usually implemented.

In **the first method** (approach), the basic relations of theory "1" (*Theory of large (finite) subcritical deformations*) are applied in the form of relations (2.5)–(2.14) and other corresponding expressions, and **the Fundamental Statement and the Consequences from it,** stated above and corresponding to the transition from the nonlinear theory of finite deformations to the nonlinear theory of small deformations. With this method, the linearized theory 2.2 is getting from the linearized theory 2.1.

In **the second method** (approach), first, all relations of the nonlinear theory of small deformations of continuum mechanics for the corresponding model are constructed, taking into account the derivation of each of the relations (in full measure, according to the authors of each of the constructed theories), all the simplifications corresponding to the above-stated **Fundamental Statement and the Consequences from it, the** nonlinear theory of small deformations.

After obtaining all the relations of nonlinear mechanics of small deformations (it can be considered that with the involvement of considerations of a physical nature), their linearization is carried out, and thus the linearized theory 2.2 is obtained, *without involving the theory of finite (large) deformations.*

It should be noted that with the involvement of the above **two methods (approaches),** the completely coinciding results are not always obtained for all relations (within the framework of the linearized theory). An example of such a situation will be considered in the final part of this section. According to the author of this monograph, out of the two discussed methods (approaches) in terms of consistency and rigor, **the first method (approach)** is preferable.

The above information and considerations refer only to the procedure for constructing theory 2.2 (*The first variant of the theory of small subcritical deformations*).

Below, in a brief form, the main relations of theory 2.2 constructed using **the first method (approach)** are presented.

In presenting the noted relations, the symmetric stress tensor σ is already used (instead of the symmetric stress tensor S in the theory of finite stresses), which is apparently already quite generally accepted in the mechanics of deformable bodies, for example [1–3, 50–52].

So, for theory 2.2 (*The first variant of the theory of small subcritical deformations*), the basic relations are as follows:

the equations of motion

$$\nabla_i \left[\left(g_n^j + \nabla_n u_0^j \right) \sigma^{in} + \sigma_0^{in} \nabla_n u^j \right] - \rho \ddot{u}^j + \rho F^j = 0; \qquad (2.15)$$

the relations for determining the contravariant components of the non-symmetric Kirchhoff stress tensor t

$$t^{ij} = \left(g_n^i + \nabla_n u_0^j \right) \sigma^{in} + \sigma_0^{in} \nabla_n u^j; \qquad (2.16)$$

the incompressibility condition remains in the form (2.7), where the notation is introduced

$$q^{nj} = g^{nm}\left(g_m^j + \nabla_m u_0^j\right);\qquad(2.17)$$

the boundary and initial conditions remain in the forms (2.10)–(2.13); the relations for the covariant components of the Green strain tensor for the subcritical state (as well as for theory 2.1)

$$2\varepsilon_{nm}^0 = \nabla_n u_m^0 + \nabla_m u_n^0 + \nabla_n u_p^0 \nabla_m u_0^p;\qquad(2.18)$$

the relations for determining the covariant components of the strain velocity tensor (only for theory 2.2)

$$2e_{nm} = \left[\left(g_n^j + \nabla_n u_0^j\right)\nabla_m + \left(g_m^j + \nabla_m u_0^j\right)\nabla_n\right]\dot{u}_j + \left[\left(\nabla_m \dot{u}_0^j\right)\nabla_n + \left(\nabla_n \dot{u}_0^j\right)\nabla_m\right]u_j;\qquad(2.19)$$

the expressions for determining the contravariant components of the vectors of the right-hand sides of the boundary conditions in stresses (2.10) (quantities P^j) under the action on S_1 of the "following" load of intensity $\tilde{P}$, which is directed along the normal to S_1,

$$P^j = \tilde{P} N_k\left[\left(g_n^j + \nabla_n u_0^j\right)\left(g_m^\alpha + \nabla_m u_0^\alpha\right)\left(g^{kn}g^{m\beta} - g^{nm}g^{\beta k} - g^{n\beta}g^{mk}\right) + g^{k\beta}g^{j\alpha}\right]$$
$$\times \nabla_\beta u_\alpha.\qquad(2.20)$$

We restrict ourselves to the above information in a very brief discussion of theory 2.2 (*The first variant of the theory of small subcritical deformations*). Additional information can be obtained from the reviews [8, 9] and the literature cited in publications.

2.2.3 The Second Variant of Theory of Small Subcritical Deformations

The main relations of this variant of the theory (theory 2.3) follow from the previous variant of the theory (theory 2.2) with the introduction of appropriate simplifications, which were considered in 1934 in publication No. 110 from the list of references to the review [8], in the article [50] (p. 360, near the figure) of 1939 and were systematically used by several authors, as noted in the review [8].

The Fundamental Statement (simplification) of the second version of the theory of small subcritical deformations (theory 2.3) *is that the subcritical state can be determined within the framework of a geometrically linear theory. In the* mathematical aspect, this statement (simplification) is equivalent to the assumption that *the derivatives of the displacements of the subcritical state are small values in comparison with unity, and they can be neglected in comparison with unity*, which corresponds to the expressions

$$g_m^j + \nabla_m u_0^j \approx g_m^j. \tag{2.21}$$

Thus, taking into account expressions (2.21), the basic relations of *the second variant of the theory of small subcritical deformations (theory "3")* can be obtained from relations (2.15)–(2.20) of *the first version of the theory of small subcritical deformations* (theory 2.2), in the following form:

the equations of motion

$$\nabla_i\left(\sigma^{ij} + \sigma_0^{in}\nabla_n u^j\right) - \rho\ddot{u}^j + \rho F^j = 0; \tag{2.22}$$

the relations for determining the contravariant comp Kirchhoff stress tensor t

$$t^{ij} = \sigma^{ij} + \sigma_0^{in}\nabla_n u^j; \tag{2.23}$$

the incompressibility condition remains in the form (2.7), where the notation is introduced

$$q^{nj} = g^{nj}; \tag{2.24}$$

the boundary and initial conditions remain in the forms (2.10)–(2.13); the relations for the covariant components of the Green strain tensor for the subcritical state

$$2\varepsilon_{nm}^0 = \nabla_n u_m^0 + \nabla_m u_n^0; \tag{2.25}$$

the relations for determining the perturbations of the covariant components of the Green strain tensor and the strain velocity tensor

$$2\varepsilon_{nm} = \nabla_n u_m + \nabla_m u_n, \; 2e_{nm} = \nabla_n \dot{u}_m + \nabla_m \dot{u}_n; \tag{2.26}$$

the expressions for determining the contravariant components of the vectors of the right-hand sides of the boundary conditions in stresses (2.10) (quantities P^j) under the action on S_1 of the "following" load intensity $\tilde{P}$, which is directed along the normal to S_1,

$$P^j = \tilde{P}\left(N^j\nabla_\alpha u^\alpha - N^\alpha g^{j\beta}\nabla_\beta u_\alpha\right). \tag{2.27}$$

We restrict ourselves to the above information in a very brief discussion of theory 2.3 (*the second variant of the theory of small subcritical deformations*). Additional information can be obtained from the reviews [8, 9] and the publications cited in the bibliography, as well as from the monographs [1, 2] and publications cited in the bibliography for these monographs.

2.2.4 On the Linearized Theory of Stability at Small Deformations and Small Averaged Angles of Rotation

In the monograph [51], the nonlinear theory for small deformations and small averaged angles of rotation is proposed, which is obtained by introducing the simplifications in the basic relations of the theory of finite (large) deformations. From this nonlinear theory, after the linearization, the three-dimensional linearized theory of deformable bodies stability under small subcritical deformations and small subcritical averaged angles of rotation is obtained. In the monograph [51], the concept of averaged angles of rotation was introduced for the first time to characterize the rotation of the entire neighborhood of a point of the body. At that, the averaging was carried out for all material fibers passing through the point of the body.

It should be noted that the averaged angles of rotation introduced in [51] *have no direct physical meaning* since they do not characterize the change of the geometric objects. At that, the elongations and shears, which are included in the **Fundamental Statement (definition, simplification)** of theory 2.2, *have a direct physical meaning*, since they characterize the change of the concrete geometric objects (change in the length of material fibers, change in the angle between two material fibers).

This consideration is related both to the construction of the nonlinear theory of the small deformations and small averaged angles of rotation and to the following from it the three-dimensional linearized theory of stability for the small subcritical deformations and small subcritical averaged angles of rotation.

When constructing the nonlinear theory of small deformations and small averaged angles of rotation, the **Fundamental Statement (definition, simplification)** of the theory 2.2 is adopted and an additional **Fundamental Statement (definition, simplification)** of the theory 2.4 is introduced, which is reduced to the fact that

the average angles of rotation are small values in comparison with the unit and these quantities in comparison with the unit can be neglected.

Naturally, when constructing the corresponding three-dimensional linearized theory of stability for the small subcritical deformations and small subcritical angles of rotation, the above two Fundamental Statements (definitions, simplifications) are used.

Taking into account the above, it seems that it is necessary to consider the discussed three-dimensional linearized theory of stability under the small subcritical deformations and small subcritical averaged angles of rotation *logically inconsistent* since the simplifications are made in it, considering a small value in comparison with unity (averaged angles of rotation) and discarding it. At that, as noted above, the indicated quantities (averaged angles of rotation) *have no direct physical meaning* and can be considered as some mathematical expressions associated with the description of the deformation process.

A more detailed discussion of this issue (theory 2.4) is presented in the review [8] of 2001. In the subsequent years, the author of this monograph did not return to the analysis of this issue.

2.2.5   On the Theory of Incremental Deformations

In the monograph of M. A. Biot [53], published in 1965, the foundations of the theory of incremental deformations and its numerous applications in the mechanics of deformable bodies are set out.

It should be noted that the monograph [53] is, apparently, *the first monograph in the world scientific literature* on the linearized mechanics of deformable bodies, although it considers only the plane (two-dimensional) problems.

Partially, the numerous publications by M. A. Biot are given in the bibliography for the review [8] and monograph [3] of 1971, which was the first monograph on the three-dimensional linearized theory of deformable bodies stability by the author of this monograph and in which, along with the plane problems, the spatial problems of this theory were considered.

The theory of incremental deformations [53] from the point of view of the positions discussed here and the classification of this monograph are the three-dimensional linearized theory of deformable bodies stability or, in a broader sense, the three-dimensional linearized mechanics of deformable bodies.

From the point of view of the analysis of the approaches that were analyzed concerning theories 2.1–2.4, the theory of incremental deformations (theory 2.5) is the theory of small subcritical deformations, since it adopts **the Fundamental Statement (definition, simplification)** of the theory 2.2. Moreover, in the theory 2.5, the subcritical state is also determined by the geometrically linear theory, since it accepts **the Fundamental Statement (definition, simplification)**of theory 2.3.

The above simplifications and constructions in theory 2.5 are sufficiently consistent and correspond to the simplifications and constructions of theories 2.2 and 2.3.

The simplifications and transformations are introduced *additionally* in the theory of incremental deformations, associated with a certain interpretation of the angles of rotation of material fibers, to determine in which the relations are applied within the framework of the *linear* theory. In fact, within the framework of the nonlinear theory of small deformations, from which, using linearization and simplifications, the main relations of the discussed theory 2.5 are obtained, and the angles of rotation of the material fibers are determined by the completely different (compared to the linear theory) expressions.

The above considerations indicate that the discussed *theory of incremental deformations* [53] *is inconsistent*. Besides, according to the author of the present monograph, the main relations of the theory 2.3 are simpler (in the structure) in comparison with the theory 2.5. Additional information can be obtained from the review [8].

We restrict ourselves to the above information in a brief discussion of the incremental theory of deformations [53].

2.2.6   Approximate Approach in the ILIDBS

The considered approximate approach in the three-dimensional theory of deformable body stability was originally proposed by L. S. Leibenson in [54] (pp. 110–121) of 1951 and became quite popular among the Russian-speaking researchers after the publication of A. Yu. Ishlinsky [55] of 1954.

The essence of the approximate approach [54, 55] is *the application* of the **linear** *equations of motion* and in **a purely approximate introduction** (*into the boundary conditions in stresses) of quantities that are associated with the curvature of the boundary surface caused by the loss of stability.*

A sufficiently detailed presentation and analysis of the approach from [54, 55] are presented in the monographs [3] of 1971 and [1] of 1986 in Russian and [2] of 1999—in English as well as in a short form in the survey [8] of 2001.

As noted in Sect. 2.1 of this monograph, the survey [8] in Russian was included in the multivolume collective monograph "Advances in Mechanics" (in six volumes, seven books), which included **174** summary review articles by the authors from **26** countries around the world, and which is published in 2000–2009 in the journal "Prikladnaya Mekhanika" in Russian and in 2001–2010 in the journal "International Applied Mechanics" in English during the campaign dedicated to **the Beginning of the III Millennium**.

Below, in presenting and analyzing the approach [54, 55], we will follow the review [8]. Following the monograph [1] (pp. 370–376), the basic equations of the approximate approach [54, 55] (the linear equations) have the following form:

$$\nabla_i \sigma^{ij} - \rho \ddot{u}^j = 0, \tag{2.28}$$

and the boundary conditions on a part of the surface S_1 [54, 55] can be represented in the form

$$N_i \left(\sigma^{ij} + \sigma_0^{mj} g^{ni} \nabla_n u_m \right) \Big|_{S_1} = P^j. \tag{2.29}$$

A further possible simplification of the boundary conditions (2.29) following [54, 55] is also presented in the monograph [1] (p. 374).

The analysis of the approximate approach [54, 55] was carried out in the monographs [1–3] and the review [8], proceeding from the strict linearized formulations of the TLTDBS problems (theories 2.1–2.3) with the involvement of the Lagrangian coordinates.

It should be noted that the attraction of the Lagrangian coordinates is generally accepted in the mechanics of deformable bodies. Without dwelling on the presentation of the analysis from [1–3], the results of this analysis in a very brief form are only presented, which can be formulated in the form of the following two positions.

1. The approach [54, 55] is purely approximate and does not follow from the strict linearized statement (theories 2.1–2.3) for any logically substantiated system of simplifications.
2. The assertion of some authors that the approach [54, 55] follows from the TLTDBS as a result of neglecting terms of the order of rotation angles in the equations of motion and saving the terms of the same order in the boundary conditions in stresses is incorrect.

The above results [1–3] refer only to the general analysis of the approximate approach [54, 55]. Undoubtedly, more impressive is the information on the accuracy of the results of solving the specific problems obtained within the framework of the approach [54, 55] (as well as within the framework of any purely approximate approach in mechanics).

To analyze the accuracy of the results obtained using the approximate approach [54, 55], the simplest problem is usually considered—the plane problem of the stability of a hinged-supported plate of the constant thickness under axial compression. In this case, it is convenient to introduce the notation:

P_{Eul}—the Euler critical force (the value of the critical load, calculated using the hypothesis of plane sections or the Kirchhoff–Love hypothesis);
P_{3D}—the value of the critical load, calculated in the framework of the three-dimensional theory of stability.

Based on the well-known and generally accepted physical considerations (with a decrease in the number of degrees of freedom, the values of the critical values of the parameters of the system should increase or, in extreme cases, not decrease), in the example under consideration, *the condition must be fulfilled*

$$P_{3D} \leq P_{Eul} \tag{2.30}$$

and the condition *must not be fulfilled*

$$P_{3D} \geq P_{Eul}. \tag{2.31}$$

In the publications in periodicals [27] of 1967, [17] of 1968, and [19] of 1969, as well as in the monographs [1–3], it is rigorously proved analytically that even for the simplest TLTDBS (theory 2.3) *the condition* (2.30) *is satisfied*. As follows from the solution [55] of the same problem within the framework of the approximate approach [54, 55], for the approximate approach *the condition* (2.31) *is satisfied*, which should not be fulfilled due to well-known physical considerations.

Fig. 2.1

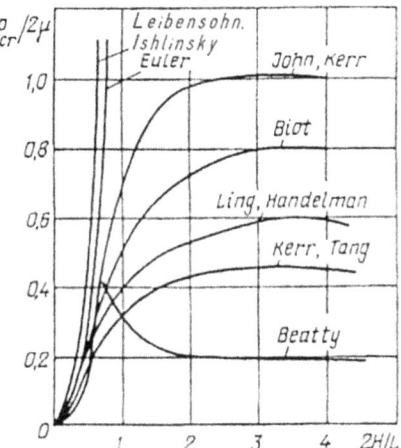

Thus, *the approximate approach* [54, 55] *leads* not only to the quantitative errors but *also to the qualitatively incorrect results*. It can be considered that in the Russian-language literature, the above situation is known and proven in an analytical form since 1967–1969.

In the English-language scientific literature, the above situation, apparently, became known after the publication of the article [56] in 1969, which presents the results of a numerical study of this problem for an incompressible material and compares the results obtained by various authors for this problem. Following [56] in Fig. 2.1 for the above problem, the dependence of the dimensionless value of the critical load on the dimensionless thin-wallness parameter is presented. At that, in Fig. 2.1 near each curve representing the above dependence, the names of the authors are indicated, based on whose solution this dependence was obtained.

Note that in Fig. 2.1 the curve associated with Euler's name was obtained based on a two-dimensional (in the considered problem, one-dimensional) applied theory, built using the Kirchhoff–Love hypothesis. All other curves shown in Fig. 2.1 [56] were obtained based on various formulations of problems of the three-dimensional linearized theory of deformable bodies stability (TLTDBS).

It follows from the results presented in Fig. 2.1 [56] that for all the indicated solutions based on TLTDBS, all curves lie *to the right* of the curve constructed based on the Kirchhoff–Love hypothesis. Therefore, for all the above solutions constructed based on the TLTDBS, the following condition is satisfied:

$$P_{3D} < P_{Eul},   \tag{2.32}$$

which is stronger than the condition (2.30).

Also, it follows from the results presented in Fig. 2.1 that only for the solution constructed based on the approximate approach [54, 55], the curve lies *to the left* of the curve constructed based on the theory using the Kirchhoff–Love hypothesis.

Therefore, the condition is fulfilled

$$P_{\text{Трex}} > P_{\text{эл}} \tag{2.33}$$

which is stronger than the condition (2.31). In this case, as noted above, condition (2.31) *should not be satisfied* due to the well-known and generally accepted considerations of a physical nature.

Thus, the results of [56], obtained based on a numerical study, and the results of [17, 19, 27] and other publications, obtained based on studies in an analytical form, qualitatively coincide and indicate that the approximate approach [54, 55] *can also lead to qualitatively incorrect results.*

Considering the above, the conclusion can be done

that the approach [54, 55] is purely approximate and does not follow from the TLTDBS. Moreover, the approach [54, 55] can lead not only to quantitative errors but also to qualitatively incorrect results.

Basing on the above-formulated conclusion, all the results obtained using the approach [54, 55] require additional research to determine the reliability of these results and their correspondence to the considered phenomena of mechanics and related fields of natural science. The brief information on the application of the approach [54, 55] to the study of problems in mechanics and related fields of natural science is presented in the review [8].

2.2.7 Notes

Thus, this section presents brief but informative information about the six three-dimensional theories of deformable body stability, which, in a certain sense, are basic and most frequently used at present. Below, the final information, which is presented in the form of Notes, is considered.

Note 2.3 In this section, the main relations are given in a brief form relative to six theories (theories 2.1–2.6), which are formulated concerning stress and displacement disturbances. If to add the linearized constitutive equations (in particular, the linearized elasticity relations) to the indicated above relations, then the closed systems of equations are obtained, which are formulated through the perturbations, for the considered models of the mechanics of deformable bodies.

Note 2.4 The above-mentioned closed systems of equations include the quantities of the stresses of the subcritical state, which are marked with the index "zero" and are unknown (they must be determined as a result of solving the stability problems) in the TLTDBS. If to consider in the closed systems of equations the stresses marked with the index "zero" as the known values, then the basic relations of the linearized mechanics of deformable bodies for the materials with initial (residual) stresses are obtained.

Note 2.5 The considered *six* theories can be characterized by the rigor of their construction (by the sequence and accuracy).

Theory 2.1 (*the theory of large (finite) subcritical deformations*) *is the most general, consistent, and exact theory.*

Theory 2.2 (*the first variant of the theory of small subcritical deformations*) is *consistent* (is obtained from theory 2.1 with generally accepted assumptions about the smallness of elongations and shears) and *not quite accurate* (these assumptions were introduced) theory.

Theory 2.3 (*the second variant of the theory of small subcritical deformations*) is *consistent* (in addition to the simplifications of the theory 2.2, the generally accepted assumption was introduced about the determination of the subcritical state according to the geometrically linear theory) and *not quite accurate* (the indicated simplifications were introduced) *theory.*

Theory 2.4 (*the linearized theory of stability at small deformations and small averaged angles of rotation*) is *inconsistent* (an assumption is made about the smallness of the quantity, which **has no direct physical meaning**—the smallness of the averaged angles of rotation) and *inaccurate* (the indicated simplification is introduced) *theory.*

Theory 2.5 (*theory of incremental deformations*) is *linearized, inconsistent* (in addition to the generally accepted assumptions of theories 2.2 and 2.3, the simplifications are introduced due to inconsistent interpretation of the angles of rotation) and *not accurate* (the indicated simplifications are introduced) *theory.*

Theory 2.6 (*the approximate approach in the three-dimensional theory of deformable body stability*) *is inconsistent* (in no way following from the rigorous linearized theories) and *inaccurate* (leads to the qualitatively different results and quantitative errors) *theory.* This is a purely approximate approach leading to the above situation.

Note 2.6 In this subsection, when describing the procedure for constructing theory 2.2 (*the first variant of the theory of small subcritical deformations*), in the introductory part of this presentation, two existing methods of constructing the basic relations of the theory under discussion were indicated.

In the *first* method, all the main relations of the discussed theory 2.2 are obtained from the corresponding relations of the theory 2.1 (*theory of large (finite) subcritical deformations*) by introducing the basic simplifications of the nonlinear theory of small deformations into the indicated relations of theory 2.1. With this method, all the basic relations of the theory 2.2 are obtained, which are given in this monograph.

In the *second* method, first, the basic relations of the nonlinear theory of small deformations are obtained, proceeding from the Fundamental Statement (simplification) of the nonlinear theory of small deformations and related physical considerations. Then, the obtained relations of the nonlinear theory of small deformations are linearized, following the procedure outlined in the introductory part of Sect. 2.1.

As already noted, the results obtained by the first and second methods do not always coincide.

In connection with this note, below an example of such a non-coincidence concerning expression (2.14) for determining the contravariant components of the vectors of the right-hand sides of the boundary conditions in stresses (2.10) (quantities P^j) when a "following" load intensity $\tilde{P}$ is applied to S_1 and is directed along the normal S_1 will be considered.

Thus, when implementing the *first* method to determine these quantities, the expression (2.20) was obtained concerning theory "2" (*the first variant of the theory of small subcritical deformations*) and expression (2.27) concerning theory 2.3 (*the second variant of the theory of small subcritical deformations*).

It should be noted that expressions (2.20) and (2.27) do not coincide with each other, which was to be expected due to the introduction into theory 2.3 of the **Fundamental Statement** of this theory. When implementing the *second* method, the following identical expression was obtained for theory 2.2 and theory 2.3

$$P^j = \tilde{P} N^\beta \nabla_\beta u^j; \qquad (2.34)$$

Moreover, the expression (2.34) was obtained and used in the publications of almost all researchers involved in the nonlinear theory of small deformations.

It is quite obvious that the expressions (2.34) do not coincide with any of the expressions (2.20) and (2.27). Apparently, it can be considered that the noted non-coincidence of the results is a consequence of the following situation existing in the procedure for constructing the theories (carrying out transformations in constructing the theories) relative to the theory 2.2 (a *first variant of the theory of small subcritical deformations*).

The linearization of the basic relations of the theory of large (finite) deformations + the introduction into the indicated relations of simplifications corresponding to the **Fundamental Statement** (definition) of the theory of small deformations $\boxed{\equiv}$ the construction of the basic relations of the nonlinear theory of small deformations based on the **Fundamental Statement** (definition) of the small deformations and accompanying considerations of a physical nature + linearization of the obtained approach of the main relations.

Since the TLTDBS, in essence, operates with the perturbations of the small quantities, the given example testifies to the necessary rigor and caution. The sufficiently exact expressions (2.20) and (2.27), as will be noted in the subsequent presentation, make it possible to consistently consider some general questions, in contrast to expressions (2.34).

We restrict ourselves to the above information when analyzing the construction and classification of the main theories and approaches in the TLTDBS. Additional information can be obtained from the monographs [1, 2] and reviews [8, 9], as well as from other publications presented in the bibliography to the present monograph.

2.3 On Stability Criteria in the TLTDBS

First of all, it is advisable to note that the stability criteria in TLTDBS are an application, development, and generalization of the corresponding stability criteria that
exist in the two-dimensional and one-dimensional applied theories of stability of
thin-walled deformable structural elements (rods, plates, and shells), constructed
using the hypotheses of plane sections, Kirchhoff–Love, Timoshenko, etc.

The noted situation makes it possible to consider in a rather brief form the various
criteria for the stability of deformable bodies relative to the various models, noting
the specificity of their application to the TLTDBS. Additional information can be
obtained from the monographs [1, 2]. This information is presented in a fairly
consistent form in the review [8].

Preliminarily, in a general form for all models of deformable bodies, let us present
some information related to the classification of problem statements following the
structure of perturbations of the components of the mass forces F^j in (2.5) and the
surface forces P^j in (2.10). In a fairly general case, the corresponding contravariant
components of the vectors of these forces can be represented in the form

$$F^j = F_{(1)}^{j\alpha} u_\alpha + F_{(2)}^{j\alpha} \dot{u}_\alpha, \ P^j = P_{(1)}^{j\alpha} u_\alpha + P_{(2)}^{j\alpha} \dot{u}_\alpha, \qquad (2.35)$$

where $F_{(k)}^{j\alpha}$ and $P_{(k)}^{j\alpha}$ denote the differential operators in the space coordinates θ^n, the
coefficients of which depend on θ^m and time τ.

The *dynamical* problems of stability theory include the problems for which at
least one of the following conditions is satisfied

$$F_{(2)}^{j\alpha} \neq 0, \ P_{(2)}^{j\alpha} \neq 0. \qquad (2.36)$$

The *statical* problems of the theory of stability include the problems for which
the following conditions are satisfied

$$F_{(2)}^{j\alpha} = 0, \ P_{(2)}^{j\alpha} = 0. \qquad (2.37)$$

As an example of the statical problems for which a more particular form of the
condition is satisfied in comparison with conditions (2.37), the statical problems can
be indicated for which the conditions are fulfilled

$$F_{(2)}^{j\alpha} = 0, \ P_{(2)}^{j\alpha} = 0, \ F_{(1)}^{j\alpha} = 0, \ P_{(1)}^{j\alpha} \neq 0. \qquad (2.38)$$

Such statical problems include the problems under the action of a "following"
load in the form of the uniform pressure. In this case, the differential operators
are determined from the expressions (2.14) for theory 2.1 (*theory of large (finite)
subcritical deformations*), from the expressions (2.20) for the theory 2.2 (*the first
variant of the theory of small subcritical deformations*), and expression (2.27) for
the theory 2.3 (*the second variant of the theory of small subcritical deformations*).

Below, following the presentation of the article [8], the information on stability criteria will be considered separately for the main models of the mechanics of deformable bodies.

2.3.1 Elastic Bodies

Consider the stability of the equilibrium state of an elastic body. In this case, it is assumed that *after the transition to a neighboring equilibrium form of equilibrium (the perturbed state) with a decrease in the loading parameter (with one-parameter loading), the elastic body returns to its original (unperturbed) state.*

This feature is a specific property of the theory of stability of elastic systems, which cannot be transferred into the inelastic bodies.

Usually, when studying the stability of the equilibrium state of elastic bodies, the dynamical or statical (bifurcation) criteria are used, corresponding to the dynamical and statical (Euler's method) research methods.

When applying *the dynamical criterion of stability*, the state of equilibrium is stable if the disturbances attenuate with time (at $\tau \to \infty$). Conditionally, in this case, the equilibrium state is also considered to be *stable* if for the perturbations only periodic by τ solutions are obtained.

When applying the dynamical criterion of stability, the state of equilibrium is considered *unstable* if the disturbances increase indefinitely at $\tau \to \infty$.

When applying the *dynamical research method*, the equations of motion (2.5), (2.15), (2.22), (2.28) and the corresponding boundary conditions are used. At that, the corresponding ones are understood as the equations of motion and boundary conditions written in a different form. In all quantities of the disturbances that are included in the above equations of motion and boundary conditions, a factor $\exp i\Omega\tau$ is allocated.

As a result, from the above equations and boundary conditions, an eigenvalue problem concerning the parameter Ω is obtained.

Denote the eigenvalues of the above-formulated problem by $\Omega_k (k = 1, 2, ..., \infty)$. Considering the above, the equilibrium state is considered as stable (with the adopted dynamical criterion) if for all eigenvalues Ω_k the condition is fulfilled

$$Im\Omega_k \geq 0; k = 1, 2, ..., \infty. \tag{2.39}$$

The boundary of the stability region in the space of loading parameters is determined in this case as a result of minimization by the various combinations of loading parameters, provided that these loading parameters correspond to the conditions for the eigenvalues in the following form

$$Im\Omega_k = 0; k = 1, 2, ..., \infty. \tag{2.40}$$

When applying *the statical research method* (Euler's method) *and the statical criterion* for the stability of the equilibrium state (in this case, it should be accepted $\Omega \equiv 0$), the eigenvalue problem is obtained concerning the loading parameters that are included in the basic relations indicated above. The boundary of the stability area in the space of loading parameters (the critical values of loading parameters) is determined as a result of minimization over the various combinations of loading parameters corresponding to the obtained eigenvalues.

It should be noted that the dynamical research method is applicable to the dynamical and statical problems. The statical research method is applicable only to the study of the statical problems and even then not to all the statical problems. Since the use of the statical method (in its implementation) is incomparably simpler in comparison with the use of the dynamical method, the problem always arises of formulating the conditions for the applicability of the statical method (Euler's method) for the statical problems. The sufficient conditions for the applicability of the Euler's method in the general form for the elastic and plastic bodies with a fairly general form of the constitutive equations will also be given in this chapter with the further presentation of the basics of TLTDBS.

We restrict ourselves to the above information in a very brief discussion of the generally accepted criteria for the stability of the equilibrium state of elastic bodies.

The above presentation of the issue under consideration coincides with the presentation of this issue in [8] since with time the publication of paper [8] in 2001, there have been no changes in the generally accepted views on the issue under discussion.

Also, it should be noted that relative to several problems in the mechanics of elastic bodies (e.g., to the theory of impact on the elastic bodies), specific more approximate criteria for the stability of elastic bodies are also used.

2.3.2 Plastic Bodies

When even sufficiently small plastic deformations appear in the process of deformation, *due to their irreversibility* (when trying to apply the concept of the theory of elastic stability) an *unstable* state of equilibrium is obtained, since a system that has received the plastic deformations, with a decrease in the loading parameter, never returns to its original position. In this regard, in the mechanics of inelastic bodies, the question of returning the system to its original state after the loss of stability is never considered.

Taking into account the above considerations, in the mechanics of plastic systems, the question of the stability of the equilibrium state is considered in terms of the behavior of perturbations that are described by the linearized theory. In this case, the question of returning the system to its original position with a decrease in the load, as already noted, is never envisaged.

It should be emphasized that when studying the stability of the equilibrium state at a point O_0 on the diagram $\sigma \sim \varepsilon$ (Fig. 2.2) with the use of a linearized system of equations, with a *rigorous* approach, it is necessary to consider for disturbances the

Fig. 2.2

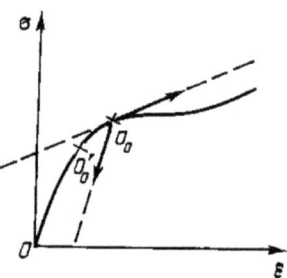

presence of active loading and unloading processes. Thus, with a rigorous consideration, a linearized system of equations for the perturbations is obtained with a varying
unloading zone, which, in essence, excludes the possibility of obtaining the solutions
of specific problems in the framework of the TLTDBS.

A similar complexity arises in the theory of stability of the thin-walled systems
based on the two-dimensional applied theories of thin-walled systems. In this case,
in the theory of stability of plastic systems for several problems, it is possible to
overcome the above-formulated complexity by using the well-known hypotheses on
the distribution of the stress–strain state over the thickness of thin-walled systems.

To overcome the above difficulties in the study of the TLTDBS problems for the
plastic bodies, **two approaches** have been formed:

The **first** approach is based on the study of the stability of the equilibrium state
using the *assumption*—the generalized concept of continuing loading.
The **second** approach is based on the study of the stability of the deformation
process using the *assumption*—the tendency of the system to the loss of stability
with an equally active bifurcation.

Without considering or analyzing at the beginning each of the two indicated
approaches, let us formulate the **main conclusion** that follows from the noted
analysis.

The main conclusion: With two different approaches (the *first* is the study of the
stability of the equilibrium state, and the *second* is the study of the stability of the
deformation process) and *under different assumptions for each of the approaches*

for the *first*—the generalized concept of continuing loading without clarifying the
type of additional load that compensates for the additional unloading zones with
loss of stability;
for the *second*—the tendency to the loss of stability with an equally active
bifurcation

we arrive at the same linearized problem of the type of Sect. 2.2 with the known unloading
zones that are determined in the analysis of the subcritical state.

Thus, taking into account the above-stated **main conclusion**, in the study of three-
dimensional linearized problems of the theory of stability for the plastic bodies, we

arrive following the above two approaches to the linearized problems, which in the structure correspond to the problems of Sect. 2.2.

At that, it is necessary to introduce in the corresponding equations and boundary conditions in Sect. 2.2 the relations of the linearized constitutive equations of the corresponding theory of plasticity.

Taking into account the criteria and approaches for the plastic bodies (based on the above-linearized systems of equations), when solving the specific problems, the dynamical and statical research methods are used in the same form as for the elastic bodies. Therefore, we can assume that the following statement is proved.

Statement. With the considered approaches, the study of problems of the TLTDBS can be carried out uniformly for elastic and plastic bodies. In this case, for the plastic bodies, the subcritical state should be determined taking into account the unloading zones changing during the loading.

Consider briefly the *first* approach, which is based on the generalized concept of continuing loading. For the first time, the concept of continuous loading as applied to the theory of stability of thin-walled plastic systems based on two-dimensional applied theories was proposed by F. Shenley (an article of 1951 with No. 104 in the list of references to the review [8]). A generalization of the above concept to the TLTDBS, called the generalized concept of continuous loading, was considered in the publications of the author of this monograph (e.g., an article of 1968 (No. 25 in the list of references to the review [8]), an article of 1969 (No. 27 in the list of references to the review [8]), and an article [23] of 1973, as well as the number of works by other authors). A comparatively detailed discussion of the generalized concept of continuing loading is presented in the monographs of the author of this monograph [1, 2, 5].

It is assumed in *the generalized concept* of continuing loading that the stability of the equilibrium state of the elastic–plastic bodies (within the framework of the three-dimensional theories) can be judged by the behavior of perturbations, which are described by the linearized equations of Sect. 2.2 with the involvement of the linearized constitutive equations of the corresponding theory of plasticity.

In this case, the unloading zones formed in the subcritical state *do not change*.

Also, it is believed that the process of the loss of stability begins *somewhat earlier* than the critical state is reached (a state of neutral equilibrium). In this regard, the process of the loss of stability occurs with *a slight but continuing loading*, and the unloading during the loss of stability does not occur. With this approach, in fact, instead of the situation shown in Fig. 2.2, the situation is studied presented in Fig. 2.3.

We restrict ourselves to the above information in a very brief discussion of the first approach. Additional information can be obtained from the monographs [1, 2, 5].

The *second* approach is also described in some detail in the monographs [1, 2, 5]. In this regard, this monograph not provides information related to the second approach.

Moreover, as noted above in the **main conclusion**, when applying the first and second approaches, we arrive at the same linearized problems.

Fig. 2.3

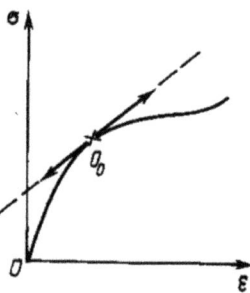

In the final part of subsection 2.3.2, following the review [8], two Notes are presented that contribute to a clearer understanding of the discussed issues concerning the TLTDBS for the plastic bodies.

Note 2.7 It is necessary to note a significant fundamental difference (both in the severity and consistency of the analysis and in the correspondence to the physical meaning of the phenomena under consideration) of the following two situations.

The *first situation* is associated with the use of a dynamic method for studying the stability of the plastic bodies based on the analysis of a linearized system obtained using the above two approaches in the TLTDBS. Naturally, in this case, the change in the unloading zones is not taken into account in the study of oscillatory movements for the disturbances corresponding to the application of the dynamic method of analysis. In this case, the values of the critical loads (finite values) are determined as a result of the analysis of the behavior of the small disturbances. At that, the logical and physically based approaches and concepts are applying.

The second situation arises in the study of dynamic problems for the plastic bodies (a determination of the stress–strain state), for example, in the study of vibrations of the plastic bodies. Several authors in this situation, referring to the use of the theory of small deformations, also do not take into account the change in the unloading zones during oscillations, i.e., in fact, instead of the phenomenon in Fig. 2.2, consider the phenomenon in Fig. 2.3. It should be emphasized that in this case the not small quantities are considered (the perturbations—as in the linearized theory of stability), but the quantities (stresses and displacements during oscillations), the values of which are determined by the level of external load. At that, with the indicated replacement of phenomena (Fig. 2.2 on Fig. 2.3), the behavior of the plastic bodies on one half-period is correctly described and incorrectly described—on the same second half-period of oscillations.

Thus, neglect of changes in the unloading zones *in the second situation* leads to results that *do not correspond* to the physical phenomenon under consideration.

Note 2.8 The concept of continuing loading (both within the framework of applied two-dimensional theories of stability of thin-walled plastic structural elements and its generalized version as applied to the TLTDBS) can be considered an approximate method for studying the stability of the equilibrium state of plastic bodies if its following interpretation is accepted.

Assume that the point O_0 in Fig. 2.2 is the bifurcation point for a specific problem for a plastic body. For an approximate determination (calculation) of the point O_0, consider on the diagram $\sigma \sim \varepsilon$ in Fig. 2.2 the point O'_0 that is below the point O_0. When loading in the neighborhood of the point O'_0, the loading will continue and unloading will not occur. In this case, it is assumed that the continuing loading in the vicinity of the point O'_0 will compensate for the appearance of additional unloading zones if in the vicinity of the point O'_0 a linearized problem like Fig. 2.3 is considered.

By approaching a point O'_0 as close as arbitrarily to a point O_0, the above considerations can be kept. As a result, the position of the point O_0 on the diagram of Fig. 2.2 can be determined as accurately as desired.

We restrict ourselves to the above information in a very brief consideration of the stability criteria for the equilibrium state of plastic bodies. Additional information on the issue under discussion is presented in the monographs [1, 2, 5].

2.3.3   Bodies with Rheological Properties

First of all, it is advisable to note that in the case of elastic (subsection 2.3.1) and plastic (subsection 2.3.2) bodies, the study of the problem of stability of the equilibrium state is based on the analysis of the behavior of small perturbations, which are described by the linearized three-dimensional equations, and as a stability criterion with the dynamic method of research, the attenuation of disturbances in time is assumed. Naturally, this approach extends to all models of deformable bodies, in which the elastic and plastic deformations are taken into account. The above approach for the elastic and plastic bodies corresponds to the generally accepted traditional approach in mechanics and allows to carry out the analysis of the corresponding various problems with a uniform, general, and traditional methodology.

In the case of bodies with the rheological properties, the developed approaches to the study of the problem of stability do not have the aforementioned generality and uniformity characteristic of the elastic and plastic bodies relative to the study of various problems, which will follow from the further presentation in this subsection.

Consider in a very brief form some approaches and stability criteria that are the characteristics of the bodies with the rheological properties (viscoelastic, hereditary elastic, viscoelastoplastic, and other bodies). Somewhat more detailed information on this issue, mainly concerning the TLTDBS, is given in the monographs [1, 2, 5].

Usually, one proceeds from the following basic assumption:

the stability can be judged by the behavior of disturbances within the framework of the linearized problem.

In this case, in the case of bodies with the plastic properties, the simplifications such as the concept of continuing loading are additionally adopted.

In what follows, the following general criterion is applied:

a state of equilibrium or motion is considered stable if the disturbances attenuate in time, and unstable if the disturbances increase in time.

The above approach and general criterion is only *a general scheme*. Further progress in this scientific direction, at least in the opinion of the author of this monograph, has not yet been achieved due to the following difficulties.

When applying the above general criterion, the very complex linearized systems of equations are obtained, since the coefficients of these equations also depend on time (in addition to the dependence on coordinates in the case of inhomogeneous subcritical states, which is observed for the elastic and plastic bodies). In connection with the noted difficulties, at present, in the study of the stability of bodies with the rheological properties, the various approximate approaches, methods, and stability criteria are mainly used.

The analysis of a number of the mentioned approximate approaches, research methods, and stability criteria (both within the framework of the applied two-dimensional theories of thin-walled elements and the framework of the TLTDBS) was considered in the monographs [1, 2, 5] and partly in the reviews indicated in [8]. The most systematized results of this analysis are presented in the survey [8], which we will follow in the further presentation in this section.

Note that below the information is given on the approximate approaches and stability criteria for the bodies with the rheological properties, the governing equations of which necessarily include the elastic deformations. Thus, information about the approximate approaches and stability criteria is excluded that is characteristic of the non-Newtonian fluids and related media.

Taking into account the above information and considerations, below, following the review [8], the five approximate approaches and the corresponding five approximate criteria for the stability of the equilibrium state for the bodies with rheological properties are briefly considered.

Approach 1. In the case of slow changes in the coefficients of the linearized problem for the indicated coefficients, as if the "own" time τ_1 is introduced (in fact, the value of the coefficients at τ_1 is fixed) and time τ is introduced for the disturbances. It is assumed at that "time τ_1 and time τ do not depend on each other." Apparently, it can be assumed that such an approach can lead to the correct results only when studying the behavior of perturbations in a small time interval, during which the coefficients of the system change insignificantly. This approach is still associated, as it were, with "freezing the time."

The above approach in the TLTDBS was applied in the monograph [3] of 1971 and the publications indicated in them periodicals, for example [21]. In connection with the slow change in the coefficients of linearized systems assumed in this case, one can also additionally introduce, following the monographs [1, 2, 5], the condition

$$\dot{u}_m^0 \approx 0. \tag{2.41}$$

Approach 2. A fairly widespread approach is obtained when the question of the stability of bodies with the rheological properties is investigated based on the analysis of limiting systems. These systems are understood as the linearized systems of equations and the corresponding boundary conditions (both within the framework of the applied two-dimensional theories of thin-walled structural elements and the framework of the TLTDBS) when $\tau \to \infty$ is accepted in the coefficients.

As an example, the article (No. 138 in the bibliography for the review [8]) and several other publications can be pointed. In this case, the asymptotic stability is investigated.

A partial survey of the results obtained using this approximate approach is given in the review articles (No. 73 and No. 74 in the list of references to the review [8]).

It is worth noting that the approximate *Approach 2* follows from the approximate *Approach 1* if the latter $\tau_1 \to \infty$ is adopted.

Approach 3. In the case of slow motions, in the study of the stability of the equilibrium state of bodies with the rheological properties, the inertial forces can be discarded.

The considered approach for the applied two-dimensional theories of thin-walled structural elements under creep was used in the monograph of 1966 (No. 102 in the bibliography for the review [8]) and was called there the quasi-static approach. This approach was applied in the TLTDBS in the monograph [3] of 1971 and the publications indicated in it in periodicals, for example [21], at the homogeneous subcritical states.

It is obvious that the discussed quasi-static approach can lead to the correct results only for a certain type of external loads. In the case of the elastic and plastic bodies, this approach is called the static method for studying the stability of the equilibrium state.

Approach 4—An approximate approach and approximate criterion of stability for the bodies and thin-walled structural elements with rheological properties. In an article by N. Hoff [57] of 1954, such an approach and criterion, called the approximate Hoff stability criterion, was proposed for studying the *creep* stability of the thin-walled systems (rods, plates, and shells) in the framework of one-dimensional and two-dimensional applied theories constructed by using appropriate hypotheses.

Following this approach [57], a small deviation $f(x)$ of the initially rectilinear shape of thin-walled systems is introduced, and subsequently (based on the solution of the corresponding problem *under creep*), the time variation $f(x, \tau)$ is analyzed under a constant or increasing external load. In this case, the critical time τ_{cr} is determined from the following condition:

$$|f(x, \tau)| \to \infty \text{at} \tau \to \tau_{cr}. \tag{2.42}$$

In the case of the layered and fibrous composite materials, to study the stability problems under *viscoelastic* deformation, it is necessary to introduce the small deviations of layers and fibers from a rectilinear shape (such as the quantity $f(x)$ in the case of [57] of the thin-walled structural elements) into the internal structure of the composite material. The above procedure for introducing the small deviations into the structure of composite material is carried out using the corresponding methods of mechanics of the curved composites, described in the monograph [58].

This approximate criterion for the stability of the layered and fibrous composite materials under the *viscoelastic* deformation was considered in the publications in periodicals [59–61] in 1997–1999. And in several other publications, the analysis of which is presented in the article [62] of 2007. In this approach, the critical time τ_{cr} is determined from conditions of the type (2.42), while the values σ_T and ε_T are also calculated from the corresponding expressions.

Approach 5—The approximate approach and approximate criterion of stability for the bodies and structural elements with the rheological properties. In the article by J. Gerard and A. Hilbert of 1959 (No. 77 in the list of references to the review [8]), an approximate approach and approximate criterion, called the critical deformation method, was proposed for studying the *creep* stability of thin-walled systems (plates and shells) within the framework of two-dimensional applied theories, constructed using the corresponding hypotheses.

Following the critical deformation method or critical deformation criterion, it is assumed that the critical time and critical forces under creep are determined from the equality of the creep deformation and the critical deformation calculated under the assumption that a thin-walled structural element operates in an elastic or elastic–plastic region. Thus, the critical time is determined as the time required to achieve the critical deformation value for an elastic or elastic–plastic body at a given strain load.

For a complex stress–strain state, the critical deformation is calculated from the value of the shear intensity (from the invariant quantity).

In the monograph [63] (pp. 604–614) and in the two-volume monograph [64] (vol. 2, pp. 636–648), as well as in publications in periodicals, for example [65], the application of the criterion of critical deformation to the construction of the theory of stability (TLTDBS) of the simplest body with the rheological properties (orthotropic linear hereditary elastic body of not aging) is considered. These results formed the basis of the continuum theory of the long-term fracture of composite materials under compression.

Note 2.9 It follows from the information given in subsection 2.3.3 above that for a body with the rheological properties, in contrast to the elastic and plastic bodies, at present, the general method and a corresponding clearly formulated stability criterion that would allow one to study various problems are absent. For the elastic and plastic bodies, such a fairly general method is the dynamic method. Besides, the above-considered approximate stability criteria for the bodies with rheological properties do not always correspond to the generally accepted traditional approach in mechanics, when the study of the stability of the equilibrium state is based on the analysis of the behavior of small perturbations described by the linearized equations, and the attenuation of perturbations in time with dynamic research method is adopted.

We restrict ourselves to the above information in a very brief discussion of the problems for the bodies with the rheological properties.

In conclusion, it should only be noted that in the theory of stability of materials (bodies, media) and structural elements with the rheological properties, other approaches and criteria are also used. Above in subsection 2.3.3, only the main, from the point of view of the author of this monograph, approaches and criteria are used most often.

2.4 General Questions of the TLTDBS

In this section, the information related to the following general issues of the TLTDBS is presented in a very brief form:

The general formulation of TLTDBS problems for various models of deformable bodies;
The sufficient conditions for the applicability of the static research method (Euler's method);
The examples of different types of variational principles of TLTDBS.

These general issues are considered quite consistently (with a possible degree of generality for the various models of the deformable bodies) in the monographs [5] of 1977, [1] of 1986, and [2] of 1999, as well as in a number of the review articles, which are included in the list of references to this monograph. In the most concise form, these issues are given in the review [8], which will be used below in this section.

The general issues of TLTDBS will be considered only for three theories, which, following the terminology of Sect. 2.2, are named:

Theory 2.1 (the theory of large (finite) subcritical deformations);
Theory 2.2 (the first variant of the theory of small subcritical deformations);
Theory 2.3 (the second variant of the theory of small subcritical deformations).

The consideration is carried out in a common general form for the above three theories, which are consistent following Note 2.5.

2.4.1 General Formulation of the TLTDBS Problems for Various Models of Deformable Bodies

The discussed results are presented separately for the compressible bodies, introducing a tensor of the fourth rank ω, and separately for the incompressible bodies, introducing a tensor of the fourth rank κ, and also some symmetry properties of the components of the introduced tensors ω and κ are separately considered.

These results are obtained by introducing the linearized constitutive equations for the various models of deformable bodies into the corresponding expressions (2.5)–(2.14) for theory 2.1, (2.15)–(2.20) for theory 2.2, and (2.22)–(2.27) for theory 2.3. Besides, the expressions (2.35) are used to determine, in a fairly general case of the covariant components of the disturbance vectors of the mass and surface forces, and all quantities of disturbances, following the dynamic research method (Sect. 2.3), the factor $(\exp i\Omega\tau)$ is selected.

Compressible bodies. In this case, the linearized equations of motion in the displacements

$$\nabla_i\left(\omega^{ij\alpha\beta}\nabla_\beta u_\alpha\right) + \rho\Omega^2 u^j + \rho\left(F_{(1)}^{j\alpha} + i\Omega F_{(2)}^{j\alpha}\right)u_\alpha = 0 \qquad (2.43)$$

and the linearized boundary conditions in stresses on a part of the surface S_1

$$\left(N_i \omega^{ij\alpha\beta} \nabla_\beta u_\alpha\right)\Big|_{S_1} = \left[\left(P^{jk}_{(1)} + i\Omega P^{jk}_{(2)}\right)u_k\right]\Big|_{S_1}. \tag{2.44}$$

are obtained.

The boundary conditions in displacements on a part of the surface S_2 remain in the form (2.11). The contravariant components of the non-symmetric Kirchhoff stress tensor t are then determined by the following relations

$$t^{ij} = \omega^{ij\alpha\beta} \nabla_\beta u_\alpha. \tag{2.45}$$

Incompressible bodies. In this case, the linearized equations of motion in the displacements

$$\nabla_i \left(\kappa^{ij\alpha\beta} \nabla_\beta u_\alpha + q^{ij} p\right) + \rho \Omega^2 u^j + \rho\left(F^{j\alpha}_{(1)} + i\Omega F^{j\alpha}_{(2)}\right)u_\alpha = 0 \tag{2.46}$$

and the linearized boundary conditions in stresses on a part of the surface S_1

$$\left[N_i\left(\kappa^{ij\alpha\beta} \nabla_\beta u_\alpha + q^{ij} p\right)\right]\Big|_{S_1} = \left[\left(P^{jk}_{(1)} + i\Omega P^{jk}_{(2)}\right)u_k\right]\Big|_{S_1}. \tag{2.47}$$

are obtained.

The boundary conditions in displacements on a part of the surface S_2 remain in the form (2.11), and the incompressibility condition remains in the form (2.7). The contravariant components of the non-symmetric Kirchhoff stress tensor t are then determined by the following relations:

$$t^{ij} = \kappa^{ij\alpha\beta} \nabla_\beta u_\alpha + q^{ij} p. \tag{2.48}$$

It should be noted that the above relations (2.43)–(2.45) for a compressible body were formulated concerning three components of the displacement perturbation vector, and relations (2.46)–(2.48) for an incompressible body were formulated concerning three components of the displacement perturbation vector and a scalar value p. For an incompressible body, the quantities q^{ij} are determined by expressions (2.8) for theory 2.1, (2.17) for theory 2.2, and (2.24) for theory 2.3.

In the monographs [1, 2, 5], the relations are given for calculating the constituent tensors ω and κ for the various models of the elastic and plastic bodies, as well as the bodies with the rheological properties as applied to theories 2.2, and 2.3. In the same monographs, for theory 2.1, the relations are given for calculating constituent tensors ω and κ for the hyperelastic anisotropic bodies with an arbitrary structure of the elastic potential. It should be noted that the monograph [4] is exclusively devoted to the study of the stability of the hyperelastic bodies with an arbitrary structure of the elastic potential under large (finite) subcritical deformations, i.e., within the framework of theory 2.1.

Some properties of tensors ω and κ—analogies—It follows from the results presented in the monographs [1, 2, 5] that for the elastic and plastic bodies, the components of the tensors ω and κ do not depend on Ω and are real

$$Im\omega^{ij\alpha\beta} = 0; \; Im\kappa^{ij\alpha\beta} = 0. \tag{2.49}$$

It also follows in the general case from the results of the monographs [1, 2, 5] for the bodies with the rheological properties that the components of tensors ω and κ depend on Ω and are complex

$$Im\omega^{ij\alpha\beta} \neq 0; \; Im\kappa^{ij\alpha\beta} \neq 0. \tag{2.50}$$

In the general case of the elastic and plastic bodies, and the bodies with the rheological properties, it has been rigorously proven that the bodies are components of tensors ω and κ do not have symmetry properties that are the characteristics of the linear theory of elasticity of an anisotropic body, i.e., the following conditions hold

$$\omega^{ijnm} \neq \omega^{jinm}, \omega^{ijnm} \neq \omega^{ijmn}, \omega^{ijnm} \neq \omega^{nmij};$$
$$\kappa^{ijnm} \neq \kappa^{jinm}, \kappa^{ijnm} \neq \kappa^{ijmn}, \kappa^{ijnm} \neq \kappa^{nmij}. \tag{2.51}$$

Also, in the general case considered in the monographs [1, 2, 5], it was proved that for the elastic and plastic bodies the following symmetry conditions hold

$$\omega^{ijnm} = \omega^{mnji}, \kappa^{ijnm} = \kappa^{mnji}. \tag{2.52}$$

Note that the symmetry conditions (2.52) are essential in the formulation of the sufficient conditions for the applicability of the static method (Euler's method).

The formulated basic relations of TLTDBS for the compressible bodies in the forms (2.43) and (2.44) and the incompressible bodies in the forms (2.46) and (2.47) resemble the corresponding relations of the linear mechanics of deformable bodies, in particular, the linear theory of elasticity of an anisotropic body. In this regard, the question arises of the existence of analogies between the linear and linearized problems of the mechanics of deformable bodies. The statement of the question about the indicated analogy and its brief study was carried out in the article [22]. Subsequently, this question in a more extended form was considered in the monographs [1, 2, 6, 69] both from the point of view of the exhaustive solution and from the point of view of the application in solving the particular problems.

2.4.2   Sufficient Conditions for Applicability of the Euler's Method (Statical Method)

As noted in subsection 2.3.1, when characterizing the methods for studying the stability of an equilibrium state and the corresponding stability criteria for the elastic bodies, the statical method (Euler's method) is simpler than the dynamical method of research, although the statical method (Euler's method) is only applicable to the statical problems (and even then not to all the static problems).

The dynamical research method is more general than the statical one, and the dynamic research method is applicable to the dynamical and statical problems.

In connection with the above, the question arises—when the results obtained by Euler's method (statical method) coincide with the results obtained by the dynamical one.

The above question is solved by formulating and proving the sufficient (in the sense under discussion) conditions for the applicability of the Euler's method (statical method, bifurcation method) for solving the statical TLTDBS problems (for the case of fulfilling conditions (2.37)).

The question discussed in this subsection, like any scientific problem, has certain stages of development in the historical aspect. The development of research on the question under discussion, along with the presentation of results on other questions, has been quite consistently considered in the monographs [1, 2, 4–6], and in the monographs [1, 2], the results of consideration are presented in the most general form, which made it possible in the survey [8] to present the results of the consideration in an annotated form.

The noted situation makes it possible to recommend the readers who wish to receive more detailed information on the issue under discussion, to refer to the above publications, in the lists of references to which there are the publications of many authors on the issue under consideration.

Considering the above, below in this subsection, the information will be only presented in the annotated form about two fairly general results that were obtained on the issue under consideration.

Sufficient conditions for nonlinear models of the elastic and plastic bodies. In several publications, for example [20, 66], and then quite consistently in the monographs [1, 2], the sufficient conditions for the applicability of the Euler method for studying the stability of an equilibrium state for the hyperelastic anisotropic bodies were formulated and proved (in a unified general form for theories 2.1–2.3) with an arbitrary structure of the elastic potential and the plastic bodies with governing equations of a fairly general form (in the latter case, of course, using the generalized concept of continuing loading).

These results were obtained for the compressible and incompressible piece-wise homogeneous bodies with allowance for the interface between the properties of bodies, which can arise in the subcritical state due to the appearance of unloading zones in the case of plastic bodies.

The main result for the above *nonlinear* models of deformable bodies can be formulated as follows:

> the sufficient conditions for the applicability of the Euler method are fulfilled if the external loads (mass and surface forces) are conservative forces, i.e., the sufficient conditions are the same as for the linear elastic bodies.

In the proof of the discussed sufficient conditions for the applicability of the Euler method, the conditions of the type (2.49) and (2.52) are essentially used. Therefore, the obtained proof refers *only to the elastic and plastic bodies* and *does not apply to the bodies with the rheological properties*, since for the latter, in the general case, the conditions (2.50) hold.

Thus, the question of formulating and proving the sufficient conditions for the applicability of the Euler method for the bodies with the rheological properties remains open. The statement of individual authors about the allegedly obtained proof for the bodies with the rheological properties is erroneous, which was noted in the monographs [1, 2, 5] and in review articles [67, 68].

Sufficient conditions under the action of a "following" load in the form of a uniform load. Concerning this situation, it is necessary to distinguish between two cases:

In the *first* case, the "following" load is applied to the entire surface of the body.
In the *second* case, the "following" load is applied to a part of the body surface.

In *the first case*, in a common general form for all versions of the TLTDBS problem formulations (theories 2.1, 2.2, and 2.3) as applied to the elastic and plastic compressible and incompressible bodies, it is rigorously proved using relations (2.14) for theory 2.1, (2.20) for theory 2.2, and (2.27) for theory 2.3 that the sufficient conditions for the applicability of the Euler method are satisfied.

These results are presented in the most complete form in the monographs [1, 2, 6] and the modern review [12].

It should be noted that the above results were obtained for the simply connected bodies (in this case, a homogeneous subcritical state arises) and for the multiply connected bodies (as applied to the particular problem statements).

In *the second case*, the situation is considered when on the part S_1 of the body surface, which (part of the body surface) is bounded by a curve L, a "following" load is set in the form of a uniform load. In this case, it is rigorously proved that the sufficient conditions for the applicability of the Euler method are satisfied if the following conditions are satisfied on the curve L

$$u_3|_L = 0 \text{ or } u_M|_L = 0, \tag{2.53}$$

where the designations are introduced:

u_3 is the displacement along the normal to the surface S_1.
u_M is the displacement along the normal to the curve L in the tangent plane to the surface S_1.

The conditions (2.53) are obtained in a unified general form for the various variants of the TLTDBS problem formulations (theories 2.1, 2.2, and 2.3) as applied to the elastic and plastic compressible and incompressible bodies using relations (2.14) for theory 2.1, (2.20) for theory 2.2, and (2.27) for theory 2.3 to determine the components of the "following" load. These results are presented in the most complete form in the monographs [1, 2].

Note 2.10 As follows from Note 2.6, the expressions (2.20) for theory 2.2 (*the first variant of the theory of small subcritical deformations*) and expressions (2.27) for theory 2.3 (*the second variant of the theory of small subcritical deformations*) were obtained consistently and strictly from the expressions (2.14) for theory 2.1 (*theory of large (finite) subcritical deformations*). In this connection, the expressions (2.20) and (2.27) can be considered sufficiently accurate for the corresponding variants of the theory.

The expressions (2.34) for theories 2.2 and 2.3 are obtained in the general form by the linearization from the nonlinear theory of small deformations, which has already been constructed with the introduction of certain simplifications, proceeding from considerations of a physical nature. Therefore, the expressions (2.34) for theories 2.2 and 2.3 can be considered insufficiently accurate for the corresponding variants of the theory.

Taking into account the introductory information presented in Note 2.10, it should be noted that with the use of relations (2.34) for theories 2.2 and 2.3, when determining the components of the "following" load, **it is no longer possible to obtain** the result (2.53) under strict consideration.

In the monograph (No. 13 from the bibliography for the survey [8]), for theory 2.3, the result (2.53) was obtained, proceeding from the expressions (2.34), within the framework of theory 2.3, but at that the Kirchhoff–Love hypothesis was used. Hence, in this case, the result (2.53) was obtained only for an applied two-dimensional theory constructed with the use of the Kirchhoff–Love hypothesis.

The result (2.53), as already noted before Note 2.10, was obtained rigorously and consistently for *the three-dimensional* theories 1, 2, and 3 with the use of the corresponding strict and consistent expressions (2.14), (2.20), and (2.27). The analysis of this problem was considered back in 1980 in the publication of the author of this monograph (No. 141 from the list of references to the review [8]).

We restrict ourselves to the above information when discussing the issue under consideration. Additional information can be obtained from the monographs [1, 2] and review [8], as well as from publications indicated in the references to [1, 2, 8].

2.4.3 Sufficient Conditions of Stability

The formulation and proof of the sufficient stability conditions in the theory of stability of deformable bodies are an actively developed problem throughout the history of the development and formation of the discussed scientific direction. In this subsection, following the review [8], the basic information is given in a very brief form on the issue under consideration. Additional information can be obtained from the monographs [1, 2], as well as from the publications indicated in the references to [1, 2, 8].

At present, the sufficient conditions of stability in a strict form are formulated concerning the stability of the state of equilibrium of an elastic body under the action of conservative loads, based on *the energy approach* usually adopted in mechanics.

Taking the generalized concept of continuing loading, the sufficient conditions for the stability of the equilibrium state for plastic bodies are formulated in the same form.

In a common general form for all the considered formulations of the TLTDBS problems (theories 2.1, 2.2, and 2.3) for the nonlinearly elastic anisotropic compressible (with the involvement of the tensor components ω) and incompressible (with the involvement of the tensor components κ) bodies, the sufficient conditions for the stability of the equilibrium state are considered in the monographs [1, 2, 6]. There, the approaches presented in the monograph (No. 19 from the bibliography to the review [8]) were applied for theory 2.1 (*the theory of large (finite) subcritical deformations*).

Below, as an example, the sufficient conditions are given for the stability of the equilibrium state for the compressible elastic bodies with an arbitrary structure of the elastic potential, when the perturbations of the external mass and surface forces are equal to zero. These sufficient conditions for the stability of the equilibrium state remain valid for the plastic bodies if the generalized concept of continuing loading is adopted. In the notation of the monographs [1, 2], this condition has the following form

$$J_1 > 0; \ J_1 = \int_V \left(\omega^{ij\alpha\beta} \nabla_i u_j \nabla_\beta u_\alpha\right) dV, \tag{2.54}$$

which must be fulfilled for all kinematically possible displacements.

The sufficient conditions for the stability of the equilibrium state in the form (2.54) for the considered compressible elastic and plastic bodies are associated with the conditions for the uniqueness of the solution of the corresponding linearized problems.

In a sufficient general form, the conditions for the uniqueness of the linearized three-dimensional problems for the elastic and plastic bodies, apparently for the first time, were obtained in 1957 and 1958 in publications (No. 153 and No. 154 from the bibliography for the review [8]). In the notation used in this monograph, these conditions are presented in the monographs [1, 2]. For the considered elastic and plastic bodies, when the perturbations of the external mass and surface forces are equal to zero, the conditions for the uniqueness of the solution of the linearized problems can be represented [1, 2] in the following form the condition:

$$\omega^{ij} \varsigma_{ji} \varsigma_{\alpha\beta} > 0 \qquad (2.55)$$

should be fulfilled for the arbitrary ς_{nm} that is not equal to zero at the same time.

The conditions (2.55) correspond to the results of publications (No. 153 and No. 154 from the list of references to the review [8]) and are the conditions of strong ellipticity that must be satisfied at each point of the deformable body. In the terminology of the above publications, the condition (2.55) is the condition of strict local convexity.

It follows from (2.54) and (2.55) that if the sufficient condition (2.55) for the uniqueness of the solution of the linearized problems is satisfied, then the sufficient condition (2.54) for the stability of the equilibrium state is also satisfied. Thus, in this case, there is no loss of stability.

We also note that the fulfillment of condition (2.55) ensures that the phenomenon of internal instability of the material does not arise. The concept of the phenomenon of internal instability of the material, apparently for the first time, was introduced in publications for 1963 (No. 111 and No. 120 from the list of references for the review [8]).

As follows from Note 2.2, a related scientific direction concerning the TLTDBS is the theory of propagation of elastic waves in the bodies with the initial (residual) stresses, set forth, for example, in the monograph [7]. This theory also applies (weaker than (2.55)) the following condition

$$\omega^{ij\alpha\beta} \varsigma_{ji} \varsigma_{\alpha\beta} \geq 0, \qquad (2.56)$$

which must be fulfilled for all ς_{nm} with non-zero values simultaneously.

The condition (2.56) is also called the Hadamard condition. In this case, the elastic bodies for which condition (2.56) is satisfied are called the Hadamard bodies. Hadamard's condition (2.56) ensures that the velocities of propagation of the plane harmonic waves (for an arbitrary wave normal) in materials with the initial (residual) stresses are not imaginary values.

We restrict ourselves to the above information when discussing the sufficient conditions for the stability of the equilibrium state within the framework of the TLTDBS and related issues.

2.5   On Variational Principles of the TLTDBS for Elastic and Plastic Bodies

In this section, the information is briefly considered about the variational principles of TLTDBS as applied to the elastic and plastic bodies. At that, for the plastic bodies, the generalized concept of continuing loading is applied, which makes it possible for the elastic and plastic bodies to carry out a unified general consideration of the issue under discussion.

The formulation and proof of the variational principles of the theory of stability of deformable bodies (both within the framework of two-dimensional applied theories for thin-walled structural elements and the framework of the TLTDBS) were considered in all the years of development and formation of the discussed scientific direction. The lists of references to the monographs [1, 2] and review [8] contain the main publications related to the issue under discussion concerning TLTDBS.

The results of the author of this monograph on this issue are presented in publications [23, 24] and several others, with the main results presented in monographs [1, 2, 5]. In these monographs, the variational principles are mainly presented in a unified general form for various statements of TLTDBS problems (theories 2.1, 2.2, and 2.3) for a nonlinear elastic anisotropic body with an arbitrary structure of an elastic potential and a plastic body with the governing equations of a fairly general form.

Note that these variational principles are presented for the compressible and incompressible bodies under the action of "dead" loads and in several cases (for certain theories) under the action of "following" loads. The variational principles are formulated for a piece-wise homogeneous body, taking into account the presence of a surface of separation of material properties, designated as a surface $\hat{S}$, which is determined by the presence of the unloading zones arising in the subcritical state.

In this regard, all designations are introduced for two homogeneous parts of the material (active loading zone and unloading zone) with signs $(\pm)$: $V^{(\pm)}$ is the volume; $S^{(\pm)}$ is the surface; $S_1^{(\pm)}$ are the parts of surfaces on which boundary conditions in stresses are given; $S_2^{(\pm)}$ are the parts of surfaces on which boundary conditions in displacements are given.

In monographs [1, 2, 5] for the TLTDBS relative to the compressible and incompressible bodies, the various variational principles are stated:

The most general principle of Hu–Vashizu type, all relations, and equations, including the boundary conditions in displacements, follow from the condition of stationarity of the corresponding functional;
The principle of the Hamilton–Ostrogradsky type for the dynamic method;
The principle of the type of possible displacements for a static method;
The separate principles for the case of the action of the "tracking" load.

Several variational principles are considered below only as examples. The more complete information can be obtained from the monographs [1, 2, 5].

2.5.1 Variational Principle of Hu–Vashizu Type in the TLTDBS for Incompressible Bodies with "Dead" External Loads. Unified General Form for Theories 2.1, 2.2, and 2.3

Consider, as an example, the incompressible bodies, taking into account the general formulation of problems for the dynamical research method in the forms (2.46)–(2.48), taking into account the incompressibility condition in the form (2.8) and boundary conditions in displacements on S_2 in the form (2.11) as applied to the case when "dead" external loads act (there are no disturbances of mass and surface loads, as well as the right-hand sides of the boundary conditions in displacements in the form (2.11) on a part S_2 of the surface). Additionally, as is generally accepted in the Hu–Vashizu principle, a kinematic tensor v is introduced, the covariant components of which are determined by the following relations

$$v_{\alpha\beta} = \nabla_\beta u_\alpha. \tag{2.57}$$

Taking into account the relations (2.8), (2.11), (2.46)–(2.48), and (2.57), for a body that has an interface $\hat{S}$ of properties, under the action of the above "dead" loads in the general formulation of problems when applying the dynamic research method, it is necessary to include the following relations:

The equations of motion

$$\nabla_i t^{ij}_{(\pm)} + \rho^{(\pm)}\Omega^2 u^j_{(\pm)} = 0, \theta^k \in V^{(\pm)}; \, . \tag{2.58}$$

The constitutive equations

$$t^{ij}_{(\pm)} = \kappa^{ij\alpha\beta}_{(\pm)} v^{(\pm)}_{\alpha\beta} + q^{ij}_{(\pm)} p^{(\pm)}; \, . \tag{2.59}$$

The expressions for determining the components of the kinematic tensor

$$v^{(\pm)}_{\alpha\beta} = \nabla_\beta u^{(\pm)}_\alpha; \tag{2.60}$$

The incompressibility condition

$$q^{ij}_{(\pm)} v^{(\pm)}_{ji} = 0, \theta^k \in V^{(\pm)}; \tag{2.61}$$

The boundary conditions in stresses on body surface parts $S^{(\pm)}_1$

$$N^{(\pm)}_i t^{ij}_{(\pm)} = 0, \theta^k \in S^{(\pm)}_1; \tag{2.62}$$

The boundary conditions in displacements on parts of the body surface $S^{(\pm)}_2$

$$u^j_{(\pm)} = 0, \theta^k \in S_2^{(\pm)}; \tag{2.63}$$

The conditions of continuity of the components of the stress vector at the interface $\hat{S}$

$$N_i^{(+)} t^{ij}_{(+)} + N_i^{(-)} t^{ij}_{(-)} = 0, \theta^k \in \hat{S} \tag{2.64}$$

and the conditions for the continuity of the components of the displacement vector at the interface

$$u^j_{(+)} = u^j_{(-)}, \theta^k \in \hat{S}. \tag{2.65}$$

The expressions (2.58)–(2.65) completely exhaust the general formulation (in general form for theories 2.1, 2.2, and 2.3) of the TLSTDBS problems for the incompressible bodies, taking into account the existence of an interface $\hat{S}$ of material properties.

For the above formulation of problems in the notation of the monographs [1, 2], the following functional is introduced

$$
\begin{aligned}
J_4 &\left(t^{(\pm)}, v^{(\pm)}, u^{(\pm)}, p^{(\pm)} \right) \\
&= \int_{V^{(+)}} \left[\frac{1}{2} \kappa^{ij\alpha\beta}_{(+)} v^{(+)}_{\alpha\beta} v^{(+)}_{ji} - t^{ij}_{(+)} \left(v^{(+)}_{ji} - \nabla_i u^{(+)}_j \right) + q^{ij}_{(+)} p^{(+)} v^{(+)}_{ji} - \frac{1}{2} \rho^{(+)} \Omega^2 u^{(+)}_j u^j_{(+)} \right] dV^{(+)} \\
&+ \int_{V^{(-)}} \left[\frac{1}{2} \kappa^{ij\alpha\beta}_{(-)} v^{(-)}_{\alpha\beta} v^{(-)}_{ji} - t^{ij}_{(-)} \left(v^{(-)}_{ji} - \nabla_i u^{(-)}_j \right) + q^{ij}_{(-)} p^{(-)} v^{(-)}_{ji} - \frac{1}{2} \rho^{(-)} \Omega^2 u^{(-)}_j u^j_{(-)} \right] dV^{(-)} \\
&- \int_{S_2^{(+)}} N_i^{(+)} t^{ij}_{(+)} u^{(+)}_j dS^{(+)} - \int_{S_2^{(-)}} N_i^{(-)} t^{ij}_{(-)} u^{(-)}_j dS^{(-)}. \tag{2.66}
\end{aligned}
$$

Statement. If the functions $u^j_{(\pm)}$ satisfy the conditions of continuity of the components of the displacement vector on the interface $\hat{S}$ in the form (2.65), then from the condition of stationarity of the functional (2.66) all equations and relations in the form (2.58)–(2.64) follow.

The proof of this statement is given in the monographs [1, 2].

It should be noted that when using the function (2.66) and when proving the above statement, the components of the tensors $t^{(\pm)}$ and $v^{(\pm)}$ and the components of the vectors $u^{(\pm)}$ and scalars $p^{(\pm)}$ are varied.

The variational principle based on the functional (2.66) is the most general variational principle for the incompressible bodies within the TLTDBS. Note that the above-discussed variational principle is formulated in a single general form for theories 2.1, 2.2, and 2.3 following the terminology of Sect. 2.2 and the introductory part of Sect. 2.4.

When applying this variational principle to solve the particular classes of problems, the fulfillment of condition (2.65) can be ensured by specifying the coordinate functions in the form of single general expressions for $V^{(+)}$ and $V^{(-)}$.

In the case of compressible bodies, a similar variational principle was also formulated and proved. The corresponding results are given in the monographs [1, 2].

If there is no interface $\hat{S}$ of the properties of the material, then, naturally, the continuity conditions on $\hat{S}$ in the forms (2.64) and (2.65) are excluded and the formulation of the basic relations of the problem under discussion is carried out for the tensors t and v, the vector u and the scalar p, and then the dynamic research method includes the following expressions:

The equations of motion

$$\nabla_i t^{ij} + \rho \Omega^2 u^j = 0, \theta^k \in V; . \tag{2.67}$$

The constitutive equations

$$t^{ij} = \kappa^{ij\alpha\beta} v_{\alpha\beta} + q^{ij} p; \tag{2.68}$$

The expressions for determining the components of the kinematic tensor

$$v_{\alpha\beta} = \nabla_\beta u_\alpha; . \tag{2.69}$$

The incompressibility condition

$$q^{ij} v_{ji} = 0, \theta^k \in V; . \tag{2.70}$$

The boundary conditions in stresses on a part S_1 of the body surface

$$N_i t^{ij} = 0, \theta^k \in S_1; \tag{2.71}$$

The boundary conditions in displacements on a part S_2 of the body surface

$$u^j = 0, \theta^k \in S_2. \tag{2.72}$$

The expressions (2.67)–(2.72) completely exhaust the general formulation (in general form for theories 2.1, 2.2, and 2.3) of the TLTDBS problems for incompressible bodies (without the interface $\hat{S}$).

For the above formulation of problems in the notation of monographs [1, 2], the following functional is introduced

$$J_4(t, v, u, p) = \int_V \left[\frac{1}{2} \kappa^{ij\alpha\beta} v_{\alpha\beta} v_{ij} - t^{ij} \left(v_{ij} - \nabla_i u_j \right) + q^{ij} p v_{ij} - \frac{1}{2} \rho \Omega^2 u_j u^j \right] dV$$

$$- \int\limits_{S_2} N_i t^{ij} u_j dS. \tag{2.73}$$

Statement. All relations and Eqs. (2.67)–(2.72) follow from the stationarity condition for the functional (2.73).

The proof of this statement corresponds to the monographs [1, 2].

In the proof, the components of the tensors t and v and the components of the vector u and the scalar p are varied.

The variational principle based on functional (2.73) is the most general variational principle for incompressible bodies (without the interface $\hat{S}$ of the material properties) within the TLTDBS.

Note that this variational principle is formulated in a common general form for theories 2.1, 2.2, and 2.3 following the terminology of Sect. 2.2 and the introductory part of Sect. 2.4. In the case of compressible bodies, a similar variational principle was also formulated and proved. The corresponding results are given in the monographs [1, 2].

2.5.2   Variational Principle of the TLTDBS for Compressible Bodies Under "Following" Load. Results for Theory 2.3

First of all, it should be noted that in the case of a "following" load in the form of uniform external pressure, the variational principles are formulated and proved in the monographs [1, 2], as well as in corresponding publications, for example, article [24] of 1979, only for theory 2.3 (*the second variant of the theory of the small subcritical deformations*), when *the expressions* (2.27)*are used* to determine the components of the external surface load vector, obtained as a result of successive simplifications from the corresponding *expressions* (2.14), originally constructed within the framework of the rigorous and exact theory 2.1 (*theory of the large (finite) subcritical deformations*).

Note that in the monographs [1, 2], the variational principles are formulated and proved [1, 2] under the action of a "following" load in the form of a uniform external pressure.

Below, as an example, these variational principles will be considered only for a compressible body.

Denote by the intensity Q of the external normal pressure. In this case, taking into account the direction of the external normal pressure ($Q = -\tilde{P}$) from (2.27) for the compressible and incompressible bodies, the expressions for the contravariant components of the disturbance of the surface load vector are obtained in the form

$$P^j = -Q\left(N^j \nabla_\alpha u^\alpha - N^\alpha g^{\beta j} \nabla_\beta u_\alpha\right). \tag{2.74}$$

Consider the case when a "following" load is applied to the entire surface S of the body in the form of a uniform external pressure. In this case, following (2.43), (2.44), and (2.49), with the general formulation of the problems, the next expressions are obtained as follows:

The equations of motion in displacements

$$\nabla_i \left(\omega^{ij\alpha\beta} \nabla_\beta u_\alpha \right) + \rho \Omega^2 u^j = 0, \theta^k \in V \tag{2.75}$$

The boundary conditions in stresses on the entire surface of the body (on the surface S)

$$N_i \omega^{ijnm} \nabla_m u_n = -Q \left(N^j \nabla_\alpha u^\alpha - N^\alpha g^{\beta j} \nabla_\beta u_\alpha \right), \theta^k \in S. \tag{2.76}$$

The expressions (2.75) and (2.76) completely exhaust the general formulation of the TLTDBS problems in displacements for the compressible bodies as applied to this case.

For the above formulation of problems in the notations of the monographs [1, 2], the following functional is introduced:

$$J_1(u) = \frac{1}{2} \int_V \left\{ \omega^{ij\alpha\beta} \left(\nabla_\beta u_\alpha \right) \left(\nabla_i u_j \right) - \rho \Omega^2 u_j u^j + Q \left[\left(\nabla_i u^i \right)^2 - \left(\nabla_i u^j \right) \left(\nabla_j u^i \right) \right] \right\} dV.$$
$$\tag{2.77}$$

Statement. The stationarity condition for the functional (2.77) implies the equations of motion (2.75) and the boundary conditions in stresses, formulated through the displacements, on the entire surface of the body.

The proof of the variational principle is given in the monographs [1, 2]. In this case, when studying the function (2.77), the components of the vector u are varied. In the case of incompressible bodies, a similar variational principle is also formulated and proved. In this case, the components of the vector u and the scalar p are already variating. The corresponding results are presented in the monographs [1, 2].

This variational principle can be applied in studying the problems of the stability of mine workings, where the effect of gas on the working surface can be modeled by a "following" load in the form of uniform pressure.

In the monographs [1, 2], the variational principles for the compressible and incompressible bodies are also formulated and proved in the framework of theory 2.3 when a uniform external pressure in the form of a "following" load is applied to a part of the body surface. In this case, the results obtained have a more complex form and are not discussed in this monograph, confining themselves only to the above information.

Note 2.11 As already noted above, the results of this section were obtained only for theory 2.3 (*the second variant of the theory of the small subcritical deformations*), when *the sufficiently accurate expressions* (2.27) were used to determine the components of the vector of the external surface load ("following" load). If the more approximate expressions (2.34) are used to determine the components of the vector of the external surface load ("following" load), then *it is no longer possible to formulate and prove these variational principles*, i.e., *the results of this section cannot be obtained*.

It should be noted that more approximate expressions (2.34) were obtained and used by almost all researchers who initially proceeded from the nonlinear theory of *small* deformations. A critical discussion of the very approximate expressions (2.34) is carried out in the final part of Sect. 2.2, starting with Note 2.6.

It is advisable to note that the above situation is already *the second* case when, in the study of mechanical phenomena associated with the action of the "following" load, applying (generally accepted among specialists dealing with the applied problems, but not being accurate enough) expressions (2.34) for theory 2.3 cannot obtain the results that are obtained for theory 2.3 using the expressions (2.27), which are more accurate than (2.34).

The first situation arose when proving the applicability of the Euler method for the case of the action of a "following" load. The analysis of this situation is carried out in the second part of subsection 2.4.2. It turned out that when applying the expressions for theory 2.3 (2.27), the results are obtained that completely coincide with the corresponding results for theories 2.1 and 2.2. It should be noted that *nothing of the kind can be obtained* by applying (2.34).

The two cases of application of the well-known expressions (2.34) considered in Note 2.11 indicate that to obtain the reliable results within the framework of theory 2.3, corresponding to the results within the framework of more accurate theories 2.2 and, especially, 2.1, *it is advisable not to apply the expressions* (2.34), although they are well known and popular.

We restrict ourselves to the above information in Sect. 2.5 when discussing general issues of the TLTDBS in their brief consideration. Additional information on the issues discussed in this section can be obtained from the monographs [1, 2] and the survey [8].

2.6   General Solutions of the TLTDBS Under Homogeneous Subcritical States

In this section, the information on the already constructed general solutions of the TLTDBS for the homogeneous subcritical states is presented in a very brief form. The more detailed information on the construction of the general solutions and their application to the study of various problems are presented in the monographs of the author of this monograph, which are included in the list of references to this monograph.

For generality, the orthotropic bodies with rectilinear orthotropic will be considered, in which the axes of symmetry of the material properties are directed along the coordinate lines $0x_n$ of the Cartesian coordinate system. The study will be carried out in a common general form for the hyperelastic bodies with an arbitrary structure of elastic potential with the above symmetry property and for plastic bodies with defining equations of a fairly general form also with the above symmetry property. At that, for the plastic bodies, the generalized concept of continuing loading will be taken.

The general solutions will be constructed for the case of triaxial compression along the axes of the above-mentioned Cartesian coordinate system by the uniformly distributed loads of constant intensity (of different intensity for different coordinate axes). In this case, a homogeneous subcritical stress state arises

$$\sigma_{ij}^0 = \text{const for } i = j; \ \sigma_{ij}^0 = 0 \text{ for } i \neq j; \ \sigma_{11}^0 \neq \sigma_{22}^0 \neq \sigma_{33}^0. \tag{2.78}$$

The displacements in the subcritical stress state (2.78) are defined as follows:

$$u_m^0 = (\lambda_m - 1)x_m; \ \lambda_1 \neq \lambda_2 \neq \lambda_3; \ \lambda_m = \text{const}, \tag{2.79}$$

where λ_m are the elongation coefficients.

Note 2.12 This monograph provides information on the construction of the general solutions for the elastic and plastic bodies. In this case, the conditions $\sigma_{ij}^0 = \text{const}$ are satisfied. The constructed general solutions remain valid for the bodies with the rheological properties, but in this case σ_{ij}^0 and u_m^0 already depend on time. The indicated generalization of the general solutions to the bodies with the rheological properties was carried out in the monographs [1, 2].

It should be noted that due to the dependence of σ_{ij}^0 and u_m^0 on time, there is no prospect of applying the general solutions for the bodies with the rheological properties when solving the specific problems. In connection with the foregoing, in this monograph, the general solutions are considered only for the elastic and plastic bodies in the general form for theories 2.1, 2.2, and 2.3.

When constructing general solutions, it is usually assumed that there are no disturbances of the mass forces. In this regard, when applying the dynamical method of studying the problems of the TLTDBS in the basic relations in displacements for the compressible bodies in the forms (2.43)–(2.45) and for the incompressible bodies in the forms (2.46)–(2.48) and the incompressibility condition in the form (2.7), it should be accepted

$$F_{(1)}^{j\alpha} = 0, \ F_{(2)}^{j\alpha} = 0. \tag{2.80}$$

In the case of the homogeneous subcritical states in the forms (2.78) and (2.79), we thus obtain the systems of partial differential equations with the constant coefficients in the form (2.43) taking into account (2.80) for the compressible bodies and in the forms (2.7) and (2.46) taking into account (2.80) for the incompressible bodies. Since these systems of equations are of a sufficiently high order, following the experience of constructing the solutions of separate classes, it is promising, in this case, to construct the general solutions, when the solution of a system of equations of a higher order is represented through the solutions of equations or systems of equations of a lower order. This approach is generally also accepted in the study of other classes of problems in continuum mechanics and mathematical physics. The discussed approach has also been implemented for TLTDBS problems. At that, the main part of the results is presented in the monographs [1, 2] for the spatial, plane, and anti-plane problems with the dynamical and statical research methods.

It is expedient to note that in the *eight* non-classical problems of fracture mechanics considered in this monograph, the situations do not arise when it is necessary to take into account the occurrence of disturbances of mass and surface forces, i.e., following the notation in Sect. 2.3, the following conditions hold

$$F_{(1)}^{j\alpha} = 0, \; F_{(2)}^{j\alpha} = 0; \; P_{(1)}^{j\alpha} = 0, \; P_{(2)}^{j\alpha} = 0. \tag{2.81}$$

In case (2.81), as shown in subsection 2.4.2, the sufficient conditions for the applicability of the Euler method for the nonlinear models of the elastic and plastic bodies are satisfied. Thus, the results of the study of the TLTDBS problems for any of the *eight* non-classical problems of the fracture mechanics coincide when using the dynamical and statical research methods.

In connection with the above, below the general solutions will be considered only concerning the statical method of research, setting in the system of Eq. (2.43) for compressible bodies and the system of Eqs. (2.7) and (2.46) for incompressible bodies $\Omega = 0$. At that, the general solutions are presented separately for the compressible bodies, for the incompressible bodies, and for the case of using the complex variables.

2.6.1 General Solutions of the TLTDBS for Compressible Bodies

These general solutions are presented for the compressible bodies in a unified general form for theories 2.1, 2.2, and 2.3 (in the terminology of the introductory part of Sect. 2.4) separately for the plane and spatial problems.

The plane problem. Consider the plane deformation in the plane $x_1 0 x_2$ for the cases (2.78) and (2.79). It should be taken in this case

$$u_1 = u_1(x_1, x_2); \; u_2 = u_2(x_1, x_2); \; u_3 = 0. \tag{2.82}$$

The displacements u_j are determined through functions $\chi^{(n)}$ ($j, n = 1, 2$) as follows:

$$u_1 = \left(\omega_{1221} \frac{\partial^2}{\partial x_1^2} + \omega_{2222} \frac{\partial^2}{\partial x_2^2} \right) \chi^{(2)} - (\omega_{1122} + \omega_{2121}) \frac{\partial^2}{\partial x_1 \partial x_2} \chi^{(1)};$$

$$u_2 = -(\omega_{2211} + \omega_{1212})\frac{\partial^2}{\partial x_1 \partial x_2}\chi^{(2)} + \left(\omega_{1111}\frac{\partial^2}{\partial x_1^2} + \omega_{2112}\frac{\partial^2}{\partial x_2^2}\right)\chi^{(1)}. \qquad (2.83)$$

The equation for determining the functions $\chi^{(j)}(j = 1, 2)$ is as follows:

$$\left(\frac{\partial^2}{\partial x_2^2} + \eta_2^2\frac{\partial^2}{\partial x_1^2}\right)\left(\frac{\partial^2}{\partial x_2^2} + \eta_3^2\frac{\partial^2}{\partial x_1^2}\right)\chi^{(j)} = 0; \; j = 1, 2. \qquad (2.84)$$

In (2.84), the quantities η_2^2 and η_3^2 are determined from the following expressions:

$$\eta_{2,3}^2 = c \pm \left(c^2 - \omega_{1111}\omega_{1221}\omega_{2222}^{-1}\omega_{2112}^{-1}\right)^{\frac{1}{2}};$$

$$2c\omega_{2222}\omega_{2112} = \omega_{1111}\omega_{2222} + \omega_{2112}\omega_{1221} - (\omega_{1122} + \omega_{2121})(\omega_{2211} + \omega_{1212}).$$
$$\qquad (2.85)$$

The construction of solutions in the concrete classes of problems is determined by the structure of the parameters η_2^2 and η_3^2.

The spatial problem. The general solution for the spatial problem is presented for the subcritical state in the forms (2.78) and (2.79), for which the conditions are fulfilled

$$\sigma_{11}^0 = \sigma_{22}^0 \neq \sigma_{33}^0; \; \lambda_1 = \lambda_2 \neq \lambda_3. \qquad (2.86)$$

Besides, it is assumed that the nonlinear elastic and plastic bodies are the transversely isotropic bodies, in which the planes $x_3 = $ const are the isotropic planes (in this case, the axis $0x_3$ is the isotropy axis).

The general solution is presented in a form that is convenient for studying the problems for a cylindrical body with an arbitrary cross-sectional contour, the axis of which is directed along the axis $0x_3$. The following designations are introduced:

N and S are the unit vectors of the normal and tangent to the contour of the cross section in the plane $x_3 = $ const.
u_N and u_S are the displacements along the normal and tangent to the contour of the cross section in the plane $x_3 = $ const.
u_3 is the displacement along the axis of the cylinder.
N_1 and N_2 are the components of the normal to the contour of the cross section in the plane $x_3 = $ const.

With the above notation, the general solution can be represented in the following form:

$$u_N = \frac{\partial}{\partial S}\psi - \frac{\partial^2}{\partial N \partial x_3}\chi; \; u_S = -\frac{\partial}{\partial N}\psi - \frac{\partial^2}{\partial S \partial x_3}\chi;$$

$$u_3 = (\omega_{1133} + \omega_{3131})^{-1}\left(\omega_{1111}\Delta_1 + \omega_{3113}\frac{\partial^2}{\partial x_3^2}\right)\chi. \qquad (2.87)$$

The equations for determining the functions ψ and χ have the following form:

$$\left(\Delta_1 + \xi_1^2 \frac{\partial^2}{\partial x_3^2}\right)\psi = 0; \quad \Delta_1 = \frac{\partial^2}{\partial x_1^2} + \frac{\partial^2}{\partial x_2^2};$$

$$\left(\Delta_1 + \xi_2^2 \frac{\partial^2}{\partial x_3^2}\right)\left(\Delta_1 + \xi_3^2 \frac{\partial^2}{\partial x_3^2}\right)\chi = 0. \tag{2.88}$$

In (2.88), the quantities $\xi_j^2 (j = 1, 2, 3)$ are determined from the following expressions:

$$\xi_1^2 = \omega_{3113}\omega_{1221}^{-1}; \quad \xi_{2,3}^2 = c \pm \left(c^2 - \omega_{3333}\omega_{3113}\omega_{1111}^{-1}\omega_{1331}^{-1}\right)^{\frac{1}{2}};$$

$$2c\omega_{1111}\omega_{1331} = \omega_{1111}\omega_{3333} + \omega_{1331}\omega_{3113} - (\omega_{1133} + \omega_{3131})(\omega_{3311} + \omega_{1313}). \tag{2.89}$$

The construction of solutions in the specific classes of problems is determined by the structure of the parameters $\xi_j^2 (j = 1, 2, 3)$ and the relationships between them.

The general solutions for the compressible bodies in the forms (2.83)–(2.85) for the plane problems and in the forms (2.87)–(2.89) for the spatial problems were widely used in the monographs [1–6, 63, 64, 69] and several other monographs, as well as in the corresponding articles published in periodicals. The most complete and consistently discussed general solutions are presented in the monographs [1, 2], which also contain the expressions for determining the components $\omega_{ij\alpha\beta}$ for general and many concrete models of the elastic, plastic, and bodies with rheological properties.

2.6.2 The General Solutions of the TLTDBS for Incompressible Bodies

These general solutions are presented for the incompressible bodies in a unified general form for theories 2.1, 2.2, and 2.3 (in the terminology of the introductory part of Sect. 2.4) separately for the plane and spatial problems. It should be noted that in the case of the incompressible bodies, the general solutions are presented for the components of the vector u and the scalar p, in contrast to the compressible body (subsection 2.5.1), for which the general solutions are presented only for the components of the vector u. Also, the basic relations for the incompressible body in the forms (2.46)–(2.48) also include the quantities q^{ij} that in Cartesian coordinates have the form

$$q_{ij} = \delta_{ij}q_j, \tag{2.90}$$

where the quantities q_j, respectively, for theories 2.1, 2.2, and 2.3 are determined by the expressions

$$q_j = \lambda_j^{-1}; q_j = \lambda_j; q_j = 1. \tag{2.91}$$

The elongation coefficients $\lambda_j (j = 1, 2, 3)$ are related to each other by the relationships that follow from the incompressibility condition for the subcritical state; the discussed relations, respectively, for theories 2.1, 2.2, and 2.3 have the following form

$$\lambda_1\lambda_2\lambda_3 = 1; \lambda_1^2 + \lambda_2^2 + \lambda_3^2 - 3 = 0; \lambda_1 + \lambda_2 + \lambda_3 - 3 = 0. \tag{2.92}$$

The plane problem. Consider the plane deformation in the plane $x_1 0 x_2$ for the cases (2.78) and (2.79). It should be taken in this case

$$u_1 = u_1(x_1, x_2); u_2 = u_2(x_1, x_2); u_3 = 0; p = p(x_1, x_2). \tag{2.93}$$

In the general case of this plane deformation, the expressions (2.92) hold. In the special case of the plane deformation, when the subcritical state is also determined under conditions of the plane deformation, it should additionally be taken $\lambda_3 = 1$. Thus, taking into account $\lambda_3 = 1$ from (2.92), respectively, for theories 2.1, 2.2, and 2.3, the conditions are obtained

$$\lambda_1\lambda_2 = 1; \lambda_1^2 + \lambda_2^2 - 2 = 0; \lambda_1 + \lambda_2 - 2 = 0. \tag{2.94}$$

In the considered general solution for the plane deformation, the displacements $u_j (j = 1, 2)$ and the scalar p are determined in terms of the functions $\chi^{(n)} (n = 1, 2, 4)$ as follows:

$$u_1 = -q_2^2 \frac{\partial^2}{\partial x_2^2} \chi^{(1)} + q_1 q_2 \frac{\partial^2}{\partial x_1 \partial x_2} \chi^{(2)}$$
$$- \left[q_1 \left(\kappa_{1221} \frac{\partial^2}{\partial x_1^2} + \kappa_{2222} \frac{\partial^2}{\partial x_2^2} \right) - q_2(\kappa_{1122} + \kappa_{2121}) \frac{\partial^2}{\partial x_2^2} \right] \frac{\partial}{\partial x_1} \chi^{(4)} ;$$

$$u_2 = q_1 q_2 \frac{\partial^2}{\partial x_1 \partial x_2} \chi^{(1)} - q_1^2 \frac{\partial^2}{\partial x_1^2} \chi^{(2)}$$
$$- \left[q_2 \left(\kappa_{1111} \frac{\partial^2}{\partial x_1^2} + \kappa_{2112} \frac{\partial^2}{\partial x_2^2} \right) - q_1(\kappa_{2211} + \kappa_{1212}) \frac{\partial^2}{\partial x_1^2} \right] \frac{\partial}{\partial x_2} \chi^{(4)} ;$$

$$p = -\left[q_1 \left(\kappa_{1221} \frac{\partial^2}{\partial x_1^2} + \kappa_{2222} \frac{\partial^2}{\partial x_2^2} \right) - q_2(\kappa_{2211} + \kappa_{1212}) \frac{\partial^2}{\partial x_2^2} \right] \frac{\partial}{\partial x_1} \chi^{(1)}$$
$$- \left[q_2 \left(\kappa_{1111} \frac{\partial^2}{\partial x_1^2} + \kappa_{2112} \frac{\partial^2}{\partial x_2^2} \right) - q_1(\kappa_{1122} + \kappa_{2121}) \frac{\partial^2}{\partial x_1^2} \right] \frac{\partial}{\partial x_2} \chi^{(2)}. \tag{2.95}$$

The equation for determining the functions $\chi^{(n)}(n = 1, 2, 4)$ is as follows:

$$\left(\frac{\partial^2}{\partial x_2^2} + \eta_2^2 \frac{\partial^2}{\partial x_1^2}\right)\left(\frac{\partial^2}{\partial x_2^2} + \eta_3^2 \frac{\partial^2}{\partial x_1^2}\right)\chi^{(j)} = 0; \ j = 1, 2, 4. \qquad (2.96)$$

In (2.96), the quantities η_2^2 and η_3^2 are determined from the following expressions:

$$\eta_{2,3}^2 = c \pm \left(c^2 - q_1^2 \kappa_{1221} q_2^{-2} \kappa_{2112}^{-1}\right)^{\frac{1}{2}};$$

$$2cq_2^2 \kappa_{2112} = q_2^2 \kappa_{1111} + q_1^2 \kappa_{2222} - q_1 q_2 (\kappa_{1122} + \kappa_{2121} + \kappa_{2211} + \kappa_{1212}). \qquad (2.97)$$

Note 2.13 The general solutions for the plane TLTDBS problems in the forms (2.83)–(2.85) for the compressible bodies and in the forms (2.95)–(2.97) for the incompressible bodies were widely used in the monographs [1–6, 63, 64, 69] and some other monographs, as well as in the corresponding articles published in periodicals. Moreover, in most studies it was enough to keep one function $\chi^{(2)}$, assuming for the compressible bodies $\chi^{(1)} = 0$ and the incompressible bodies $\chi^{(1)} = 0$ and $\chi^{(4)} = 0$. There are situations (the classes of problems or particular problems) when it is expedient to keep in these general solutions two functions for the compressible and incompressible bodies, assuming for the compressible bodies $\chi^{(1)} \neq 0$, $\chi^{(2)} \neq 0$ and for the incompressible bodies $\chi^{(1)} \neq 0$, $\chi^{(2)} \neq 0$, and $\chi^{(4)} = 0$. This situation is considered in the monograph [69] and volume 2 of [70] when introducing the representation in terms of complex potentials to ensure their transition at $\sigma_{ij}^0 \to 0$ to the classical representations of Kolosov–Muskhelishvili [71] and Lehnitskiy [72] for the linear theory.

The spatial problem. The general solution of the spatial problem for the incompressible bodies is presented for the subcritical state in the forms (2.78) and (2.79), for which condition (2.86) is satisfied. In this case, from expressions (2.86) and (2.91) in the general form for theories 2.1, 2.2, and 2.3, the following condition can be obtained for the quantities q_j (2.91):

$$q_1 = q_2. \qquad (2.98)$$

Besides, from conditions (2.86) and expressions (2.92), the relations can be obtained for determining the quantity λ_3 in terms of quantity λ_1, respectively, for theories 2.1, 2.2, and 2.3 in the following form.

$$\lambda_3 = \lambda_1^{-2}; \ \lambda_3 = \left(3 - 2\lambda_1^2\right)^{\frac{1}{2}}; \ \lambda_3 = 3 - 2\lambda_1. \qquad (2.99)$$

Both in the case of the compressible bodies and the discussed case of the incompressible bodies for the spatial problem, it is assumed that the nonlinear elastic and plastic bodies are the transversely isotropic bodies, in which the planes $x_3 = $ const are the isotropy planes (in this case, the axis $0x_3$ is the isotropy axis).

The general solution is presented in a form that is convenient for studying the problems for a cylindrical body with an arbitrary cross-sectional contour, the axis of which is directed along the axis $0x_3$. In this regard, all the notations introduced for the compressible bodies before expressions (2.87) are applied. With the above notation, the general solution for the incompressible body (for the displacements u_N, u_S, u_3 and the scalar p) can be represented in the following form:

$$u_N = \frac{\partial}{\partial S}\psi - \frac{\partial^2}{\partial N \partial x_3}\chi; \; u_S = -\frac{\partial}{\partial N}\psi - \frac{\partial^2}{\partial S \partial x_3}\chi;$$

$$u_3 = q_1 q_3^{-1}\Delta_1\chi; \; \Delta_1 = \frac{\partial^2}{\partial x_1^2} + \frac{\partial^2}{\partial x_2^2}; \quad\quad\quad (2.100)$$

$$p = q_1^{-1}q_3^{-1}\left\{[q_3\kappa_{1111} - q_1(\kappa_{1133} + \kappa_{3131})]\Delta_1 + q_3\kappa_{3113}\frac{\partial^2}{\partial x_3^2}\right\}\frac{\partial}{\partial x_3}\chi.$$

The functions ψ and χ are determined from Eq. (2.88) with the following notation

$$\xi_1^2 = \kappa_{3113}\kappa_{1221}^{-1}; \; \xi_{2,3}^2 = c \pm \left(c^2 - q_3^2\kappa_{3113}q_1^{-2}\kappa_{1331}\right)^{\frac{1}{2}};$$

$$2cq_1^2\kappa_{1331} = q_1^2\kappa_{3333} + q_3^2\kappa_{1111} - q_1 q_3(\kappa_{1133} + \kappa_{3311} + \kappa_{1313} + \kappa_{3131}). \quad (2.101)$$

The construction of solutions in the specific classes of problems is determined by the structure of parameters $\xi_j^2 (j = 1, 2, 3)$ in the form (2.101) and the relationships between them.

The general solutions for the incompressible bodies in the forms (2.95)–(2.97) for the plane problems and in the forms (2.88), (2.100), and (2.101) for the spatial problems were widely used in the monographs [1–6, 63, 64, 69] and several other monographs, as well as in corresponding articles published in periodicals. The most complete and consistently discussed general solutions are presented in the monographs [1, 2], which also contain the expressions for determining the components $\kappa_{ij\alpha\beta}$ of the tensor κ for the general and many specific models of the elastic and plastic bodies as well as the bodies with rheological properties.

2.6.3 The Complex Potentials in Plane Problems of the TLTDBS. Preliminary Discussion

The complex potentials in the plane TLTDBS problems for the statical research method (Euler method) were consistently introduced in the monograph [69] in 1983. Preliminarily, in a more reduced form, these results were presented in the articles listed in the references to [69]. In subsequent years, the discussed results were presented in full in volume 2 of the collective monograph [70] in 1991, in the supplement to the monograph [2] in 1999, and in volume 2 of the two-volume monograph [64] in 2008. Since the monograph [2] was published in English by Springer, the results discussed are readily available to the English-speaking researchers.

These complex potentials in the plane TLTDBS problems are introduced for the statical (Euler method) method of investigation in the absence of disturbances of the mass and surface forces (when conditions (2.81) are satisfied), as in all other results of Sect. 2.5.

In this case, according to subsection 2.4.2, the sufficient conditions for the applicability of the static method (Euler method) are satisfied. Consequently, when using these complex potentials, the results of solving the problems with the dynamical and statical methods coincide.

Below, in this section, following the notation of [64], the construction of the complex potentials of the plane TLTDBS problems will be discussed for the linearized mechanics of materials with the initial (residual) stresses, since the latter, following Note 2.4, coincides with the TLTDBS if the values in the TLTDBS marked by index "0" (in TLTDBS, the quantities of the subcritical state), consider them known and call them the initial (residual) stresses. Also, of the *eight* non-classical problems of fracture mechanics considered in this monograph, following Sect. 1.3 in **Problem 4** (*brittle fracture of materials with cracks, taking into account the action of the initial (residual) stresses along the cracks*), the complex potentials of the plane problems of the linearized mechanics of materials with the initial (residual) stresses are used.

When constructing the complex potentials of the plane problems of the linearized mechanics of materials with the initial (residual) stresses, it is necessary to take into account the following situation.

In this chapter, when considering the basic relations of the TLTDBS following Sect. 2.1, the Lagrangian coordinate system is used, which in the reference state (the first state) coincided with a certain curvilinear coordinate system. At that, the components of the stress tensor were used, also referred to the dimensions of the areas in the reference state (the first state), which is generally accepted in the mechanics of deformable bodies. In the case of the plane TLTDBS problems in subsection 2.6.1 as applied to the compressible bodies and in subsection 2.6.2 as applied to the incompressible bodies, the Lagrangian coordinate system was used, introduced in the reference state (the first state) and coinciding in this reference state (the first state) with the Cartesian system coordinates.

When introducing the complex potentials of plane problems of the linearized mechanics of materials with the initial (residual) stresses, as in any other problem of this mechanics, the initial state should be considered as the reference state (the second, unperturbed state in the terminology of Sect. 2.1). Thus, in the second state, one should introduce the Lagrangian coordinates, which in the second state coincide with certain curvilinear coordinate systems. Besides, in this case, it is advisable to introduce the stress tensors, the components of which are measured per the unit area also in the second (unperturbed) state.

Below, following the above approach, the basic relations and general solutions of linearized mechanics of materials with initial (residual) stresses are considered, when the initial state is determined by relations (2.78) and (2.79) and is a homogeneous state. Also, the materials will be considered with the symmetry properties specified in the introductory part of Sect. 2.6.

2.6.4 Basic Relations and General Solutions of the TLTDBS in Coordinates of Initial State

It is worth noting that the main relations of the mechanics for the cases (2.78) and (2.79) are presented in a compact form in Sect. 1.4 of volume 1 of the two-volume monograph [64], which will be also followed below.

Following the above, in this case, the Lagrangian coordinates $y_m (m = 1, 2, 3)$ are introduced, which in the initial state (the second, the unperturbed; in the present case—the reference state) coincide with the Cartesian coordinates. In this case, the relationship between the introduced Cartesian coordinates y_n and the Cartesian coordinates x_n that were used in Sects. 2.5.1 and 2.5.2, for the same material point, has the following form:

$$y_n = \lambda_n x_n; \ \lambda_n = \text{const.} \tag{2.102}$$

The components of the displacement vector (2.79) for the initial state in coordinates y_n can be represented in the following form:

$$u_m^0 = \lambda_m^{-1}(\lambda_m - 1)y_m. \tag{2.103}$$

Later on, for all quantities that are considered in the initial (the second, the unperturbed, in this case—the reference) state and referred to the geometric objects in the same state, designations with the "prime" index are introduced. The exception is the complex potentials, in which the "prime" index denotes the derivative concerning the complex variable.

Thus, the following are introduced:

Q'_{nm} are the components of the asymmetric stress tensor.
P'_m are the components of the surface load vector on the area with the normal unit vector N.
ρ' is the density of the material.

For the introduced quantities, the following expressions take place

$$Q'_{nm} = (\lambda_1\lambda_2\lambda_3)^{-1}\lambda_n t_{nm}; \ P'_m = (\lambda_1\lambda_2\lambda_3)^{-1}\lambda_n N_n^{-1} P_m; \ \rho' = (\lambda_1\lambda_2\lambda_3)^{-1}\rho. \tag{2.104}$$

Note that in all monographs of the author of this monograph, when using the Lagrangian method for describing the motion of a continuum, N denotes the unit vector of the normal at a certain point of the material surface in the reference state, which is chosen as the undeformed (first) state. The components N_m of the unit vector of the normal in (2.104) also refer to the above unit vector of the normal. With the indicated designations, a surface load P acts at the same point of the material surface in a disturbed state (the third state, in the actual configuration).

For the formulation of the basic relations in the coordinates of the initial state, the unit vector N^0 of the normal to the same material surface at the same definite point in the initial (the second, the undisturbed, the subcritical state) is introduced in this section. In this case, the following relation holds [64] (vol. 1, p. 153):

$$N_j^0 = \left(\lambda_\alpha N_\alpha^{-1}\right)\lambda_j^{-1}N_j. \tag{2.105}$$

Taking into account expressions (2.102)–(2.105), the basic relations of linearized mechanics of materials with initial (residual) stresses are obtained in Cartesian coordinates $y_n(n = 1, 2, 3)$, introduced in the initial state, in the form of the following relations:

The equations of motion

$$Q'_{ij,i} - \rho'\ddot{u}_j = 0, \, y_n \in V'; \tag{2.106}$$

The incompressibility equations

$$u_{n,n} = 0, \, y_m \in V'; \tag{2.107}$$

The boundary conditions in stresses on a part S'_1 of the body surface

$$Q'_j = P'_j, \, y_n \in S'_1; \, Q'_j = N_i^0 Q'_{ij}; \tag{2.108}$$

The boundary conditions in displacements on a part S'_2 of the body surface

$$u_j = 0, \, y_n \in S'_2; \tag{2.109}$$

The linearized constitutive relations for compressible bodies

$$Q'_{ij} = \omega'_{ij\alpha\beta}u_{\alpha,\beta}, \tag{2.110}$$

where the tensor components ω' have the following form:

$$\omega'_{ij\alpha\beta} = (\lambda_1\lambda_2\lambda_3)^{-1}\lambda_i\lambda_\beta\omega_{ij\alpha\beta}; \tag{2.111}$$

The linearized constitutive relations for incompressible bodies

$$Q'_{ij} = \kappa'_{ij\alpha\beta}u_{\alpha,\beta} + \delta_{ij}p', \tag{2.112}$$

where the tensor components κ' have the following form:

$$\kappa'_{ij\alpha\beta} = \lambda_i\lambda_\beta\kappa_{ij\alpha\beta}. \tag{2.113}$$

It should be noted that in the monographs [1–6, 63, 64, 69], for the general and specific models of the elastic and plastic bodies and bodies with the rheological properties, the expressions are shown for determining the components of the tensors ω and κ under the homogeneous initial (residual) states or the subcritical states in the forms (2.78) and (2.79).

Thus, the expressions (2.106)–(2.113), together with the expressions for determining the components of the tensors ω and κ in the above monographs, represent the basic relations of linearized mechanics of materials with the initial (residual) stresses in the forms (2.78) and (2.79) or the TLTDBS (which is identical owing Note 2.2 for a homogeneous subcritical state in the forms (2.78) and (2.79)).

The above basic relations are constructed in the Lagrangian coordinates, which coincide with the Cartesian coordinates introduced in the initial (the second, the subcritical, and the unperturbed) state. In this case, all quantities are referred to as the geometric objects in the same initial state.

Note that the above-discussed basic relations are given in the notation that corresponds to the notation in Sect. 1.4 of the first volume of the monograph [64]. The above basic relations are presented in a unified general form for theories 2.1, 2.2, and 2.3 (according to the terminology of the introductory part of Sect. 2.4).

It is possible to construct for the above basic relations in the coordinates of the initial state the general solutions in a form similar to the general solutions in the coordinates of the first (the undeformed, the reference) state, which is described in subsections 2.5.1 and 2.5.2. These general solutions are presented in sufficient detail in Sect. 1.4 of the first volume of the monograph [64].

Below, as an example, the general solutions are given for the plane deformation when using the statical research method (Euler method) by analogy with the results of subsections 2.6.1 and 2.6.2. Since for the considered case (2.81) of the perturbations of the mass and surface forces, the sufficient conditions for the applicability of the statical method (Euler method) are satisfied, and the results obtained based on the discussed general solutions for the statical method will coincide with the results obtained in the framework of the dynamical method.

Thus, the plane deformation in the plane $y_1 0 y_2$ is considered. In this case, by analogy with (2.82) and (2.93), the following relations hold:

for the compressible bodies

$$u_1 = u_1(y_1, y_2); \ u_2 = u_2(y_1, y_2); \ u_3 = 0 \qquad (2.114)$$

and for incompressible bodies

$$u_1 = u_1(y_1, y_2); \ u_2 = u_2(y_1, y_2); \ u_3 = 0; \ p = p(y_1, y_2). \qquad (2.115)$$

In the considered case of the plane deformation, all quantities are determined through the functions $\chi'^{(j)}$ ($j = 1, 2$ for the compressible bodies and $j = 1, 2, 4$ the incompressible bodies) or through their linear combinations, which are solutions of the equation

$$\left(\frac{\partial^2}{\partial y_2^2} + \eta_2'^2\frac{\partial^2}{\partial y_1^2}\right)\left(\frac{\partial^2}{\partial y_2^2} + \eta_3'^2\frac{\partial^2}{\partial y_1^2}\right)\chi'^{(j)} = 0; \tag{2.116}$$

At that, the parameters $\eta_2'^2$ and $\eta_3'^2$ can be represented in the following form:

$$\eta_{2,3}'^2 = A' \pm \sqrt{A'^2 - A_1'}. \tag{2.117}$$

In (2.117), the following notation was introduced:

for the compressible bodies

$$2\omega_{2222}'\omega_{2112}'A' = \omega_{1111}'\omega_{2222}' + \omega_{2112}'\omega_{1221}' - \left(\omega_{1122}' + \omega_{1212}'\right)^2;$$

$$\omega_{2222}'\omega_{2112}'A_1' = \omega_{1111}'\omega_{1221}' \tag{2.118}$$

and for the incompressible bodies

$$2\kappa_{2112}'A' = \kappa_{1111}' + \kappa_{2222}' - 2\left(\kappa_{1122}' + \kappa_{1212}'\right); \ \kappa_{2222}'A_1' = \kappa_{1221}'. \tag{2.119}$$

Taking into account the above results and notation, the general solution can be represented in the following form:

for the compressible bodies

$$u_1 = \left(\omega_{1221}'\frac{\partial^2}{\partial y_1^2} + \omega_{2222}'\frac{\partial^2}{\partial y_2^2}\right)\chi'^{(2)} - \left(\omega_{1122}' + \omega_{1212}'\right)\frac{\partial^2}{\partial y_1\partial y_2}\chi'^{(2)};$$

$$u_2 = -\left(\omega_{1122}' + \omega_{1212}'\right)\frac{\partial^2}{\partial y_1\partial y_2}\chi'^{(2)} + \left(\omega_{1111}'\frac{\partial^2}{\partial y_1^2} + \omega_{2112}'\frac{\partial^2}{\partial y_2^2}\right)\chi'^{(1)} \tag{2.120}$$

and for the incompressible bodies

$$u_1 = -\frac{\partial^2}{\partial y_2^2}\chi'^{(1)} + \frac{\partial^2}{\partial y_1\partial y_2}\chi'^{(2)} - \left[\kappa_{1221}'\frac{\partial^2}{\partial y_1^2} + \left(\kappa_{2222}' - \kappa_{1122}' - \kappa_{1212}'\right)\frac{\partial^2}{\partial y_2^2}\right]\frac{\partial}{\partial y_1}\chi'^{(4)};$$

$$u_2 = \frac{\partial^2}{\partial y_1\partial y_2}\chi'^{(1)} - \frac{\partial^2}{\partial y_1^2}\chi'^{(2)} - \left[\left(\kappa_{1111}' - \kappa_{1122}' - \kappa_{1212}'\right)\frac{\partial^2}{\partial y_1^2} + \kappa_{2112}'\frac{\partial^2}{\partial y_2^2}\right]\frac{\partial}{\partial y_2}\chi'^{(4)};$$

$$p' = -\left[\kappa_{1221}'\frac{\partial^2}{\partial y_1^2} + \left(\kappa_{2222}' - \kappa_{1122}' - \kappa_{1212}'\right)\frac{\partial^2}{\partial y_2^2}\right]\frac{\partial}{\partial y_1}\chi'^{(1)}$$

$$- \left[\left(\kappa_{1111}' - \kappa_{1122}' - \kappa_{1212}'\right)\frac{\partial^2}{\partial y_1^2} + \kappa_{2112}'\frac{\partial^2}{\partial y_2^2}\right]\frac{\partial}{\partial y_2}\chi'^{(2)}. \tag{2.121}$$

The general solutions for the plane problem (plane deformation) in the forms (2.114)–(2.121) are used to introduce the complex potentials. The brief information and main results on this issue are given in the next subsection.

2.6.5   Complex Potentials in Plane Linearized Problems in Coordinates of Initial State

As already noted in subsection 2.5.2, the complex potentials in the plane problems for the statical research method (Euler method) were consistently and fully introduced in the monographs [69] in 1983. Preliminary, in a more reduced form, these results were presented in the articles listed in the bibliography to [69]. In subsequent years, the discussed results were presented in full in volume 2 of the collective monograph [70] in 1991, in the supplement to the monograph [2] in 1999, and in volume 2 of the two-volume monograph [64] in 2008.

Below, brief information is given about the complex potentials in the plane linearized problems, which are introduced in the coordinates of the initial (the second, the unperturbed) state. These results in more detail are presented in Sect. 7.2 of the second volume of the two-volume monograph [64], the notation and style of presentation of which is also applied below concerning the elastic and plastic materials.

Along with the parameters η'^2_2 and η'^2_3, which are introduced by the expressions (2.117)–(2.119) and which completely characterize the considered compressible or incompressible material and the acting initial (residual) stresses, the complex parameters μ'_1 and μ'_2 are also introduced by the following relations:

$$\mu'^2_{1,2} = -\eta'^2_{2,3} = -\left(A' \pm \sqrt{A'^2 - A'} \right); \; \mu'_{1,2} = i\sqrt{A' \pm \sqrt{A'^2 - A'_1}}. \quad (2.122)$$

The complex variables in the considered case of the plane deformation are introduced as follows:

$$z_k = y_1 + \mu'_k y_2; \; \bar{z}_k = y_1 + \bar{\mu}'_k y_2; \; k = 1, 2. \quad (2.123)$$

By (2.117), (2.118), and (2.122), the complex variables z_k (2.123) depend on the properties of the considered compressible or incompressible material and the acting initial (residual) or subcritical stresses.

It is strictly proved that there are the cases of the equal and unequal complex parameters or roots μ'_1 and μ'_2, for which certain conditions are satisfied:

in the case of unequal roots

$$\mu'_1 \neq \mu'_2; \; \mu'_1 = -\bar{\mu}'_2 \quad (2.124)$$

in the case of equal roots

$$\mu'_1 = \mu'_2; \; \bar{\mu}'_1 = -\mu'_1; \; \text{Re} \, \mu'_1 = 0. \tag{2.125}$$

After introducing the complex variables in the form (2.123), taking into account the notation (2.122), the representation of stresses Q'_{ij} and displacements u_m for the plane problem in terms of the complex potentials is constructed using Eq. (2.116) and representations of the general solutions in the form (2.120) for the compressible bodies and the form (2.121) for the incompressible bodies.

In this construction, certain linear combinations of functions $\chi'^{(1)}$, $\chi'^{(2)}$ and $\chi'^{(4)}$ are used and to provide a transition to the classical complex representations of Kolosov–Muskhelishvili [71] and Lekhnitsky [72] when the initial (residual, subcritical) stresses tend to zero. The corresponding transformations are presented in sufficient detail in the monograph [69], in volume 2 of the collective monograph [70], and volume 2 of the two-volume monograph [64].

Below the complex representations are given for stresses and displacements separately for the cases of unequal (2.124) and equal (2.125) roots, following Sect. 7.2 of the second volume of the two-volume monograph [64].

So, for the case of unequal roots (2.124), the representation holds

$$Q'_{22} = 2\text{Re}\left[\Phi'_1(z_1) + \Phi'_2(z_2)\right];$$
$$Q'_{21} = -2\text{Re}\left[\gamma^{(1)}_{21}\mu'_1\Phi'_1(z_1) + \gamma^{(2)}_{21}\mu'_2\Phi'_2(z_2)\right];$$
$$Q'_{12} = -2\text{Re}\left[\mu'_1\Phi'_1(z_1) + \mu'_2\Phi'_2(z_2)\right]; \tag{2.126}$$

$$Q'_{11} = 2\text{Re}\left[\gamma^{(1)}_{11}\mu'^2_1\Phi'_1(z_1) + \gamma^{(2)}_{11}\mu'^2_2\Phi'_2(z_2)\right];$$
$$u_k = 2\text{Re}\left[\gamma^{(1)}_k\Phi_1(z_1) + \gamma^{(2)}_k\Phi_2(z_2)\right]; \; k = 1, 2.$$

In (2.126) and below, as noted after expressions (2.103), the "prime" index near the functions of complex variables denotes the derivative of functions concerning a complex variable.

Thus, the expressions (2.126) determine the stresses and displacements of the plane linearized problem in the case (2.124) of the unequal roots through the complex potentials $\Phi_1(z_1)$ and $\Phi_2(z_2)$. The functions $\Phi_1(z_1)$ and $\Phi_2(z_2)$ are the analytical functions in the area occupied by the material under consideration.

For the compressible and incompressible materials with models of a fairly general form, the expressions are given in [64] (Sect. 2, Chap. 7, vol. 2) for determining the quantities $\gamma^{(j)}_{21}, \gamma^{(j)}_{11}, \gamma^{(j)}_1$, and $\gamma^{(j)}_2$ (for $j = 1, 2$) through $\mu'_1, \mu'_2, \omega'_{nm\alpha\beta}$, and $\kappa'_{nm\alpha\beta}$.

For the case of equal roots (2.125), the following representation holds:

$$Q'_{22} = \text{Re}\left\{\left[\Psi(z_1) + \bar{z}_1\Phi'(z_1)\right] + \gamma^{(2)}_{22}\Phi(z_1)\right\};$$
$$Q'_{21} = \text{Re}\left\{\mu'_1\gamma^{(1)}_{21}\left[\Psi(z_1) + \bar{z}_1\Phi'(z_1)\right] + \gamma^{(2)}_{21}\Phi(z_1)\right\};$$

$$Q'_{12} = \mathrm{Re}\left\{-\mu'_1[\Psi(z_1) + \bar{z}_1\Phi'(z_1)] + \gamma^{(2)}_{12}\Phi(z_1)\right\};$$

$$Q'_{11} = \mathrm{Re}\left\{\mu'^2_1\gamma^{(1)}_{11}[\Psi(z_1) + \bar{z}_1\Phi'(z_1)] + \gamma^{(2)}_{11}\Phi(z_1)\right\};$$

$$u_k = \mathrm{Re}\left\{\gamma^{(1)}_k[\psi(z_1) + \bar{z}_1\varphi'(z_1)] + \gamma^{(2)}_k\varphi(z_1)\right\}; \quad k = 1, 2;$$

$$\Phi(z_1) = \varphi'(z_1), \quad \Psi(z_1) = \psi'(z_1). \tag{2.127}$$

In (2.127) and below, as noted after expressions (2.103) and (2.126), the "prime" index near the functions of complex variables denotes the derivative of functions concerning a complex variable.

Thus, the expressions (2.127) determine the stresses and displacements of the plane linearized problem in the case (2.125) of equal roots through the complex potentials $\Psi(z_1)$ and $\Phi(z_1)$. The functions $\Psi(z_1)$ and $\Phi(z_1)$ are the analytical functions in the area occupied by the material under consideration. For the compressible and incompressible materials with models of a sufficient general form, the expressions are given in [64] (Sect. 2.2, Chap. 7, vol. 2) for determining the quantities $\gamma^{(2)}_{22}$, $\gamma^{(j)}_{21}$, $\gamma^{(2)}_{12}$, $\gamma^{(j)}_{11}$, $\gamma^{(j)}_1$, and $\gamma^{(j)}_2$ (at $j = 1, 2$) through μ'_1, $\omega'_{nm\alpha\beta}$, and $\kappa'_{nm\alpha\beta}$.

It should be noted that representations (2.126) and (2.127) of displacements and stresses of the plane linearized problems in terms of the analytical functions of complex variables refer to the materials with the *homogeneous* mechanical properties in the case of constitutive equations of a sufficiently general form, having the considered symmetry properties, and the *homogeneous* initial (residual, subcritical) states in the forms (2.78) and (2.79).

The discussed representations (2.126) and (2.127) for the above generality of the problem statement make it possible, using the methods of the theory of analytic functions of the complex variables, to obtain the exact solutions of various mixed problems when the boundary conditions are given at $y_2 = \mathrm{const}$.

In particular, using the Keldysh–Sedov formula [73] and representation in terms of the complex potentials in the form (2.126) for the unequal roots (2.124) and the form (2.127) for the equal roots (2.125) with the considered generality of the problem statement (in a unified general form for theories 2.1, 2.2, and 2.3, according to the terminology of the introductory part of Sect. 2.4), the exact solutions are obtained for a crack located in a plane $y_2 = \mathrm{const}$.

For the simplest case of an *inhomogeneous* material (piece-wise homogeneous material, the crack is located at the interface at $y_2 = \mathrm{const}$), it is no longer possible to obtain the exact solutions using the above methods. In this case (the piece-wise homogeneous material, the mixed boundary conditions are given on the line of separation of material properties at $y_2 = \mathrm{const}$) for the statical problems within the framework of the classical linear theory of elasticity of the isotropic body, the method for solving these problems is described in the Muskhelishvili's monograph [71], which is based on reducing to the conjugation problem if two holomorphic functions are defined in the whole plane.

Below, in a very brief form, the results are presented on the reduction of the plane linearized problems with the considered generality of the statement of the problems for the case under discussion (the piece-wise homogeneous material, the interface $y_2 = 0$, crack in the interface) to the problem of the conjugation of two holomorphic functions given in the entire plane.

These results in more detail are presented in the monograph [64] (Sect. 2, Chap. 8, vol. 2).

Denote:

L_1 as the set of segments on the line $y_2 = 0$ corresponding to cracks;
L_2 as the set of segments on the line $y_2 = 0$ corresponding to the complete connection of two components of the composite material (continuity of stress and displacement vectors).

In this case, the boundary conditions on the connection line (at $y_2 = 0$) following [64] (vol. 2, p. 171) can be represented in the following form:

$$Q_{22}^{\prime(\pm)} = 0, \ Q_{21}^{\prime(\pm)} = 0 \text{ for } y_1 \in L_1 \text{ and } y_2 = 0;$$

$$Q_{22}^{\prime(+)} = Q_{22}^{\prime(-)}, \ Q_{21}^{\prime(+)} = Q_{21}^{\prime(-)}, \ \frac{\partial}{\partial y_1}\left(u_1^{(+)} - u_1^{(-)}\right) = 0,$$

$$\frac{\partial}{\partial y_1}\left(u_2^{(+)} - u_2^{(-)}\right) = 0 \text{ for } y_1 \in L_2 \text{ and } y_2 = 0; \qquad (2.128)$$

At that, the corresponding damping conditions "at infinity" $y_2 \to \pm\infty$ are also obtained, since composite material is considered composed of two half-planes. The subscripts $(\pm)$ in (2.128) and below all quantities are indicated related to the above two half-planes.

The belonging of the complex potentials in the representations (2.126) and (2.127) to a certain concrete material occupying the corresponding half-plane is achieved by using the complex parameters μ_1' and μ_2' (2.122) calculated for the same material.

Thus, in the discussed case of a piece-wise homogeneous material consisting of two half-planes, two complex potentials (in representations (2.126) and (2.127)) for each of the half-planes are obtained. The above complex potentials are the analytical functions *only* for the corresponding half-planes.

To develop a method for constructing the exact solutions for the connected two half-planes in the framework of the plane linearized problem, similar to Muskhel-ishvili's method [71] in the framework of the plane problem of the classical linear theory of elasticity of the isotropic body, it is necessary, instead of representations (2.126) and (2.127), to construct the representations for *the half-plane* in terms of the analytical functions complex variables when these functions are defined for the entire plane.

Representations of this type are constructed in the articles [42–45, 48] and are presented in sufficient detail in the monograph [64] (Sect. 2, Chap. 8, vol. 2).

In this regard, below the corresponding representations are only given for the quantities that enter the boundary conditions (2.128) on the connection line (at $y_2 = 0$) separately for the case of the unequal roots (2.124) and the case of the equal roots (2.125).

Thus, for the case of the unequal roots (2.124) for the quantities included in the boundary conditions (2.128), the representation is obtained

$$Q'_{22} = 2\text{Re}\{C_{11}[\Phi(z_1) - \Phi(\bar{z}_2)] + C_{12}[\overline{\Phi(z_1)} - \overline{\Phi(z_2)}]\};$$
$$Q'_{21} = 2\text{Re}\{C_{21}[\Phi_1(z_1) - \Phi(\bar{z}_2)] + C_{22}[\Phi(z_1) - \Phi(z_2)]\};$$
$$\frac{\partial u_1}{\partial y_1} = 2\text{Re}[C_{31}\Phi(z_1) + C_{32}\Phi(\bar{z}_2) + C_{33}\Phi(z_2)];$$
$$\frac{\partial u_2}{\partial y_1} = 2\text{Re}[C_{41}\Phi(z_1) + C_{42}\Phi(\bar{z}_2) + C_{43}\Phi(z_2)]. \tag{2.129}$$

The expressions (2.129) refer equally to the upper and lower half-planes. To obtain the expressions relating only to the upper or lower half-planes, it is necessary to put indices $(\pm)$ in all quantities and functions included in (2.129).

In the monograph [64] (vol. 2, p. 175), the expressions are presented for the determination in (2.129) through μ'_1, μ'_2 and the coefficients in the representation (2.126) for the case of the unequal roots (2.124). It should be noted that in (2.129), the function $\Phi(z)$ is holomorphic in the entire plane.

Similarly, for the case of the equal roots (2.125) for the quantities included in the boundary conditions (2.128), the following representation is obtained:

$$Q'_{22} = 2\text{Re}\{C_{11}[\Phi(z_1) - \Phi(\bar{z}_1)]\} + C_{12}(z_1 - \bar{z}_1)[\Phi'(z_1) - \overline{\Phi'(z_1)}];$$
$$Q'_{21} = 2\text{Re}\{C_{21}[\Phi(z_1) - \Phi(\bar{z}_1)]\} + C_{22}(z_1 - \bar{z}_1)[\Phi'(z_1) + \overline{\Phi'(z_1)}];$$
$$\frac{\partial u_1}{\partial y_1} = 2\text{Re}[C_{31}\Phi(z_1) + C_{32}\Phi(\bar{z}_1)] + C_{33}(z_1 - \bar{z}_1)[\Phi'(z_1) - \overline{\Phi'(z_1)}];$$
$$\frac{\partial u_2}{\partial y_1} = 2\text{Re}[C_{41}\Phi(z_1) + C_{42}\Phi(\bar{z}_1)] + C_{43}(z_1 - \bar{z}_1)[\Phi'(z_1) + \overline{\Phi'(z_1)}]. \tag{2.130}$$

The expressions (2.130) refer equally to the upper and lower half-planes. To obtain the expressions relating only to the upper or lower half-planes, it is necessary to put the indices $(\pm)$ in all values and functions included in (2.130).

In the monograph [64] (vol. 2, p. 178), the expressions are presented for determining C_{nm} from (2.130), through μ'_1 and the coefficients entering into the representation (2.127) for the case of the equal roots (2.125). It should be noted that in (2.130), the function $\Phi(z)$ is holomorphic in the entire plane.

Using the representations (2.129) in the case of the unequal roots (2.124) and (2.130) in the case of the equal roots (2.125), the *mixed* plane linearized problems or the TLTDBS problems for a piece-wise homogeneous material (two half-planes connected to each other) with an interface $y_2 = 0$ are reduced to the problem of the conjugation of two functions holomorphic in the whole plane.

In many cases, it is possible to obtain an exact solution of the discussed problem of the conjugation of two functions within the framework of the mathematical apparatus of plane linearized problems. In particular, in the articles [8, 9, 11, 12, 35–40, 48, 74–102] of 2000–2001 and in monograph [64] (vol. 2, Chap. 8, Sect. 2) of 2008, the exact solution of the problem of stability of a piece-wise homogeneous material (two half-planes connected to each other) under compression along the interface containing the plane cracks is presented with the General Statement of the problems considered in Sect. 2.6.

It is expedient to note that the monograph [64] (vol. 2) also presents a representation in terms of the complex potentials of the type (2.126) and (2.127) for the anti-plane linearized problem in the coordinates of the initial state. To reduce the volume of the presentation, these results are not discussed in this monograph.

2.6.6 Complex Potentials in Dynamical Plane Linearized Problems in Coordinates of Initial State for Moving Cracks and Loads

In this subsection, in a very short form, short even in comparison with subsections 2.6.4 and 2.6.5, the main results are presented on the introduction of the complex potentials in the dynamic plane linearized problems in the coordinates of the initial state, when the cracks are located in the plane $y_2 = \text{const}$ and move uniformly and rectilinearly with a constant velocity $v = \text{const}$ along the axis $0y_1$. The plane cracks in the plane $y_2 = \text{const}$ are considered, which are infinite in the direction of the axis $0y_3$ and have a constant width along the axis $0y_3$. Since the above cracks are located in the plane $y_2 = \text{const}$, then in addition to (2.78) and (2.79) for the initial (residual, unperturbed, subcritical) state, the following condition is accepted:

$$\sigma_{22}^0 = 0. \tag{2.131}$$

It is also assumed that the materials have the same symmetry properties as in Sects. 2.6.3 and 2.6.4. The studies are carried out for the subsonic regime of the crack motion, i.e., the following conditions are met:

$$c_{l1} > v, c_{S12} > v, c_{S13} > v. \tag{2.132}$$

In (2.132), the designations of the monograph [7] are used for the "true" velocities of motion of the plane waves in bodies with the initial stresses (2.78), (2.79), and (2.131):

c_{lm}—the velocity of longitudinal waves propagating along the axis $0y_m$;
c_{smn}—the velocity of transverse waves (shear waves) propagating along the axis $0y_m$ and polarized in the plane y_m0y_n;
c_{smk}—the velocity of transverse waves (shear waves) propagating along the axis $0y_m$ and polarized in the plane y_m0y_k; $m \neq n \neq k \neq m$.

Under the conditions mentioned in subsection 2.6.6, the process of introducing and applying the complex potentials in the dynamical plane linearized problems in coordinates of the initial state for the moving cracks and loads is described in sufficient detail in the monograph [69] of 1983. Preliminarily, the results obtained before 1983 were published in periodicals in the scientific articles, which are indicated in the list of references to the monograph [69]. The most fully these results for the dynamical problems, including the results for piece-wise homogeneous materials (composite materials) such as those in the final part of subsection 2.6.5, are presented in the monograph [64] (vol. 2, Chap. 10) of 2008. Preliminarily, the corresponding results obtained in 1983–2007 were published in the scientific articles, which are indicated in the list of references to this monograph.

Additionally, a modern review [10] should be mentioned, which is devoted to the analysis of the results on the mechanics of moving cracks in the materials with initial (residual) stresses, which were obtained before 2011.

In the case under discussion, to introduce the complex potentials for the plane linearized problems as applied to the moving cracks and loads, one should proceed from the basic relations in the forms (2.106)–(2.113) of the dynamics of materials with the considered initial (residual) stresses formulated in the coordinates of the initial state. Along with the Cartesian coordinate system with coordinates $y_j (j = 1, 2, 3)$, a movable Cartesian coordinate system with coordinates $\eta_j (j = 1, 2, 3)$ is introduced as follows:

$$\eta_1 = y_1 - v\tau; \ \eta_2 = y_2; \ \eta_3 = y_3. \tag{2.133}$$

For the plane problem in a plane $\eta_1 O \eta_2$ in the moving coordinates $\eta_j (j = 1, 2, 3)$, the complex variables $z_k (k = 1, 2)$ are introduced by the following expressions:

$$
\begin{aligned}
z_k &= \eta_1 + \mu'_k \eta_2 \equiv y_1 - v\tau + \mu'_k y_2; \\
\bar{z}_k &= \eta_1 + \bar{\mu}'_k \eta_2 \equiv y_1 - v\tau + \bar{\mu}'_k y_2; k = 1, 2,
\end{aligned} \tag{2.134}
$$

where the complex parameters μ'_1 and μ'_2 are determined by the second expression (2.122), in which the following designations are introduced:

for the compressible materials

$$
\begin{aligned}
2A'\omega'_{2222}\omega'_{2112} &= (\omega'_{1111} - \rho'v^2)\omega'_{2222} + \omega'_{2112}(\omega'_{1221} - \rho'v^2) - (\omega'_{1122} + \omega'_{1212})^2; \\
A'\omega'_{2222}\omega'_{2112} &= (\omega'_{1111} - \rho'v^2)(\omega'_{1221} - \rho'v^2)
\end{aligned} \tag{2.135}
$$

for the incompressible materials

$$
\begin{aligned}
2A'\kappa'_{2112} &= \kappa'_{1111} - \rho v^2 + \kappa'_{2222} - 2(\kappa'_{1122} + \kappa'_{1212}); \\
A'_1\kappa'_{2112} &= \kappa'_{1212} - \rho v^2.
\end{aligned} \tag{2.136}
$$

With the notation for A' and A'_1 in the form (2.135) for the compressible materials and in the form (2.136) for the incompressible materials and the second expression

(2.122) for determining the roots μ'_1 and μ'_2, it is rigorously proved that two cases have the place—the case of the unequal roots in the form (2.124) and the case of the equal roots in the form (2.125).

The further procedure for introducing the complex potentials corresponds to the procedure described in subsection 2.6.5 for the statical problems.

These results are presented in detail in the monograph [64] (vol. 2, Chap. 10). In this regard, below the final results only are presented, taking into account that for dynamical problems the complex variables are introduced by relations (2.134) and the complex roots μ'_1 and μ'_2 are determined by the second expression (2.122) with notation (2.135) and (2.136).

So, in the case of the unequal roots (2.124), the stresses and displacements of the dynamical plane linearized problems are determined by the following relations:

$$Q'_{22} = 2\mathrm{Re}\big[\Phi'_1(z_1) + \Phi'_2(z_2)\big];$$
$$Q'_{21} = -2\mathrm{Re}\big[\mu'_1\gamma^{(1)}_{21}\Phi'_1(z_1) + \mu'_2\gamma^{(2)}_{21}\Phi'_2(z_2)\big];$$
$$Q'_{12} = -2\mathrm{Re}\big[\mu'_1\gamma^{(1)}_{12}\Phi'_1(z_1) + \mu'_2\gamma^{(2)}_{12}\Phi'_2(z_2)\big];$$
$$Q'_{11} = 2\mathrm{Re}\big[\mu'^2_1\gamma^{(1)}_{11}\Phi'_1(z_1) + \mu'^2_2\gamma^{(2)}_{11}\Phi'_2(z_2)\big];$$
$$u_k = 2\mathrm{Re}\big[\gamma^{(1)}_k\Phi_1(z_1) + \gamma^{(2)}_k\Phi_2(z_2)\big]; \, k = 1, 2 \qquad (2.137)$$

through the complex potentials $\Phi_1(z_1)$ and $\Phi_2(z_2)$, which are the analytical functions in the area occupied by the material in question.

It should be noted that the representations (2.137) for the dynamic problems with the unequal roots *differ* from the corresponding representations (2.126) for the statical problems with the unequal roots not only by different expressions for determining the quantities μ'_1, μ'_2, A', A'_1, ..., $\gamma^{(2)}_k$, but also by the fact that in (2.137) the quantities $\gamma^{(j)}_{12} \neq 1$ at $j = 1, 2$.

In the monograph [64] (vol. 2, p. 339) for the compressible and incompressible materials with the models of a sufficiently general form, the expressions are given for determining the quantities $\gamma^{(j)}_{21}$, $\gamma^{(j)}_{12}$, $\gamma^{(j)}_{11}$, $\gamma^{(j)}_1$, $\gamma^{(j)}_2$ (at $j = 1, 2$) through μ'_1, μ'_2, $\omega'_{nma\beta}$, and $\kappa'_{nma\beta}$.

In the case of the equal roots (2.125), the stresses and displacements of the dynamical plane linearized problems are determined by the following relations:

$$Q'_{22} = \mathrm{Re}\Big\{\big[\Psi(z_1) + \bar{z}_1\Phi'(z_1)\big] + \gamma^{(2)}_{22}\Phi(z_1)\Big\};$$
$$Q'_{12} = \mathrm{Re}\Big\{-\mu'_1\gamma^{(1)}_{12}\big[\Psi(z_1) + \bar{z}_1\Phi'(z_1)\big] + \gamma^{(2)}_{12}\Phi(z_1)\Big\};$$
$$Q'_{11} = \mathrm{Re}\Big\{\mu'^2_1\gamma^{(1)}_{11}\big[\Psi(z_1) + \bar{z}_1\Phi'(z_1)\big] + \gamma^{(2)}_{11}\Phi(z_1)\Big\};$$
$$u_k = \mathrm{Re}\Big\{\gamma^{(1)}_k[\psi(z_1) + \bar{z}_1\varphi(z_1)] + \gamma^{(2)}_k\varphi(z_1)\Big\}; \, k = 1, 2;$$
$$\Phi(z_1) = \varphi'(z_1), \, \Psi(z_1) = \psi'(z_1) \qquad (2.138)$$

through the complex potentials $\Phi(z_1)$ and $\Psi(z_1)$, which are the analytical functions in the area occupied by the material in question.

It should be noted that the representations (2.138) for the dynamical problems with the equal roots *differ* from the corresponding representations (2.127) for the statical problems with equal roots not only by different expressions for determining the quantities μ_1', A', A_1', $\gamma_{22}^{(2)}$, …, $\gamma_k^{(2)}$, but also by the fact that in (2.138) the quantity $\gamma_{12}^{(1)} \neq 1$.

In the monograph [64] (vol. 2, pp. 342–344), for the compressible and incompressible materials with the models of a sufficiently general form, the expressions are given for determining the quantities $\gamma_{22}^{(2)}$, $\gamma_{21}^{(j)}$, $\gamma_{12}^{(j)}$, $\gamma_{11}^{(j)}$, $\gamma_1^{(j)}$, and $\gamma_2^{(j)}$ (at $j = 1, 2$) in terms of μ_1', $\omega_{nm\alpha\beta}'$, and $\kappa_{nm\alpha\beta}'$.

Note also that the complex representations of stresses and displacements of the dynamical plane linearized problems in the form (2.137) for the unequal roots (2.124) and the form (2.138) for the equal roots (2.125) as the initial (residual, subcritical) stresses tend to zero, pass into the known Galin complex representations [103] of the dynamical plane problem of the classical linear theory of elasticity.

It should be noted that the representations (2.137) and (2.138) of displacements and stresses of the dynamical plane linearized problems in terms of the analytical functions of complex variables, as well as the corresponding representations (2.126) and (2.127) for the statical plane linearized problems, refer to the materials with the *homogeneous* mechanical properties in the case of the constitutive equations of a sufficiently general form with the considered symmetry properties and the *homogeneous* initial (residual, subcritical) states in the forms (2.78) and (2.79).

The representations (2.137) and (2.138) for the above generality of the problem statement make it possible (using the methods of the theory of the analytical functions of complex variables) to obtain the exact solutions of various mixed problems when the moving mixed boundary conditions are given for $y_2 = $ const. In particular, using the Keldysh–Sedov formula [736] and representations for the dynamical plane linearized problems in terms of the complex potentials in the form (2.137) for the case of the unequal roots (2.124) and in the form (2.138) for the equal roots (2.125) with the considered generality of the statement of the problems (in a unified general form for theories 2.1, 2.2, and 2.3 in the terminology of the introductory part of Sect. 2.4), the exact solutions for a crack moving in a plane $y_2 = $ const are obtained.

For the simplest case of an *inhomogeneous* material (piece-wise homogeneous material, a moving crack is located at the interface $y_2 = $ const), it is no longer possible to obtain the exact solutions using the above methods. In the case of the dynamical plane linearized problems for two connected half-planes of different materials, along the dividing line of which (interface) the plane cracks move, it is necessary to construct a representation of stresses and displacements for each of the half-planes through the analytical functions that are defined for the entire plane.

A similar situation takes place for the statical plane linearized problems, which was studied in the second half of subsection 2.6.5.

It should be noted that in the above complex representations (2.137) and (2.138) for the dynamical plane linearized problems, the complex potentials are presented, which are the analytical functions only for the areas occupied by the material under consideration. In this case (two connected half-planes) in the representations (2.137) and (2.138), the complex potentials are the analytical functions for the corresponding half-planes.

After obtaining the representation of the stresses and displacements of the dynamical plane linearized problems for a *half-plane* in terms of the analytical functions that are defined for the *entire plane*, the exact solutions in several cases can be obtained by reducing to the Riemann–Hilbert problem or, in the terminology of the monograph [71], to the problem of the conjugation of two holomorphic functions, given in the entire plane.

The corresponding results are presented in a fairly complete form in the monograph [64] (vol. 2, Chap. 10, Sects. 5 and 6). Therefore, below some of the main results are only presented.

The boundary conditions at the interface (at $y_2 = 0$) for the moving cracks at the interface in a moving coordinate system, by analogy with (2.128), can be represented in the following form:

$$Q_{22}^{\prime(\pm)} = 0, \; Q_{21}^{\prime(\pm)} = 0 \; \text{for} \eta_1 \in L_1 \text{and} \eta_2 = 0;$$

$$Q_{22}^{\prime(+)} = Q_{22}^{\prime(-)}, \; Q_{21}^{\prime(+)} = Q_{21}^{\prime(-)}, \; \frac{\partial}{\partial \eta_1}\left(u_1^{(+)} - u_1^{(-)}\right) = 0, \; \frac{\partial}{\partial \eta_1}\left(u_2^{(+)} - u_2^{(-)}\right) = 0;$$

$$\tag{2.139}$$

for $\eta_1 \in L_2$ and $\eta_2 = 0$.

The representation of the quantities included in (2.139) for each of the half-planes is given for the unequal and equal roots μ_1' and μ_2', which are determined by the second expression in (2.122) under the notation (2.135) and (2.136) for the dynamical plane linearized problems.

So, in the case of the unequal roots (2.124), the quantities included in (2.139) for each of the half-planes are determined by the following expressions:

$$Q_{22}' = 2\mathrm{Re}\{C_{11}[\Phi(z_1) - \Phi(\bar{z}_2)] + C_{12}[\overline{\Phi(z_1)} - \overline{\Phi(z_2)}]\};$$
$$Q_{21}' = 2\mathrm{Re}\{C_{21}[\Phi(z_1) - \Phi(\bar{z}_2)] + C_{22}[\Phi(z_1) - \Phi(z_2)]\};$$
$$\frac{\partial u_1}{\partial \eta_1} = 2\mathrm{Re}[C_{31}\Phi(z_1) + C_{32}\Phi(\bar{z}_2) + C_{33}\Phi(z_2)];$$
$$\frac{\partial u_2}{\partial \eta_1} = 2\mathrm{Re}[C_{41}\Phi(z_1) + C_{42}\Phi(\bar{z}_2) + C_{43}\Phi(z_2)] \tag{2.140}$$

in terms of the function $\Phi(z_j)$ at $j = 1, 2$, which is holomorphic in the entire plane.

The monograph [64] (vol. 2, p. 401) presents the expressions for the determination C_{nm} from (2.140) through μ_1', μ_2' and the coefficients in the representation (2.137) for the case of the unequal roots (2.124), as applied to the dynamical plane linearized problem. The expressions (2.140) relate equally to the upper and lower half-planes. To obtain the expressions relating only to the upper or lower half-planes, it is necessary to put the index ($\pm$) in all quantities and functions included in (2.140).

In the case of the equal roots (2.125), the quantities in (2.139) for each of the half-planes are determined by the following expressions:

$$Q'_{22} = 2\mathrm{Re}\{C_{11}[\Phi(z_1) - \Phi(\bar{z}_1)]\} + C_{12}(z_1 - \bar{z}_1)\big[\Phi'(z_1) - \overline{\Phi'(z_1)}\big];$$

$$Q'_{21} = 2\mathrm{Re}\{C_{21}[\Phi(z_1) - \Phi(\bar{z}_1)]\} + C_{22}(z_1 - \bar{z}_1)\big[\Phi'(z_1) + \overline{\Phi'(z_1)}\big];$$

$$\frac{\partial u_1}{\partial \eta_1} = 2\mathrm{Re}[C_{31}\Phi(z_1) + C_{32}\Phi(\bar{z}_1)] + C_{33}(z_1 - \bar{z}_1)\big[\Phi'(z_1) - \overline{\Phi'(z_1)}\big];$$

$$\frac{\partial u_2}{\partial \eta_1} = 2\mathrm{Re}[C_{41}\Phi(z_1) + C_{42}\Phi(\bar{z}_1)] + C_{43}(z_1 - \bar{z}_1)\big[\Phi'(z_1) + \overline{\Phi'(z_1)}\big] \quad (2.141)$$

through a function $\Phi(z_1)$ that is holomorphic in the entire plane.

The monograph [64] (vol. 2, p. 405) presents the expressions for the determination C_{nm} from (2.141) through μ'_1 and the coefficients in the representation (2.138) for the case of the equal roots (2.125), as applied to the dynamical plane linearized problem. The expressions (2.141) refer equally to the upper and lower half-planes. To obtain the expressions relating to the upper or lower half-planes, it is necessary to put the index ($\pm$) in all quantities and functions included in (2.141).

Using the representations (2.140) and (2.141) through one function, which is holomorphic in the entire plane, and with the subsequent reduction to the Riemann–Hilbert problem, a sequential analysis of a number of the dynamical plane linearized problems for a composite material consisting of two connected half-planes of different materials is carried out in the interface of which a plane crack moves. The corresponding rigorous results were originally published in articles [74–77] of 2002 and presented in a fairly complete form in the monograph [64] (vol. 2, Chap. 10, Sects. 2.5 and 2.6) of 2008. An analysis of the results on the construction of the mechanics of moving cracks in materials with the initial (residual) stresses is presented in a modern review [10] of 2011.

It is advisable to note that as applied to the dynamical linearized problems in the coordinates of the initial state for the moving cracks and loads, the representation of stresses and displacements through the complex potentials is obtained not only for the plane problem, for which they have the form (2.137) for the case of the unequal roots (2.124) and the form (2.138) for the case of the equal roots (2.125) but also for the corresponding anti-plane problem.

For this situation (the anti-plane problem), the representations of stresses and displacements have a simpler form in comparison with the representations (2.137) and (2.138). The corresponding results are given in the monograph [64] (vol. 2, Chap. 10, Sect. 2) to reduce the volume of presentation, and the above results are not presented here.

Note 2.14 In subsections 2.6.5 and 2.6.6 for the statical and dynamical plane linearized problems in the coordinates of the initial state, the presentation of stresses and displacements through the complex potentials in a unified general form for theories 2.1, 2.2, and 2.3 (according to the terminology of the introductory part of Sect. 2.4) is presented. In these results, when the Lagrangian coordinates are introduced, which in the initial (residual, subcritical, unperturbed, second) state coincide with the Cartesian coordinates, the **change** in the geometric objects during the transition from the reference (the first) state to the second (the unperturbed) state is taken into account.

It follows from the results of Sect. 2.2 that in theory 2.2 (*the first variant of the theory of small subcritical deformations*) and theory 2.3 (*the second variant of the theory of small subcritical deformations*), the above **changes** *should not be taken into account* due to the principles and accuracy of constructing the theories discussed.

Thus, it follows from the foregoing that in a sequential consideration of these theories, the noted change must be taken into account only in theory 2.1 (*the theory of large (finite) subcritical deformations*).

It should be noted in conclusion to this chapter that here the mathematical foundations of the TLTDBS are presented rather consistently, which makes it possible in the subsequent chapters not to cite the considered results concerning each corresponding problem.

References

1. A.N. Guz, Osnovy trekhmernoi teorii ustoichivosti deformiruemykh tel (Fundamentals of the three-dimensional theory of stability of deformable bodies). (Vyshcha Shkola, Kyiv, 1986)
2. Guz, A.N.: Fundamentals of the Three-Dimensional Theory of Stability of Deformable Bodies. Springer, Berlin Hiedelberg New York (1999)
3. Guz, A.N.: Ustoichivost trekhmernykh deformiruemykh tel (Stability of Three-Dimensional Deformable Bodies). Naukova Dumka, Kyiv (1971)
4. Guz, A.N.: Ustoichivost uprugikh tel pri konechnykh deformatsiiakh (Stability of Elastic Bodies under Finite Deformations). Naukova Dumka, Kyiv (1973)
5. Guz, A.N.: Osnovy teorii ustoichivosti gornykh vyrabotok (Fundamentals of the Theory of Stability of Mine Workings). Naukova Dumka, Kyiv (1977)
6. Guz, A.N.: Ustoichivost uprugikh tel pri vsestoronnem szhatii (Stability of Elastic Bodies under All-Round Compression). Naukova Dumka, Kyiv (1979)
7. Guz, A.N.: Uprugie volny v telakh s nachalnymi (ostatochnymi) napriazheniiami. V 2-kh chastiakh (Elastic Waves in Bodies with Initial (Residual) Stresses. In 2 parts). LAP LAMBERT Academic Publishing, Saarbrücken, Deutschland (2016)
8. Guz, A.N.: Construction of the three-dimensional theory of stability of deformable bodies. Int. Appl. Mech. **37**(1), 1–37 (2001)
9. Guz, A.N.: Elastic waves in bodies with initial (residual) stresses. Int. Appl. Mech. **38**(1), 23–59 (2002)
10. Guz, A.N.: Mekhanika dvizhushchikhsia treshchin v materialakh s nachalnymi (ostatochnymi) napriazheniiami (obzor) (Mechanics of moving cracks in materials with initial (residual) stresses (review)). Prikladnaya Mekhanika **47**(2), 3–75 (2011)
11. Guz, A.N.: Establishing the fundamentals of the theory of stability of mine working. Int. Appl. Mech. **39**(1), 20–48 (2003)
12. Guz, A.N.: Stability of elastic bodies under omnidirectional compression. Review. Int. Appl. Mech. **48**(3), 241–293 (2012)

13. Guz, A.N., Guz, I.A.: Mixed plane problems of linearized solids mechanics. Exact solutions. Int. Appl. Mech. **40**(1), 1–29 (2004)
14. Southwell, R.V.: On the general theory of elastic stability. Philos. Trans. R. Soc. Lond. Ser. A. **213**, 187–244 (1913)
15. Green, A.E., Rivlin, R.S., Shield, R.T.: General theory of small elastic deformations superposed on finite elastic deformations. Proc. Roy. Soc. Ser. A. **211**(1104), 128–154 (1952)
16. Guz, A.N.: O postroenii osnov mekhaniki razrusheniia materialov pri szhatii vdol treshchin (obzor) (On the construction of the foundations of the fracture mechanics of materials in compression along cracks (review)). Prikladnaya Mekhanika **50**(1), 5–89 (2014)
17. Guz, A.N.: O tochnosti gipotezy Kirkhgoffa-Liava pri opredelenii kriticheskikh sil v teorii uprugoi ustoichivosti (On the accuracy of the Kirchhoff-Love hypothesis in determining critical forces in the theory of elastic stability). Dokl. Akad. Nauk SSSR **179**(3), 552–554 (1968)
18. Guz, A.N.: Ob ustoichivosti trekhmernykh uprugikh tel (On stability of three-dimensional elastic bodies). Prikladna Matematika Mekhanika **32**(5), 930–935 (1968)
19. Guz, A.N.: Ob ustoichivosti polosy (On stability of a strip). Izvestiya Akademii Nauk SSSR Mekhanika tverdogo tela. **6**, 111–113 (1969)
20. Guz, A.N.: Ob uslovii primeneniia metoda Eilera issledovaniia ustoichivosti deformirovaniia nelineino-uprugikh tel pri konechnykh dokriticheskikh deformatsiiakh (On condition of Euler's method applicability to investigate the stability of deformation of nonlinear elastic bodies under finite subcritical deformations). Dokl. Akad. Nauk SSSR **194**(3), 38–40 (1970)
21. Guz, A.N.: O trekhmernoi teorii ustoichivosti deformirovaniia materialov s reologicheskimi svoistvami (On three-dimensional theory of stability of deformation of materials with rheological properties). Izvestiya Akademii Nauk SSSR Mekhanika tverdogo tela **6**, 104–107 (1970)
22. Guz, A.N.: Ob analogiiakh mezhdu linearizirovannymi i lineinymi zadachami teorii uprugosti (Analogies between linearized and linear problems of elasticity theory). Dokl. Akad. Nauk SSSR **212**(5), 1089–1091 (1973)
23. Guz, O.M.: Pro variatsiini printsipi trivimirnikh zadach stiikosti nepruzhnikh til (On variational principles of three-dimensional stability problems of inelastic bodies). Dopovidi Akad. Nauk URSR Ser. A. **11**, 1008–1012 (1973)
24. Guz, A.N.: O variatsionnykh printsipakh trekhmernoi teorii ustoichivosti deformiruemykh tel pri deistvii «slediashchikh» nagruzok (On variational principles of the three-dimensional theory of stability of deformable bodies under the action of "tracking" loads). Dokl. Akad. Nauk SSSR **246**(6), 1314–1316 (1979)
25. Guz, A.N., Guz, I.A.: O kontinualnom priblizhenii v teorii ustoichivosti sloistykh kompozitnykh materialov (Continual approximation in the theory of stability of layered composite materials). Dokl. Akad. Nauk SSSR **305**(5), 1073–1076 (1989)
26. Guz, A.N., Guz, I.A.: O lokalnoi neustoichivosti sloistykh kompozitnykh materialov (Local instabilities of layered composite materials). Dokl. Akad. Nauk SSSR **311**(4), 812–814 (1990)
27. Guz, A.N.: Investigation the stability of elastic systems by means of linearized equations of elasticity theory. Sov. Appl. Mech. **3**(2), 13–18 (1967)
28. Guz, A.N.: The stability of orthotropic bodies. Sov. Appl. Mech. **3**(5), 17–22 (1967)
29. Guz, A.N.: Three-dimensional theory of stability of elastic-viscous-plastic bodies. Sov. Appl. Mech. **20**(6), 512–516 (1984)
30. Guz, A.N.: Three-dimensional stability theory of deformed bodies. Internal instability. Sov. Appl. Mech. **21**(11), 1023–1034 (1985)
31. Guz, A.N.: Three-dimensional stability theory of deformable bodies. Surface instability. Sov. Appl. Mech. **22**(1), 17–26 (1986)
32. Guz, A.N.: Three-dimensional stability theory of deformable bodies. Stability of construction elements. Sov. Appl. Mech. **22**(2), 97–107 (1986)
33. Guz, A.N.: Construction of a theory of the local instability of unidirectional fiber composites. Int. Appl. Mech. **28**(1), 18–24 (1992)

34. Guz, A.N.: Stability theory for unidirectional fiber reinforced composites. Int. Appl. Mech. **32**(8), 577–586 (1996)
35. Guz, A.N.: Three-dimensional theory of stability of a carbon nanotube in a matrix. Int. Appl. Mech. **42**(1), 19–31 (2006)
36. Guz, A.N., Chekhov, V.N.: Linearized theory of folding in the interior of the earth's crust. Sov. Appl. Mech. **11**(1), 1–10 (1975)
37. Guz, A.N., Chekhov, V.N.: Variational method of investigating the stability of laminar semiinfinite media. Sov. Appl. Mech. **21**(7), 639–646 (1985)
38. Guz, A.N., Chekhov, V.N.: Investigation of surface instability of stratified bodies in three-dimensional formulation. Sov. Appl. Mech. **26**(2), 107–125 (1990)
39. Guz, A.N., Chekhov, V.N.: Problems of folding in the earth's stratified crust. Int. Appl. Mech. **43**(2), 127–159 (2007)
40. Guz, A.N., Dekret, V.A.: On two models in the three-dimensional theory of stability of composites. Int. Appl. Mech. **44**(8), 839–854 (2008)
41. Guz, A.N., Guz, I.A.: On the theory of stability of laminated composites. Int. Appl. Mech. **35**(4), 323–329 (1999)
42. Guz, A.N., Guz, I.A.: Analytical solution of stability problem for two composite half-plane compressed along interfacial cracks. Compos. B **31**(5), 405–418 (2000)
43. Guz, A.N., Guz, I.A.: The stability of the interface between two bodies compressed along interface cracks. 1. Exact solution for the case of unequal roots. Int. Appl. Mech. **36**(4), 482–491 (2000)
44. Guz, A.N., Guz, I.A.: The stability of the interface between two bodies compressed along interface cracks. 2. Exact solution for the case of equal roots. Int. Appl. Mech. **36**(5), 615–622 (2000)
45. Guz, A.N., Guz, I.A.: The stability of the interface between two bodies compressed along interface cracks. 3. Exact solution for the case of equal and unequal roots. Int. Appl. Mech. **36**(6), 759–768 (2000)
46. Guz, A.N., Rushchitskii, J.J., Guz, I.A.: Establishing fundamentals of the mechanics of nanocomposites. Int. Appl. Mech. **43**(3), 247–271 (2007)
47. Guz, A.N., Samborskaya, A.N.: General stability problem of a series of fibers in an elastic matrix. Sov. Appl. Mech. **27**(3), 223–230 (1991)
48. Guz, I.A., Guz, A.N.: Stability of two different half-planes in compression along interfacial cracks: analytical solutions. Int. Appl. Mech. **37**(7), 906–912 (2001)
49. Guz, I.A.: Investigation of the stability of a composite in compression along two parallel structural cracks at the layer interface. Int. Appl. Mech. **30**(11), 841–847 (1994)
50. Kappus, R.: Zur Elastizitatstheorie endlicher Verschiebungen. ZAMM **19**(6), 344–361 (1939)
51. Novozhilov, V.V.: Osnovy nelineynoy teorii uprugosti (Foundations of the Nonlinear Theory of Elasticity). Gostekhizdat, Moscow (1948)
52. Kappus, R.: Zur Elastizitatstheorie endlicher Verschiebungen. ZAMM **19**(5), 271–285 (1939)
53. Biot, M.A.: Mechanics of Incremental Deformations. Willey, New York (1965)
54. Leibenson, L.S.: O primenenii garmonicheskikh funkciy k voprosu ob ustoychivosti sfericheskoy i cilindricheskoy obolochek (On application of harmonic functions to stability of spherical and cylindrical shells). In: Sobranie Trudov, vol. 1, pp. 110–121. Moscow (1951)
55. Ishlinskyi, AYu.: Rassmotrenie voprosov ob ustoychivosti sostoyaniya ravnovesiya uprugikh tel s tochki zreniya matematicheskoy teorii uprugosti (Consideration of questions of stability of the equilibrium state of elastic bodies in term of mathematical theory of elasticity). Ukr. Matem. Zhurnal. **6**(2), 140–146 (1954)
56. Novinski, J.L.: On the elastic stability of thick columns. Acta Mech. **7**(4), 279–286 (1969)
57. Hoff, N.J.: Buckling and stability. J. R. Aeronaut. Soc. **58**(1), 1–11 (1954)
58. Akbarov, S.D., Guz, A.N.: Mechanics of Curved Composites. Kluwer Academic Publisher, Dordrecht Boston London (2000)
59. Akbarov, S.D., Sisman, T., Yahnioglu, N.: On the fracture of the unidirectional composites in compression. Int. J. Eng. Sci. **35**(12/13), 1115–1136 (1997)

60. Akbarov, S.D.: On the three-dimensional stability loss problems of elements of constructions fabricated from the viscoelastic composite materials. Mech. Compos. Mater. **34**(6), 537–544 (1998)
61. Akbarov, S.D., Cilli, A., Guz, A.N.: The theoretical strength limit in compression of viscoelastic layered composite materials. Compos. B Eng. **30**(5), 365–372 (1999)
62. Akbarov, S.D.: Three-dimensional stability loss problems of viscoelastic composite materials and structural members. Int. Appl. Mech. **43**(10), 1069–1089 (2007)
63. Guz, A.N.: Mekhanika razrusheniia kompozitnykh materialov pri szhatii (Fracture Mechanics of Composite Materials under Compression). Naukova Dumka, Kyiv (1990)
64. Guz, A.N.: Osnovy mekhaniki razrusheniia kompozitov pri szhatii: V 2-kh tomakh (Fundamentals of the Fracture Mechanics of Composites under Compression: In 2 volumes). Litera, Kyiv, (2008), T. 1. Razrushenie v strukture materiala. (Fracture in Structure of Materials), T. 2. Rodstvennye mekhanizmy razrusheniia. (Related Mechanisms of Fracture)
65. Guz, A.N.: Theory of delayed fracture of composite in compression. Sov. Appl. Mech. **24**(5), 431–438 (1988)
66. Babich, IYu., Guz, A.N.: O primenimosti podkhoda Eilera k issledovaniiu ustoichivosti deformirovaniia anizotropnykh nelineino-uprugikh tel pri konechnykh dokriticheskikh deformatsiiakh (On applicability of Euler's approach to the study of anisotropic nonlinear elastic bodies deformation stability under finite subcritical deformations). Dokl. Akad. Nauk SSSR **202**(4), 795–796 (1972)
67. Guz, A.N., Sporykhin, A.N.: Three-dimensional theory of inelastic stability (General questions). Sov. Appl. Mech. **18**(7), 581–596 (1982)
68. Guz, A.N., Sporykhin, A.N.: Three-dimensional theory of inelastic stability. Specific results. Sov. Appl. Mech. **18**(8), 671–692 (1982)
69. Guz, A.N.: Mekhanika khrupkogo razrusheniia materialov s nachalnymi napriazheniiami (Mechanics of Brittle Fracture of Materials with Initial Stresses). Naukova Dumka, Kyiv (1983)
70. Guz, A.N., (ed.): Neklassicheskie problemy mekhaniki razrusheniya, v 4 tomah, 5 knigah (Non-Classical Problems of Fracture Mechanics, in 4 volumes, 5 books). Naukova Dumka, Kyiv (1990–1993)
71. Muskhelishvili, N.I.: Nekotorye osnovnye zadachi matematicheskoy teorii uprugosti (Some Basic Problems of the Mathematical Theory of Elasticity). Nauka, Moscow (1966)
72. Lekhnitskyi, S.G.: Teoriya uprugosti anizotropnogo tela (Theory of Elasticity of Anisotropic Body). Nauka, Moscow (1977)
73. Keldysh, M.V., Sedov, L.I.: Effektivnoe reshenie nekotorykh kraevykh zadach dlya garmonicheskikh funkciy (Efficient solution of some boundary problems for harmonic functions). Dokl. Akad. Nauk SSSR **16**(1), 7–10 (1937)
74. Guz, A.N.: Critical phenomena in cracking of the interface between two prestressed materials. 1. Problem formulation and basic relations. Int. Appl. Mech. **38**(4), 423–431 (2002)
75. Guz, A.N.: Critical phenomena in cracking of the interface between two prestressed materials. 2. Exact solution. The case of unequal roots. Int. Appl. Mech. **38**(5), 548–555 (2002)
76. Guz, A.N.: Critical phenomena in cracking of the interface between two prestressed materials. 3. Exact solution. The case of equal roots. Int. Appl. Mech. **38**(6), 693–700 (2002)
77. Guz, A.N.: Critical phenomena in cracking of the interface between two prestressed materials. 4. Exact solution. The combined case of unequal and equal roots. Int. Appl. Mech. **38**(7), 806–814 (2002)
78. Guz, A.N.: Comments on effects of prestress on crack-tip fields in elastic incompressible solids. Int. J. Solids Struct. **40**(5), 1333–1334 (2003)
79. Guz, A.N.: On one two-level model in the mesomechanics of compression fracture of cracked composites. Int. Appl. Mech. **39**(3), 274–285 (2003)
80. Guz, A.N.: On some nonclassical problems of fracture mechanics taking into account the stresses along cracks. Int. Appl. Mech. **40**(8), 937–942 (2004)
81. Guz, A.N.: On study of Nonclassical Problems of Fracture and Failure Mechanics and Related Mechanisms, pp. 35–68. ANNALS of the European Academy of Sciences. Liège, Belgium, (2006–2007)

82. Guz, A.N.: Pascal medals lecture (written presentation). Int. Appl. Mech. **44**(1), 6–11 (2008)
83. Guz, A.N.: On study of nonclassical problems of fracture and failure mechanics and related mechanisms. Int. Appl. Mech. **45**(1), 1–31 (2009)
84. Guz, A.N.: On physical incorrect results in fracture mechanics. Int. Appl. Mech. **45**(10), 1041–1051 (2009)
85. Guz, A.N.: On the activity of the S. P. Timoshenko Institute of mechanics in 1991–2011. Int. Appl. Mech. **47**(6), 607–626 (2011)
86. Guz, A.N., Babich, IYu.: Three-dimensional stability problems of composite materials and composite construction components. Rozpr. Inz. **27**(4), 613–631 (1979)
87. Guz, A.N., Chekhov, V.N., Stukotilov, V.S.: Effect of anisotropy in the physicomechanical properties of a material on the surface instability of layered semiinfinite media. Int. Appl. Mech. **33**(2), 87–92 (1997)
88. Guz, A.N., Cherevko, M.A.: Fracture mechanics of unidirectional fibrous composites with metal matrix under compression. Theor. Appl. Fract. Mech. **3**(2), 151–155 (1985)
89. Guz, A.N., Cherevko, M.A.: Stability of a biperiodic system of fibers in a matrix with finite deformations. Sov. Appl. Mech. **22**(6), 514–518 (1986)
90. Guz, A.N., Dekret, V.A.: Interaction of two parallel short fibers in the matrix at loss of stability. Comput. Model. Eng. Sci. **13**(3), 165–170 (2006)
91. Guz, A.N., Dekret, V.A.: Stability loss in nanotube reinforced composites. Comput. Model. Eng. Sci. **49**(1), 69–80 (2009)
92. Guz, A.N., Dekret, V.A.: Stability problem of composite material reinforced by periodic row of short fibers. Comput. Model. Eng. Sci. **42**(3), 179–186 (2009)
93. Guz, A.N., Dekret, V.A., Kokhanenko, Yu.V.: Solution of plane problems of the three-dimensional stability of a ribbon-reinforced composite. Int. Appl. Mech. **36**(10), 1317–1328 (2000)
94. Guz, A.N., Dekret, V.A., Kokhanenko, Yu.V.: Two-dimensional stability problem for interacting short fibers in a composite: in-line arrangement. Int. Appl. Mech. **40**(9), 994–1001 (2004)
95. Guz, A.N., Dekret, V.A., Kokhanenko, Yu.V.: Planar stability problem of composite weakly reinforced by short fibers. Mech. Adv. Mater. Struct. **12**, 313–317 (2005)
96. Guz, A.N., Dovzhik, M.V., Nazarenko, V.M.: Fracture of a material compressed along a crack located at a short distance from the free surface. Int. Appl. Mech. **47**(6), 627–635 (2011)
97. Guz, A.N., Dyshel, MSh.: Fracture of cylindrical shells with cracks in tension. Theor. Appl. Fract. Mech. **4**, 123–126 (1985)
98. Guz, A.N., Dyshel, MSh.: Fracture and stability of notched thin-walled bodies in tension (Survey). Sov. Appl. Mech. **26**(11), 1023–1040 (1990)
99. Guz, A.N., Dyshel, MSh., Kuliev, G.G., Milovanova, O.B.: Fracture and local instability of thin-walled bodies with notches. Sov. Appl. Mech. **17**(8), 707–721 (1981)
100. Guz, A.N., Dyshel, MSh., Nazarenko, V.M.: Fracture and stability of materials and structural members with cracks: approaches and results. Int. Appl. Mech. **40**(12), 1323–1359 (2004)
101. Guz, A.N., Guz, I.A.: Substantiation of a continuum theory of the fracture of laminated composite in compression. Sov. Appl. Mech. **24**(7), 648–657 (1988)
102. Guz, A.N., Guz, I.A.: Foundation for the continual theory of fracture during compression of laminar composites with a metal matrix. Sov. Appl. Mech. **24**(11), 1041–1047 (1988)
103. Galin, L.A.: Kontaktnye zadachi teorii uprugosti (Contact Problems of Elasticity Theory). Fizmatgiz, Moscow (1953)

Part II
Fracture in Composite Materials Under Compression

Chapter 3
Problem 1. Fracture in Composite Materials Under Compression Along the Reinforcing Elements

In this chapter, in a very brief form, the main results on the issue are discussed which are obtained starting with 1967–1968 in the Department of Dynamics and Stability of the Continua of the S. P. Timoshenko Institute of Mechanics of NAS of Ukraine. At that, the presentation of the results under consideration is presented in a style announced in the Introduction to this monograph (without involving the aspects of a mathematical nature). The information on a historical character is also provided, including information on some of the results of experimental studies corresponding to the development of the scientific and technological problem under discussion.

The main results of scientific workers of the Department of Dynamics and Stability of Continua on this scientific problem are presented in monographs (30, 31, 54, 57) and some other monographic publications. These results were originally published in the scientific articles, which are listed in the literature to monographs [1–4]. The articles [4–133] and reports at international conferences [134–157] from the list of publications in this monograph are related to the scientific direction of this department.

Five dissertations for the Scientific Degree of the Doctor of Science in Physical–Mathematical or Engineering Sciences have been prepared and defended in this scientific direction of the department: I. I. Babich, V. N. Chekhov, Yu. N. Lapusta, I. A. Guz, and E. A. Tkachenko. It should be noted that the results on **Problem 6** (fracture at compression along the parallel cracks) are also included in the dissertation of I. A. Guz.

3.1 General Concept and Key Directions of Research

At present, it is considered in the scientific literature on the mechanics of the fracture of composite materials that *for the first time* microbuckling of fibers as a mechanism for the fracture of fibrous composite material *under compression* was described in the

© The Author(s), under exclusive license to Springer Nature Switzerland AG 2022
A. N. Guz, *Eight Non-Classical Problems of Fracture Mechanics*,
Advanced Structured Materials 159,
https://doi.org/10.1007/978-3-030-77501-8_3

work [158] of 1960. This phenomenon was repeatedly confirmed in several scientific centers in the corresponding experimental studies, which is considered in Sect. 3.2 of the present chapter. The mentioned information provides an opportunity to adopt the following General Concept when analyzing the particular problems of Problem 1.

3.1.1 General Concept

In the composite materials, which are modeled in a continuum approximation by the orthotropic materials, when they are compressed along the axes of symmetry of the material properties, *the initial stage (start) of fracture* is *the loss of stability in the internal structure of composites.* At that, the propagation of fracture is determined by *the behavior of perturbations* within the framework of the applied theory of stability. The propagation of fracture begins from the macro- and micro-inhomogeneities (holes, inclusions, micro- and macrocracks). *The theoretical ultimate strength at compression and the theoretical value of the limit shortening are the values of the critical load and the critical shortening calculated within the framework of the applied theory of stability.*

Note 3.1 The formulated above **General Concept** in the mechanics of fracture of the composites under compression along the axes of symmetry of properties of composites is a complete analog of the situation in the mechanics of the structural elements when the initial stage of exhaustion of the bearing capacity of the structural elements (rods, plates, shells, etc.) under compression along the axes of symmetry is the loss of stability.

Fig. 3.1

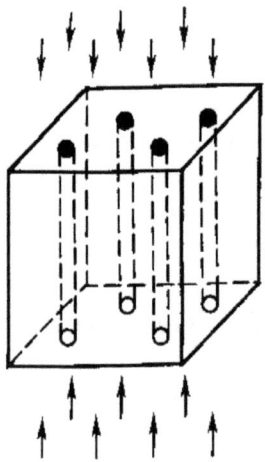

Fig. 3.2

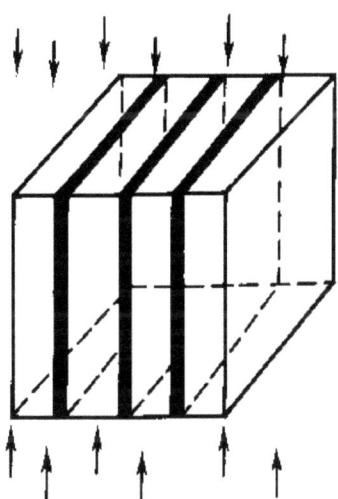

Thus, the fibrous one-directional (Fig. 3.1), the layered (Fig. 3.2) composites, and the cross-winding or placing composites are considered, as well as other composites with axes of symmetry of properties along which compression is carried out.

To apply the General Concept more consistently and rigorously, it is necessary to formulate the condition for the existence of a phenomenon of loss of stability in the internal structure *(further, the term "internal instability" will be used) relative to the structural element made of a particular composite.* Let us introduce the designations:

p_{cr} is the critical load corresponding to the (internal instability) of composite.
p_{cr}^{se} is the critical load corresponding to the loss of stability of the entire structural element.
L is the characteristic (minimal) size of the structural element.
l_{cr} is the length of the half-wave mode of the internal instability of the composite.

With allowance for the introduced above designations, the conditions of the internal instability *relative to the structural element made of this composite* can be presented in the following form

$$p_{cr} < p_{cr}^{se}; \ l_{cr} \ll L. \tag{3.1}$$

Non-fulfillment of one of the conditions (3.1) means that *in the particular case considered in this chapter* (the shape of the structural element + the composite properties) *there is no mechanism of fracture* determined by the **General Concept** when the load changes continuously.

Apparently, *for the first time,* the concept of internal instability of the material was introduced in the article [159] of 1963, and a study of internal instability within the plane problem of **theory 5** *(theory of incremental deformations)* was carried out, the analysis of which (**theory 5**) is performed in Sect. 2.2 of this monograph.

In the following years, the study of internal instability was carried out for the spatial and plane problems in the case of compressible and incompressible materials with constitutive equations of a sufficiently general form in the unified general form for **theories 1, 2, and 3** (according to the terminology of Sect. 2.2).

At that, the concrete results are obtained for both the piece-wise homogeneous medium model and the continual approach. The above results for the study of internal instability are presented in monographs [1–4, 160, 161]. These results were previously published in the articles that were partially included in the lists of literature of the monographs [1–4, 160, 161] and in the list of literature of the present monograph. For the case of the continuum composite model, the various cases of transition to the internal instability (internal fracture) are considered in the monograph [4] (volume 2, Chap. 2, Sect. 4).

Within the piece-wise homogeneous medium model, when determining the value of critical load and modes of the internal instability for the specific composites, usually the second inequality (3.1) is taking into account and the analysis is carried out for a composite of the specific structure *that occupies the infinite space*. After solving the corresponding boundary value problems for the "infinite" composite, the dependence of the load parameter p on the wave formation parameter α is determined in the following form

$$p = p(\alpha); \alpha = \pi \frac{h}{l}. \tag{3.2}$$

In (3.2) and below the following notations are introduced:

h is the characteristic geometric parameter of the composite structure (the minimum thickness of the layers in the layered composite (Fig. 3.2), the fiber radius in a one-directional fibrous composite (Fig. 3.1)).

l is the length of the half wave (along the layers or fibers) of the mode of the internal instability of the composite.

When to take into account the above considerations and definitions, then it can conclude that the phenomenon of the internal instability of the composite does not exist with arbitrary dependence (3.2). For clarity, it can imagine two types of dependency (3.2) corresponding to curves **A** and **B** in Fig. 3.3.

The curve **A** has a clear minimum. Therefore, the value p_{cri} is determined by minimizing the first expression (3.2), and the value α_{cri} is determined from the expression $p_{cri} = p(\alpha_{cri})$.

Thus, for the case of the formula (3.2) in the form of a curve **A** the value of the critical load p_{cri} and the mode of loss of stability corresponding to the wave formation parameter $\alpha_{cri} = \pi h l_{cri}^{-1}$ are determined. It is necessary to note that for the case of curve **A** the relations

$$\alpha_{cri} \neq 0; l_{cri} \neq \infty. \tag{3.3}$$

are valid.

Fig. 3.3

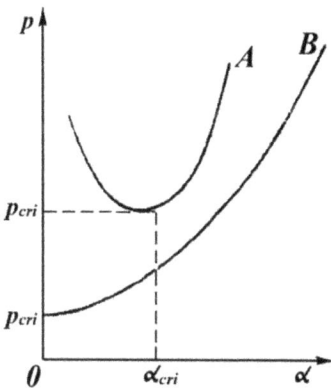

Due to the presence of relations (3.3) in the case of dependency (3.2) in the form of curve **A**, the previously noted conditions (3.1) are used *to determine those elements of structures (their sizes)* from a specific composite *for which the internal instability of the composite is realized.*

The curve **B** is a monotonous curve. Therefore, the critical value for this curve is determined as a result of the minimization of the expression (3.2) by the relation $p'_{\text{cri}} = p(0)$. Therefore, in the case of curve **B,** the relations are valid

$$\alpha_{\text{cri}} = 0; \quad l_{\text{cri}} = \infty. \tag{3.4}$$

It follows from the expressions (3.4) that it is impossible in the case of dependence (3.2) in the form of a curve **B** to determine the form of the internal instability of the composite. Thus, according to the accepted definition, the phenomenon of the internal instability of the composite *does not exist* in the case of dependency (3.4) in the form of a type **B**. Note also that in accordance with the second expression (3.4) in the case of a type **B** curve, the second condition (3.1) cannot be fulfilled *for any part of structural element.* Therefore, in this case, *only the loss of stability of the entire structural element can be realized.*

Thus, it can be considered that the phenomenon of internal instability does not occur if the dependence of the loading parameter p on the wave-forming parameter α (first expression (3.2)) *appears to be a type* **B***curve.* In the case of type **A** curves, *when α_{cri} is slightly different from zero,* the phenomenon of internal instability in the elements of the structures of a particular form also practically does not arise due to the second condition (3.1), because $l_{\text{cri}} \to \infty$. The noted situation concerning dependency (3.2) in the form of a curve **A** should be taken into account when analyzing the phenomenon in question for the specific composites set out in Sect. 3.1.1.

The **General Concept** and methodology of its application enable the receiving and analysis of results related to **Problem 1** (fracture in composite materials under compression along the reinforcing elements) and other non-classical problems of the fracture mechanics of composite obtained by three-dimensional linearized theory of the deformable bodies stability (TLTDBS). In particular, this approach is fully applicable to the "infinitely long fibers" model in the mechanics of composite materials, when the periodic along reinforcing element (fibers, layers) modes of loss of stability are studied. By the style of presentation of the material, Sect. 3.1.1 corresponds to Sect. 1.2.1 of the review article [162] and the review of the results in the monograph [163].

In the research on **Problem 1**, two scientific directions have been formed with the use of the **General Concept**. The brief information on the characteristics of these directions is presented in Sects. 3.1.2 and 3.1.3.

3.1.2 The First Direction (Very Approximate Approaches)

The first direction is based on the introduction of *various very approximate design schemes and assumptions* in the study of the phenomenon of internal instability of composites, determining the initial stage (start) of the fracture of composites under compression.

At present, it appears that such an approach has not yet been properly analyzed in the main problems, if it is at all possible. Therefore, when considering the results obtained within the framework of the first approach, the caution cannot be taken lightly, since the results under discussion are the results of theoretical studies in the introduction of *approximate design schemes and assumptions* without proper analysis and discussion.

The author of this monograph, as well as the authors of the review [162] of 2016, did not set out their goal—to provide a systematic review and analysis of the corresponding publications in **the first** direction, because a relatively large number of articles are published within the framework of this direction in the scientific and technical literature on the mechanics of composites and the fracture mechanics.

The purpose of Sect. 3.1.2 as well as Sect. 1.2.2.1 is:

1. Classification of the results discussed as **the first** direction in the studies under consideration;
2. The formulation of the main (in the opinion of the author of this monograph) approximate assumption characteristic of the first direction;
3. Consideration of the first historically corresponding publications related to the first direction.

The characteristic *approximate assumptions*, widely used in research in the first direction, can be conditionally combined into the following five groups.

1. In the study of regularities in the *fibrous* composite, a model of the layered composite is used, for which research is carried out within the framework of a plane problem.
2. In the analysis of the stability of the fillers—the reinforcing elements (fibers, layers) the widely used *one-dimensional and two-dimensional applied theories*

of the stability of thin-walled systems (rods, plates), built with the involvement of hypotheses of plane sections, Kirchhoff–Love, etc.; as is common knowledge, this type of theories is used *only* to describe the relatively long-wave modes of loss of stability.

3.   As a rule, the fact that the matrix (binding) also perceives a compressive load is not taken into account. This assumption is due to that the material of the matrix has much less stiffness as compared to the material of the filler. As a result, the study assumes that *the matrix is unloaded*. For several composites, at least for composites with a low volume fraction of filler, this assumption is sufficiently approximate.

4.   In studies, the interaction of the matrix (binding) and filler (fibers, layers) is taken into account approximately. Sufficiently often, the interaction of fiber with the matrix is modeled by the interaction of fiber with the coaxial cylinder (part of the matrix). Also, quite often the modeling of the matrix by the one-dimensional model is used when being analyzed the interaction of filler (fiber, layer) and matrix (binding).

5.   The boundary conditions at the interface of the filler and the matrix are *approximately* satisfied, quite often not even commenting on the situation.

It should be noted that, when one of these groups of assumptions is introduced, then the assumptions are automatically made on the other group. To illustrate the above situation, let us look at the following example.

So, accepting the assumption that the stresses do not arise in the matrix when the composite is compressed and the stresses in the filler–fibers exist (the third group of assumptions), the assumptions about approximate boundary conditions at the interface of the matrix and filler (the fifth group of assumptions) are automatically introduced. The fact is that the initial assumption automatically leads to the fact that in the precritical state (before the loss of stability) the matrix and filler freely slip relative to each other along the fibers, and at the moment of loss of stability for the matrix and filler the conditions of the full contact are fulfilled.

Of course, the analysis of the assumptions of the five groups and other assumptions of **the first** direction can be continued, but such an analysis is not the purpose of this monograph.

Consider some historical points related to theoretical research in the framework of the first direction. It is now generally accepted that *the first* theoretical results related to the study of the phenomenon noted in the publication [158] of 1960 were presented in the article [164] of 1965 which was published in English. The article in Russian was published in 1967 as an article [165] of 1970. Note that the articles [164] and [165] actually use in one or another form the approximate assumptions, which are the part of the above five groups of the approximations, characteristic of **the first** direction.

The articles [164] and [165] are well known and widely accepted in the world scientific and technical literature on the mechanics of composites and the fracture mechanics, despite the noted above very approximate nature of the results presented in them. So, the results of [164] and [165] were included in the seven-volume treatise on fracture of the [166] of encyclopedic nature (in the form of the article [167] of the first part of the seventh volume). At that, these results were included in the eight-volume treatise of an encyclopedic nature on composite materials [168] in the form of the material to the corresponding articles. Also, these results are also included in the well-known collective monograph [169], published in Russian in 1970.

It seems appropriate to note that the results of publications [164] and [165] were no longer intended to be included in the collective multivolume monographs [170] (in 6 volumes; Editors in Chief: A. Kelly, C. Zweben) on composite materials and [171] (in 10 vols; Int. Advisory Board: Ian Milne, R. O. Ritchie, B. Karihaloo) on fracture, which was planned for publication in Elsevier in 2006. This information can be formed from the promotional materials that were available by the end of 2006 to the author of this monograph.

Note additionally that the results of publications [164] and [165] were not included in the 4-volume edition (under the general ed. V. V. Panasyuk) on the fracture mechanics, which was published in 1988–1990.

At present, a rather large number of articles have been published in the framework of the first direction, except for the article [164] of 1965. It seems appropriate to include articles published in the first years after 1965: [172] in 1966, [173] in 1967, [174] in 1966, [175] in 1970, and [167] in 1976. In the monographs [3, 4, 163] and in the review article [162] of 2016, the above-mentioned and other publications are listed in the lists of literature.

The discussed results, referring to the article [167], are named in the monograph [176] the theory of Dou–Grundfest–Rosen–Schuertz (as the authors of the publications [158] of 1960, [164] of 1965, and [172] of 1966). A review of a number of results in the first direction is presented in the review article [177] of 1996, which is published in English.

The above information will be limited to a very brief discussion of the results obtained in the first direction (*very approximate approaches*). As is already noted, in the introductory part of Sect. 3.1.2, the author of this monograph did not seek to provide a systematic review and analysis of the corresponding publications in **the first** direction.

3.1.3 The Second Direction (Strict Sequential Approaches Based on the TLTDBS)

The second direction is characterized by the research of *the first phenomenon*—the internal instability (loss of stability in the internal structure) of composites and *the second phenomenon*—surface or near-the-surface instability (loss of stability in surface or near-the-surface layers) composites, which determine the initial stage (start) of the fracture of composites under compression in accordance with the **General Concept.**

This concept is formulated for the first phenomenon in Sect. 3.1.1 and for the second phenomenon will be formulated below in Sect. 3.1.3. Within the framework of the second direction, the study of the above first and second phenomena is carried out basing on the three-dimensional linearized theory of the deformable bodies stability (TLTDBS), the mathematical apparatus of which is summarized in Chap. 2 of this monograph.

It is appropriate to note that the study of *the second phenomenon* (surface or near-the-surface instability in the structure of composite materials) *was not carried out at all* in the framework of the first direction. It seems that taking into account the approximate schemes and assumptions of 1–5, specified in Sect. 3.1.2 and applied in the first direction, it is very difficult to construct any theory that adequately describes such a subtle phenomenon as the surface or near-the-surface instability in the structure of the composite material under compression.

3.1.3.1 Internal Fracture (Loss of Stability in the Internal Structure)

Here, the strict and consistent approaches within the framework of 3D theory (TLTDBS) are applied. Thus, the approximate design schemes and assumptions 1–5, which are characteristic of the first direction and which are specified in the first part of Sect. 3.1.2, *are not used in the second direction.* Therefore, the specific results obtained in *the second* approach can also be used to assess the accuracy of the corresponding results from the first approach. In the second direction, the specific results are obtained for several classes of spatial and plane problems of the fracture mechanics of composite under compression for different models of deformable bodies for the filler and matrix in the case of the model of a piece-wise homogeneous medium and a continuum model corresponding to the application of the principle of continualization.

The commentary to the first approach (the model of a piece-wise homogeneous medium) and the second approach (the model of a homogeneous medium with the averaged parameters—as a result of the principle of homogenization) is stated briefly in the second part of Sect. 1.4.1.

In the first approach (the model of piece-wise homogeneous medium), the basic relations of TLTDBS are applied separately for each element of the filler and the matrix and at the interface of these materials the continuity of stress and displacement vectors is ensured. In the case under discussion, the study carried out and obtained numerous results for the uniaxial compression (along fibers) of a one-directional fibrous composite (Fig. 3.1) and a layered composite (Fig. 3.2) for the one-axial or two-axial compression along layers. These studies are carried out in accordance with the methodology, which is stated briefly in Sect. 3.1.1 after the formulation of the **General Concept**.

When the second approach (model of homogeneous medium with averaged parameters) is applied, the results refer to the composites of different structures, which have one plane of symmetry, two mutual-perpendicular, or three mutual-perpendicular planes of symmetry of material properties. It is accepted that along the normals to the mentioned planes of the symmetry the one-axial, two-axial, or three-axial compression of the material is realized.

Under the above conditions (material properties + loading conditions), the continuum theory of the fracture of composites is built, based on the detection in the material of disturbances that are not of a local nature. A method of determining the theoretical ultimate strengths under one-axial loading and the method of constructing the surfaces of theoretical ultimate strengths under the two-axial and three-axial loading are developed.

3.1.3.2   Surface Fracture (Loss of Stability in the Near-The-Surface Layers of the Composite)

Apparently, for the first time, the notion of the surface instability of the material was introduced in the article [178] of 1963 and the study of near-the-surface instability within the framework of the plane problem of **theory 5** (*theory of incremental deformations* [179]) was carried out. The analysis of **theory 5** is performed in Sect. 2.2 of this monograph.

The phenomenon of surface instability [178] of the half-plane is that when the half-plane is compressed along its boundary, then the loss of stability near the boundary of the equilibrium state arises, and the modes of loss of stability attenuate with moving away from the boundary. In the author's opinion of this monograph, the phenomenon under discussion should be called "near-the-surface instability" instead of the name "surface instability" because the loss of stability occurs in layers of material near the surface and the amplitude of modes of loss of stability attenuates when moving away from the boundary of material.

Therefore, below the name "surface fracture" (surface instability, loss of stability in the structure of near-the-surface layers of material) will be applied.

In the following years, the study of near-the-surface instability was carried out for the spatial and plane problems in the case of compressible and incompressible materials (with constitutive equations of a sufficiently general form) in a common general form for the **theories 1, 2, and 3** (according to the terminology of Sect. 2.2). At that, the concrete results are obtained for both the piece-wise homogeneous medium model and the continuum approximation. The above results of the study of the near-the-surface instability are presented in monographs [1–4, 160, 161]. These results were previously published in the articles that were, in part, included in the lists of the literature of these monographs and into the list of literature for the present monograph.

In the general case, the composite materials of different structures are considered, which are modeled by orthotropic materials in a continuum approximation. It is accepted there that these materials have *a free surface*, which is parallel to one of the orthotropy planes. In the case of compression parallel to the free surface along one or two mutually perpendicular axes, the loss of stability can arise in the composite's near-the-surface layers, when the amplitude of stability loss modes attenuates when moving away from *a free surface*.

This phenomenon was called the surface instability or near-the-surface instability.

The results obtained within the above-mentioned statement provide an opportunity to formulate the **General Concept** in the study of *near-the-surface fracture* (taking into account a presence of the free surface) in analogy to the **General Concept** stated in Sect. 3.1.1 in the case of composite compression (excluding the influence of *the free surface*).

General Concept. In composite materials, which in the continuum approximation are modeled by the orthotropic materials and which have the free surface (parallel to one of the orthotropy planes), when they are compressed parallel to the "free surface" in one or two mutually perpendicular directions, *the initial stage (start) of near-the-surface fracture is the near-the-surface loss of stability*. In analyzing the further progress of this fracture mechanism, it is necessary to take into account its possible interaction with other fracture mechanisms.

The theoretical limit of strength and the theoretical value of the limit shortening in the near-the-surface fracture under compression are the values of the critical load and critical shortening calculated within the framework of the applied version of TLTDBS.

The above-formulated **General Concept** makes it possible to develop the mechanics of near-the-surface fracture of composites (in the near-the-surface layers of the material near the free boundary) under compression, when exploring the near-the-surface instability near the free boundary of the composite with the use of TLTDBS (the second direction from Sect. 3.1.3). It seems that in the first scientific direction (Sect. 3.1.2), this type of research cannot be considered promising. The above information is sufficient for the brief description of the second direction (strict sequential approaches based on TLTDBS). The additional information is presented only in the form of the following two notes.

Note 3.2 The information presented in Sect. 3.1.2 for **the first** direction of research (*very approximate approaches*) and in Sect. 3.1.3 for **the second** direction of research (*strict sequential approaches based on TLTDBS*) *cannot in any way be considered* as even very brief reviews of the results in the two directions discussed here. The information in Sects. 3.1.2 and 3.1.3 should be considered as the brief characteristics of the subject and approaches of the corresponding two directions of research. A brief overview of the main results obtained in the second direction will be presented in Sect. 3.3 of this chapter.

The following Sect. 3.2 provides the main results of the experimental studies related to the confirmation of the existence of the phenomenon discussed here (loss of stability in the structure of the composite under compression) and to the study of the fracture of composites under compression along the reinforcing elements (**Problem 1** in the terminology of Sect. 1.3 of this monograph).

Note 3.3 In the monographs [180, 181] and in the corresponding articles, for example, [182, 183] as well as other articles from the lists of literature to the aforementioned monographs, the foundations of mechanics of nanocomposites with the polymeric matrix are stated. These foundations include the principles of constructing the mechanics of composites with the polymeric matrix and a number of sections related to the dynamics, stability, and statics of these nanocomposites. Thus, the foundations of the fracture mechanics of nanocomposites with the polymer matrix under compression, which are based on TLTDBS and which can be considered as the development of the results corresponding to **the second** direction (Sect. 3.1.3), are presented in [180, 181].

3.2 Analysis of Experimental Results on Compression of Composites

At present, a sufficiently large number of scientific and technical articles are already published relative to the problem of the experimental studies of composites under compression, as well as on the theoretical studies. This in particular can be seen from the list of publications for this monograph. As an example, the review article [184] of 1996 is published in the well-known journal "Progress in Aerospace Sciences," whose list of literature includes 133 publications. This review article is devoted exclusively to analyzing the results of the experimental study of fracture of composites under compression.

Note that this article emphasizes the pioneering nature of the publication [158] in studies on the effect of the mechanism of the internal instability in the structure of the composite on the fracture of the composite under compression.

In this section, when analyzing publications with results on the experimental studies on the fracture of composites under compression, the specific goal is formulated:

First, consider a number of publications, the experimental results of which confirm the existence of the phenomenon discussed here (loss of stability in the internal structure of the composite under compression).

Second, consider a number of publications whose experimental results characterize the specifics of fracture when composites are compressed along the reinforcing elements.

The above analysis was carried out in the monographs [1] of 1971, [3] of 1990, [4] of 2008, and [163] of 2015, as well as in the review article [162] of 2016 for a different coverage of analyzing publications. It seems that the monograph [4] includes the fullest volume (compared to the other above-mentioned monographs) of information on the two problems discussed here.

It is worth to note that the monographs [180, 181], which, as indicated in Note 3.3, are devoted to the statement of the principles of building the foundations of mechanics of nanocomposites with a polymer matrix, also provide the experimental results on the study of stability nanotubes in the polymer matrix under compression.

By the results considered above, this Sect. 3.2 corresponds to a slight reduction of Sect. 1.1 of the review article [162] of 2016. At that, here the results of experimental studies showing the existence of a phenomenon of loss of stability in the internal structure of composites under compression (Sect. 3.2.1) and the results of experimental studies, characterizing the specifics of the fracture of composites when compressed along the reinforcing elements (Sect. 3.2.2), are separately analyzed.

3.2.1 Experimental Results on the Loss of Stability in the Internal Structure of Composites Under Compression

First of all, note that the phenomenon of loss of stability in the internal structure is not observed for the homogeneous materials. It is typically only for composite materials (as for structural and heterogeneous materials, in which the presence of internal structure is taken into account at different levels in their analysis).

It is worth noting that the structural homogeneity or inhomogeneity of a particular material is largely determined by the level of consideration (research) of processes, which is mainly determined by the variability of the fields of mechanical quantities (stresses, deformations) by the spatial variables.

The following situation should be taken into account when analyzing the experimental results for the loss of stability in the internal structure of composite materials. When compressing along the reinforcing elements (fibers, filler) of composite materials in the case of experimental studies, then it is very difficult to observe (fix) the loss of stability in the internal structure "in its pure form" because slight or significant fracture exists from the beginning of the process of loss of stability. In this regard, to prove the possibility of the phenomenon of loss of stability in the internal structure of the composite material under compression usually the results of the specially set experiments are given, which are the following. The fibers (composite filler) are placed in the epoxy (or other) resin and polymerized at a certain temperature, then cooling to a certain temperature and curing.

In almost all cases of experimental studies under consideration, the compression is done by shrinkage of the matrix (resin, binder) when it is curing or the block of the composite material is cooled. In this case, due to the difference in the coefficients of thermal expansion of fibers and matrix which are connected to each other, the compressive loads act in the fibers.

The above experimental studies were carried out in different scientific centers at different times. The results of such experiments are presented in the corresponding publications. Apparently, the results of the pilot studies are presented in the Russian language for the first time in a publication [165] of 1967, which is a translation of the English-language publication [164] of 1965.

In this regard, some of the results corresponding to [164, 165] are shown below.

So, Fig. 3.4 shows a photo-elastic picture for three separate fibers from the E-glass (diameter 0.13; 0.09; and 0.013 mm) in the epoxy resin matrix polymerized at the temperature of 120 °C. The periodic (with a large number of periods) picture

Fig. 3.4

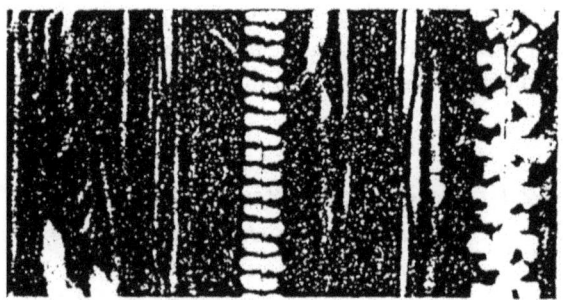

indicates for all three fibers the sinusoidal mode of the loss of stability (along the direction of fibers).

Note that Fig. 3.4 corresponds to [165, Fig. 3.20]. These results, as already noted, are obtained by the method of photo-elasticity. In the following years, the related results of experimental studies were obtained in many scientific centers, including other methods of Fig. 3.4 curing the resin (binding). Below, as an example, the results of experimental studies in the thermochemical curing of the resin (binding) with up to thermochemical curing of rosin (glass fibers and their strands floated freely in the binder) are shown.

The results of these experimental studies are published in the article [59] of 1982. These studies used fibers made of glass with a diameter of 0.01 mm. In Fig. 3.5 which corresponds to the publication [59], the results are presented (with an increase of 50 times) for the individual fibers and strands of fibers after the curing of resin (binding) by the thermochemical method.

It can be seen from Fig. 3.5 that the entire strand of fibers and individual fibers after the curing of the resin (binding) acquires a pronounced periodic sinusoidal (along the direction of fibers) form of loss of stability.

It should be noted that the results presented in Figs. 3.4 and 3.5 relate to the loss of stability in the internal structure of composite materials, in which the reinforcing elements are the glass fibers with a diameter of 0.13; 0,09; 0.013; and 0.01 mm.

The above results of experimental studies were published in the second half of the twentieth century.

Fig. 3.5

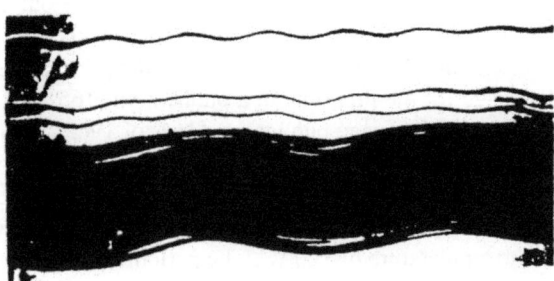

Fig. 3.6

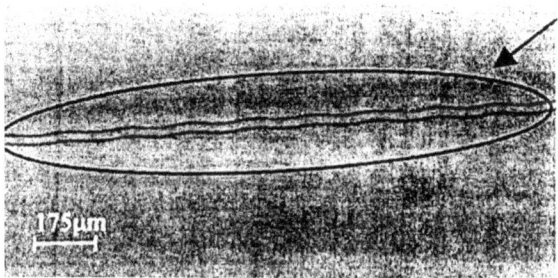

The related experimental studies are also carried out recently (at the beginning of the twenty-first century) relative to the cases when reinforcing elements (filler) are the fibers made of other materials. At that, in all cases, the compression is realized by shrinkage of the resin (binding) when it is curing or cooling.

An example of the results of experimental studies published at the beginning of the twenty-first century is the article [185] published in 2004. The results of this article, which are obtained by the above technique, are related to the study of the stability of carbon fiber in the polymer matrix (epoxy rosin). The corresponding article [185] Fig. 3.6 presents a periodic sinusoidal mode of loss of stability with a large number of periods.

This mode of loss of stability is obtained by cooling the polymer matrix and fixed according to the [185] at the 68th second after the start of the cooling process. At the bottom left corner of Fig. 3.6, the scale of the image in the microns is shown.

Along with the results presented in Figs. 3.4, 3.5, and 3.6, relating to the glass and carbon fibers, several articles are published recently with the results of experimental studies on the phenomenon under consideration for the various composite materials.

Thus, the shown above and related *results of experimental studies*, relating to the sufficiently long reinforcing elements (fibers, filler) in the matrix (binding), *confirm the existence of a phenomenon of loss of stability in the internal structure of the composite material. The modes of loss of stability found in the experimental studies* (Figs. 3.4, 3.5, and 3.6)*are periodic (along the reinforcing elements, along fibers) sinusoidal modes of loss of stability in the internal structure of a composite with a large number of periods.*

Due to the above, the boundary conditions on the ends of reinforcing elements (fibers) cannot have a significant impact on modes of loss of stability and the values of the critical loads and truncation. *The above information is in fact an experimental justification for the "infinitely long fibers" model.*

In the final part of this paragraph 3.2.1, some considerations in the form of the following two notes will be formulated.

Note 3.4 The experimental studies, the results of which are presented in the pictures of Figs. 3.4, 3.5, and 3.6, are the specially set ones, and it can be said that they are of the "model nature." They allowed proving the possibility of the existence of the phenomenon of loss of stability in the internal structure of composite materials. It follows from the careful analysis of the pictures in Figs. 3.4, 3.5, and 3.6 that the resulting loss of stability in the internal structure of composites *is fixed "in its pure form"—without signs of the phenomenon of fracture, as these images do not visually record the separation of the matrix (binding) from the fibers (filler).*

This situation appears to be quite significant because the appropriate attention should be paid to the analysis of the results of experimental studies on the fracture of composite materials at compression, which is carried out in the next subsection.

Note 3.5 It can be considered that this chapter examines the various processes in the composite materials under compression; mainly. Under compression along the reinforcing elements (fibers)—along the direction of the preferred reinforcement in relation, mainly, to the one-directional composite materials. These composite materials in the continuum approximation are modeled by the orthotropic homogeneous materials. The composite materials with reinforcement in the mutually perpendicular directions are also included in this class of materials. Thus, in the case of the model of orthotropic materials, the compression along the axes of the symmetry of the material's properties is considered.

A related situation also occurs in other types of loading. For example, it occurs in the compressed zones when bending different structural elements and in other cases. When different elements of structures (rods, plates, and shells) are compressed along the directions of symmetry (geometric shape and properties of the material), the main mechanism of exhaustion of the ultimate capacity of the structures is the loss of stability.

3.2.2   Experimental Results on Fracture of Composites Under Compression Along the Reinforcing Elements

First of all, it should be noted that this subsection examines the analysis of experimental studies on the fracture of composite materials *under compression* when the composite is already fractured. Thus, in essence, the images of the already fractured material are analyzed, and the initial stage (start) of fracture is not, of course, recorded in such experiments. Consequently, there are currently no experimental studies of the fracture of composites under compression, which record the process of fracture, starting with the initial stage (start) of fracture, corresponding to the loss of stability in the internal structure of the composite, and ending with the final stage of the structure corresponding to the division of the block of material into separate parts.

It is worth noting that there are no experimental studies of the fracture of the above type for most processes of fracture for other materials under other loads.

Thus, this paragraph examines the analysis of the nature of the fracture of composite materials *under compression*, focusing on the images (with different increases), essentially, the already destroyed composite material. The experimental studies are related to the compression of composite materials along the axes of

Fig. 3.7

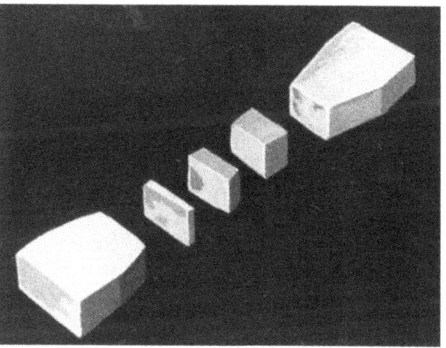

symmetry of material properties (along the reinforcing elements—along the fibers in the case of unidirectional composites, along the direction of preferred reinforcement in the case of composites with reinforcement in the mutually perpendicular directions, in the perpendicular direction to the plane of preferred reinforcement in the case of composites with reinforcement in the mutually perpendicular directions). In the analysis of the above-mentioned experimental studies, the effect of specific features of the nature of fracture in the studied kind of loading is considered.

First , note that the specific features of the nature of fracture are recorded not only when compression along the direction of preferential reinforcement, but also when compressed in a perpendicular direction.

To illustrate the above situation, let us present the results of the article [186] published in 1968.

So, in Fig. 3.7 the corresponding article picture of the fractured specimen is shown. The article [186] considered the uniaxial compression of the glass textolite perpendicular to the reinforcing elements (perpendicular to the reinforcement plane). The picture in Fig. 3.7 shows that the fracture occurred on planes perpendicular to the direction of the load, and the material is divided into parts. The fracture of this composite by planes, almost perpendicular to the action of a uniaxial compressive load, is a characteristic feature of the type of fracture in question.

The fracture of the composites under consideration by planes, almost perpendicular to the action of a uniaxial compressive load, is a characteristic feature of *this type of fracture*.

In the paper [84] published in 1969, the results of the study of the fracture of the samples from unidirectional fiberglass under compression along the reinforcing elements (fibers) are presented.

These results were obtained at the Institute of Mechanics of the Academy of Sciences of Ukrainian SSR (now the S.P. Timoshenko Institute of Mechanics of the National Academy of Sciences of Ukraine).

The cylindrical samples with a diameter of 10 mm and a height of 45 mm and the prismatic samples measuring $15 \times 15 \times 70$ mm (Fig. 3.8) were cut from the fiberglass plates made by winding on a metal fixture with the following curing under the press with the specific pressure 1 MP. As the reinforcing elements (filler), the

Fig. 3.8

alkalic glass thread of the brand NS55/6 with the oiling—paraffin emulsion, and as a matrix (binding) the epoxy-phenolic binder EFB-4 were used. The fiberglass binding content was 26.6% by weight with a polymerization level of 89.9%.

More information on the sample manufacturing technology and sample testing is given in the article [84] and the monograph [4], vol. 1, pp. 189–191.

The following situation should only be noted. To avoid crumpling the ends of the sample under compression, the metal frames were put on to its ends, which were filled with the epoxy resin of cold curing. As a result, the length of the open part of the sample was 1.5–2 of the linear cross section size.

The image in Fig. 3.9 shows the nature of the fracture of the sample with the square cross section, and the image in Fig. 3.10 shows the nature of the fracture of the sample of circular cross section. After the fracture, the samples were easily divided into two parts.

Note that the fracture, as a rule, took place near the metal ring; the observed situation is likely to indicate an occurrence of the initial local fracture at these locations caused by a cut of outermost fibers.

Note that fracture under compression of samples (Fig. 3.8) from unidirectional fiberglass when compressed along reinforcing elements (fibers) propagates across planes almost perpendicular to fibers and the direction of the compression load.

Fig. 3.9

Fig. 3.10

Thus, both under compression of unidirectional fiberglass along the fibers and when the glass textolite is compressed in a perpendicular direction to the reinforcement planes, the fracture of samples spreads over planes, almost perpendicular to the action of the uniaxial compression load.

The above situation *is a characteristic of the type of fracture in question.* Note that in the above two cases (Figs. 3.7, 3.8, 3.9, and 3.10), the compression is done along the axes of symmetry of the material properties. The additional considerations relating to the process of fracture under discussion are presented in the monograph [4], vol. 1, p. 191.

It is worth noting that the results of the experimental studies, the information of which is presented in Figs. 3.7, 3.8, 3.9, and 3.10, refer to the fiberglasses with a polymeric matrix. The related experimental studies were also carried out for other composite materials.

In the article [187] of 1985, the results of experimental studies are presented for a composite material with a metallic matrix, in which the filler (reinforcing elements) is the unidirectional sapphire fiber and the binder (matrix) is the aluminum.

The experimental studies in the [187] were carried out with uniaxial compression along the system of one-directional sapphire fibers (along the system of reinforcing elements).

The corresponding [187] image is presented in Fig. 3.11 (with some increase) of fracture in the metallic composite (sapphire fibers + aluminum matrix) with uniaxial compression along the one-directional sapphire fibers. The fracture is localized in a relatively narrow area. If to draw the plane through the middle of the destroyed area in Fig. 3.11, this plane will be almost perpendicular to the action direction of the compressible load.

Thus, when compressing one-directional composites along reinforcing elements (along fibers) than both in the case of the fiberglass with a polymer matrix (Figs. 3.8, 3.9 and 3.10) and in the case of the metallic composites (Fig. 3.11) there is a characteristic feature of this type of fracture—the fracture occurs or spreads almost perpendicular to the action of the load. However, this type of fracture should not occur instantly in the entire thickness of the material, although its origin (start) is the loss of stability in the internal structure of the composite.

Fig. 3.11

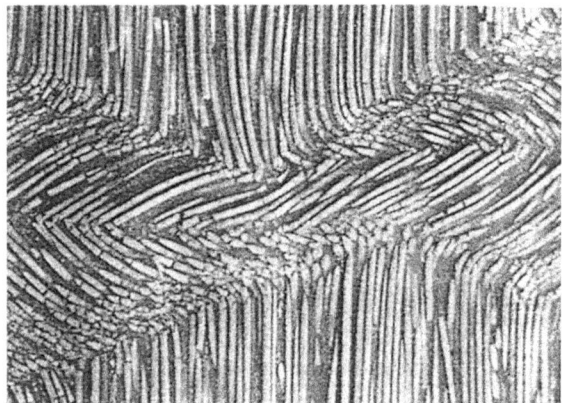

Naturally, this type of fracture may occur near any micro-inhomogeneity (breach including the continuity) in the internal structure.

Further propagation corresponds to the above characteristic feature. In this regard, it is interesting to study the regularities of the spread of fracture near the macro-inhomogeneity in the composite under compression along the reinforcing elements (fibers).

Consider this type of experimental research.

The article [188] of 1991 presents the results of experimental studies on the spread of fracture from a circular hole in the plate of composite material when compressed in the direction of reinforcement, and the results of studies are shown in the pictures of Figs. 3.12, 3.13, 3.14, and 3.15, corresponding to [188].

Figure 3.14 shows the design scheme with the denotation of the direction of the coordinate axes. According to these denotations, the compression was carried out along the vertical axis (along the $0y$-axis). The plates were made of a layered composite, each layer of which consisted of a one-directional fibrous material (filler—carbon fibers, matrix—epoxy rosin).

The layers were stacked by the thickness (along the axis $0z$ in Fig. 3.14) in such a way that the axes $0x$ and $0y$ (Fig. 3.14) were the axes of symmetry of the properties

Fig. 3.12

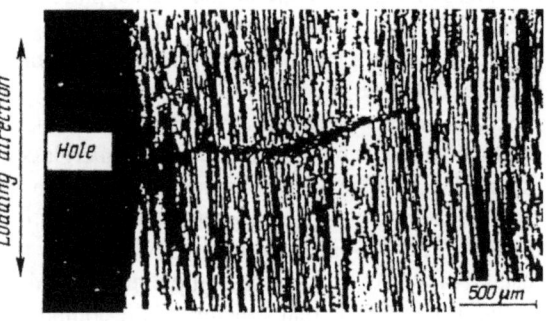

Fig. 3.13

Fig. 3.14

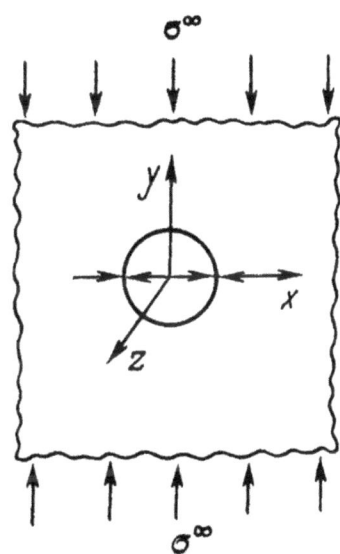

of the composite layer material (the longitudinal laying of layers along the axes Ox and Oy).

Therefore, in the continuum approximation, the material can be considered orthotropic, where the axes Ox, Oy, and Oz (Fig. 3.14) are the axes of symmetry of the material properties. At that, the compression was carried out along the axis Oy (along the axis of the symmetry of the material properties). In the longitudinal–transverse arranging, the unidirectional fibers were oriented in most layers along the axis Oy (Fig. 3.14). Therefore, the resulting layered plates can be considered the plates with preferring reinforcement along the axis Oy, along which the compression was realized. In these experimental studies, the fracture began with two points on the contour of the hole along the horizontal line in Fig. 3.14, i.e., from points with the maximum coefficient of concentration of compressive macrostresses (stresses within the framework of the continuum orthotropic model). The further development of fracture was carried out in the form of the formation of two almost straight cracks, which come out of the contour of the hole from the points with the maximum

Fig. 3.15

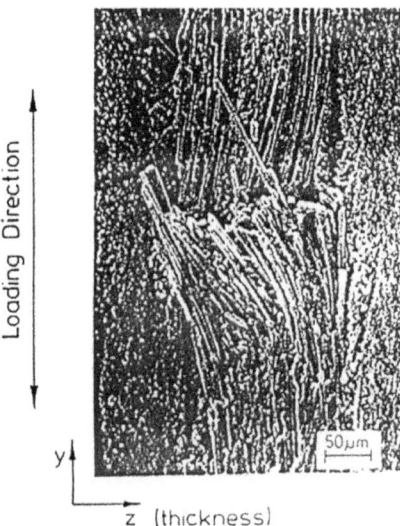

coefficients of concentration of compressing stresses (Fig. 3.14) and which spread almost perpendicular to the direction of the compression load. These cracks are filled with destructed material. Information on the nature of the motion of cracks (fracture) is presented on the microphotographs in Figs. 3.12, 3.13, and 3.15 obtained from an electron microscope and corresponding [187]. The scale of the image (in microns) of each image (Figs. 3.12, 3.13, and 3.15) is shown in the bottom right corner.

The pictures in Figs. 3.12, 3.13, and 3.15 show the fracture spreading from the right point on the contour of the hole on the horizontal axis on Fig. 3.14. So, Fig. 3.12 shows a crack spreading from the contour of the hole in a direction that is almost perpendicular to the action of the compressing load.

Figure 3.13 shows with a much larger increase (almost 20 times) the fractured part of the material inside the spreading narrow strip, which can be modeled by a filled crack. As shown in Figs. 3.12 and 3.13, results correspond to a compression load of about 95% of values of the total fracture load for the entire plate with the hole.

Figure 3.15 presents the structure of the fractured part of the material on the edge of the hole over the thickness of the plate (along the axis $0z$ in Fig. 3.14). In this image, the following types of fracture are clearly visible: the fracture of fibers, the bending of *the* destroyed fibers toward the hole, and the delamination of *the* layered material.

The results in Fig. 3.15 correspond to a compression load of about 80–85% of the values of the total fracture load for the entire plate with the hole.

It is worth noting that many authors, when analyzing the fracture in the internal structure of composites (of the type of fracture in the pictures of Figs. 3.8, 3.9, 3.10, 3.11, 3.12, 3.13, and 3.15), note only microbuckling (microswelling at the local loss of stability) and delamination. In fact, as *it is* shown in Fig. 3.13, there are significantly more mechanisms of fracture in the microstructure of the composite material under compression. Additionally, the following mechanisms of fracture can be noted: *the fracture of the fiber within the crack limit; bending of destructed fiber; fracture of fiber outside the crack; separation of fiber from the matrix delamination; fracture of the matrix,* etc.

Nevertheless, the above-mentioned and similar mechanisms of fracture in the microstructure of the composite under compression along the axes of symmetry of properties appear only in the later stages of fracture.

The initial stage of fracture (start) appears to arise in this situation only due to the loss of stability in the internal structure of the composite. The above start of fracture, of course, can occur both near the local inhomogeneities in the internal structure of the composite (when cutting, for example, the extreme fibers by the metal ring shown in Figs. 3.8, 3.9, and 3.10) and near the macro-inhomogeneities (near the hole, for example, shown in Figs. 3.12, 3.13, 3.14, and 3.15).

The appeared local fracture (shown in all the pictures of Figs. 3.7, 3.8, 3.9, 3.10, 3.11, 3.12, 3.13, 3.14, and 3.15) **then spreads over planes and surfaces that are almost perpendicular to the direction of the compression load. As has been repeatedly noted, the above situation is a characteristic of the type of fracture in question.**

The results of the experimental studies presented in 3.7, 3.8, 3.9, 3.10, 3.11, 3.12, 3.13, 3.14, and 3.15 and related ones were published in the second half of the twentieth century in 1968–1991. The similar studies are continuing in the present twenty-first century.

As an example, consider below the experimental results of the article [189] of 2004. Here, the results of experimental studies for compression along layers of the layered composite material consisting of 628 layers are presented.

The presence of so many layers makes it possible to expect that the experimental results seem to refer to the material consisting of an "infinite" number of layers. In this regard, it can be conventionally considered that the results of [189] seem to refer to the phenomena that occur in the internal structure of the layered composite and do not depend on the boundary conditions on the boundary surfaces of the entire package. However, it seems impossible to completely exclude the impact on all phenomena of the boundary conditions, especially the boundary conditions on the ends of the package. Since the compression was carried out in [189] along the layers, it can be considered in the continuum approximation that the compression was realized along the axes of the symmetry of the orthotropic material.

In the corresponding to [189, p. 1074, Fig. 2] Fig. 3.16, the shapes of the layers are shown which they acquire at the appropriate value of the compression load.

It follows from an analysis of the results presented in Fig. 3.16 that there are as if narrow strips of fractured material (conditionally in Fig. 3.16 shown by sloping solid lines). These bands seem to be periodically repeated along the horizontal, along which the single-axis compression was carried out. It should be also noted that narrow bands of the destroyed material are tilted toward the vertical axis at angles 19°; 18.5°, and 18°. They differ little from perpendicular to the horizontal axis.

Fig. 3.16

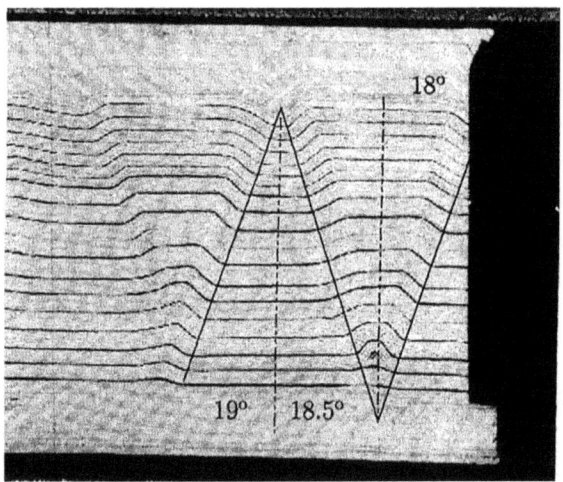

Thus, the results of experimental studies in [189] that are presented in Fig. 3.16 confirm the conclusion, formulated based on the results of experimental studies presented in 3.7, 3.8, 3.9, 3.10, 3.11, 3.12, 3.13, 3.14, and 3.15. This conclusion is that **in uniaxial compression, the fracture spreads across the planes and surfaces that are almost perpendicular to the direction of the compression load. As it was repeatedly noted, the above situation is a characteristic of the type of fracture in question.**

The above-mentioned information, considerations, and conclusions will be limited to the analysis of the results of experimental studies relating to the fracture of composite materials under compression along the reinforcing elements (or, more broadly, when compressed along the axes of the symmetry of composite properties), following the specific goal formulated in the introductory part of Sect. 3.2.

3.2.3 About the Study of the Phenomenon of "Kinking"

In addition to Sects. 3.2.1 and 3.2.2, this subsection provides a very brief account of the study of the "kinking" phenomenon, because this study arose from an analysis of experimental results under compression of composite materials.

The "kinking" phenomenon was originally considered in the article [190] of 1983. In the following years, the concept of "kinking" became quite popular, especially among English-speaking researchers, and the numerous articles have already been published within the framework of this concept, among which it is advisable to include the review [191] of 1997 published in the well-known series of publications "Advances in Applied Mechanics" (USA).

Fig. 3.17

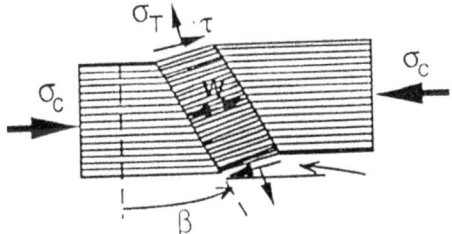

In general terms, the phenomenon of "kinking" lies in the appearance (when the composite is compressed along the reinforcing elements) rather than narrow bands of "kinking" (kink bands) of *the* already fractured material.

Schematically, the phenomenon of "kinking" is shown in Fig. 3.17, which is included in most publications on the study of this phenomenon, where W is the width of the "kinking" zone.

First of all, it is worth noting that the "kinking" zone is analyzed based on *the very approximate* design scheme and relations.

The author of this monograph also participated in the discussion (in a very brief form) of studies on the "kinking" phenomenon, which was reflected in the article [192] of 2006, monograph [4] (vol. 1, p. 73,74) of 2008, monograph [163] (p. 94–96) of 2015, and review [162] (p. 46,47) of 2016.

The thoughts and conclusions shown in the above publications are also outlined in a very brief manner. At that, the typos and inaccuracies observed are corrected.

First of all, it should be noted that the scheme of fracture presented in Fig. 3.17 refers to the already fractured sample. Therefore, the beginning (start) fracture is not determined in this approach.

Thus, it cannot be excluded that the beginning (start) of the process of fracture, which in the final phase of the process of fracture led to the appearance of the "kinking" band, was determined by the loss of stability in the internal structure of the composite following the General Concept stated in Sect. 3.1.1 of this monograph.

The above situation gives some reason to attribute the discussed study of the "kinking" phenomenon to the first direction (very approximate approaches) (Sect. 3.1.2 of this monograph). It should also be noted that for the "kinking" phenomenon under compression along the horizontal axis (Fig. 3.17) the shift is observed along the vertical axis of parts of the material located on the left and right side of the "kinking" band (Fig. 3.17). This situation seemed to be typical of the "kinking" phenomenon.

Following the above publications [4, 162, 163, 192], the three considerations on the "kinking" phenomenon are presented below.

1. The phenomenon of "kinking" in samples (Fig. 3.17) may occur if the boundary conditions on the ends of the samples allow the presence of the displacement along the vertical axis when the samples are compressed along the horizontal axis.

2. In the case of internal fracture (the model of "infinite" material), the "kinking" phenomenon *cannot* occur as a separate isolated band of "kinking" (in the form

of a separate kink band) when compressed along the axes by the symmetry of the material properties. Inside the composite, when compressed along the axes of symmetry of material properties, the phenomenon can only exist in the form of alternating "kinking" bands, so that the resulting disturbance of the stress–strain state is self-balanced.

3. The study of the "kinking" phenomenon does not seem to be able to establish unequivocally the mechanisms that determine the beginning (start) of the process of fracture. One of the mechanisms that determine the beginning (start) of the process of fracture can be in this case the mechanism of loss of stability in the internal structure (internal instability) of the composite following the General Concept (Sect. 3.1.1), which can be quite exactly researched within the framework of **the** second direction (strict sequential approaches based on TLTDBS) (Sect. 3.1.3) of this monograph.

As noted in the above publications, these considerations only reflect the views of the authors of these publications. Other considerations reflecting the views of other authors are also possible. In some sense, the second of the above considerations is confirmed by the results of the experimental studies, which are published in the article [189] and presented in Fig. 3.17. The results of the experimental studies are described and discussed in a very brief form in the text of this monograph near Fig. 3.17.

Taking into account the additional nature of Sect. 3.2.3, the above information in Sects. 3.2.1, 3.2.2, and 3.2.3 will be limited when discussing the results of experimental studies on the fracture of composites under compression along the axes of symmetry of the material properties.

3.3   Main Results of the Second Direction (Strict Sequential Approaches Based on the TLTDBS)

The short characteristics of the second direction of the theoretical research in the construction of the fracture mechanics of composites under compression along the direction of symmetry of the material properties are presented in Sect. 3.1.3. The main characteristics of researches within the framework of the second direction can be formulated as follows.

1. The characteristic of the first direction *approximate design* schemes and assumptions (such as those specified in Sect. 3.1.2) are not used.
2. The research is carried out in a three-dimensional statement based on TLTDBS (in a brief form the apparatus of which is outlined in Chap. 2), which leads to results exactness or which corresponds to other directions of the mechanics of deformed bodies.
3. By taking into account the above characteristics 2, the results of the second direction can be used to estimate the more approximate theories, including the results of the first direction.

4. The foundations of the mechanics of near-the-surface fracture are developed only within the framework of the second direction.
5. The results of the second direction strictly and consistently determine the beginning (start) of the process of fracture for the composites under compression, both in internal fracture and in the case of near-the-surface fracture.

3.3.1 Introductory Information

In Sect. 1.3, when describing Problem 1 (fracture in composite materials under compression along the *reinforcing* elements) it is noted that the results in the second direction (approach) belong mainly to the author of the monograph and his pupils.

It is worth noting that the first articles in the second direction (approach) were articles [9, 10] of 1969. List of publications (monographs, articles, and reports at international conferences), which were included in the list of literature for this monograph and which are prepared based on studies of the Department of Dynamics and Stability of Continua of the S.P. Timoshenko Institute of Mechanics of NAS of Ukraine, is shown in the introductory part of this chapter .

Below, the main results obtained in the second direction are stated, which were originally presented in the above articles and at international conferences and then entered the above and other monographs.

The discussed studies were carried out for elastic and elasto-plastic models of compressible and incompressible isotropic, transversally isotropic, and orthotropic materials under compressing along the axes of symmetry of material properties (in the case of transversally isotropic and orthotropic materials). The results of a general character are obtained in the case of elastic models—for the hyperelastic materials with an arbitrary structure of the elastic potential—and in the case of elasto-plastic models—for materials with constitutive equations of a sufficiently general form.

The results of a specific character are obtained for the elastic and elasto-plastic models of materials with the constitutive relationships of the simplest structure. For the elasto-plastic materials (as applied to the case of binder and filler), the generalized concept of the continued loading is applied, which is stated in Sect. 2.3.2. In this regard, the study is carried out in the common general form for the elastic and elasto-plastic materials.

The study of the phenomenon of loss of stability in the internal structure (internal instability) and in the near-the-surface layers (near-surface instability) for the composite materials is carried out under the action of the external "dead" load, which is typical for almost all publications on the fracture mechanics.

As applied to the studies of the second direction in the case of elastic and elasto-plastic models (with allowance for the generalized concept of continued loading), the fulfillment of the sufficient conditions of applicability *of* the static method of stability research is strongly proven (Sect. 2.4.2—the first result). Thus, these problems are *reduced* to the *boundary values* problems; i.e., the Euler method is applied.

The above proof also refers to *the* near-the-surface instability under compression along the *reinforcing* elements and near-the-surface instability near *the* loaded ends.

Thus, the studies **with**in the framework of the second approach are fully consistent with the generally **accepted** and rigorous method of **the** research of the phenomenon of loss of stability—the analysis of the behavior of small perturbations within the framework of the linearized **three-dimensional** dynamic problems.

The above conclusion, proof, and approach are valid for the elastic and elasto-plastic models (with allowance for the generalized concept of the continued loading) and are not valid for the models with the rheological properties.

Note 3.6 The results for the models with the rheological properties are not included in this monograph (in Sect. 3.3.1 "Main results of the second direction (strict successive approaches based on TLTDBS)" by the following reasons:

1. As noted in Note 2.9, there is no common method and strict criterion for bodies with rheological properties, allowing to research with such a degree of generality and strictness as for the elastic and elasto-plastic bodies.
2. The results, based on three-dimensional equations for bodies with rheological properties, are now built on *a* static method involving one of the approximate criteria of stability shown in Sect. 2.3.3. Due to the approximate character of the criterion, the commonality and validity of these results are not clear.
3. Formulated before this Note 3.6, conclusion, proof, and approach are not valid for the bodies with rheological properties.

Note 3.7 The results of the second direction are intended for the composite materials with polymer and metal matrix. For composites with the polymeric matrix, the brittle fracture is analyzed. In this case, the matrix is modeled by an elastic body, what is typical for the composites at moderate temperatures and relatively short-time load action, because in this case the viscosity effects are not taken into account. For the composites with a metallic matrix, *the* plastic fracture is analyzed (taking into account the generalized concept of continued loading). In this case, the loading stage is considered, when the entire matrix is in *a* state of plastic deformation.

Note 3.8 Within the framework of the second direction, in the study internal instability of the composite and the surface instability of composites, *the* identical shortening *of* the filler and matrix along the direction of compression is provided (along the fibers in Fig. 3.1 for a unidirectional fibrous composite, along the layers in Fig. 3.2 for a layered composite). This condition seems to be the only possible condition which allows to analyze the phenomena inside the composite material. In the experimental studies, the above conditions are met under compression with the sufficiently rigid disks (along the vertical axis in Fig. 3.2), when the minimal friction along the horizontal axis (at the ends in Fig. 3.2) is provided. In the theoretical studies along the vertical axis, the identical movement in the filler and matrix is given, and the zero tangential stresses are given along the horizontal axis.

Taking into account the stated above information of an introductory nature in this subsection, below in the following subsections the main results obtained in the second direction and stated in the publications which are shown in the introductory part of this chapter will be summarized.

Besides, note that these main results were obtained by the author and his pupils in the Department of Dynamics and Stability of Continua of the S.P. Timoshenko Institute of Mechanics of the NAS of Ukraine.

In the second direction, the next theories are developed:

The continuum theory of the fracture of composites, based on the model of a homogeneous medium with the averaged parameters and mathematical apparatus TLTDBS (Chap. 2).

The three-dimensional theory of fracture of fibrous unidirectional and layered composites, based on the model of a piece-wise homogeneous medium at the exact boundary conditions on the interface and the mathematical apparatus of TLTDBS (Chap. 2).

The monographs of [3, 4] are used significantly in the presentation of the published results, and the experience of the short presentation of these results in the review article [162] (pp. 27–45) of 2016 is used.

3.3.2 Continuum Theory of Fracture

In this paragraph, in a brief form, the main elements of the continuum theory of fracture of composites under compression are considered. This theory is based on the continuum model of composites with averaged parameters and TLTDBS, as applied to the brittle and plastic fracture, taking into account Note 3.7 for the plastic fracture. The main results are obtained for the internal fracture (loss of stability in the internal structure and internal instability) and near-the-surface fracture (the near-the-surface loss of stability in the internal structure and near-the-surface instability).

3.3.2.1 Internal Fracture

When researching the internal fracture, the composite should be considered that occupies the infinite space. The fracture and its spreading are described by the system of static equations TLTDBS (Chap. 2). In an unloaded state, this system of equations for the compressed material is a system of equations of elliptical type. The compressibility of the composite in the continuum approximation is ensured by the compressibility of the filler or binder. Due to the consideration of the infinite material, the analysis of internal fracture examined is studied as if the loss of stability in the microvolume. Thus, the changes in microvolume in the used mechanism of fracture should be manifested in some way also in the macrovolume. In the latter case, these changes should not be of a local nature, because only in this case the fracture of the entire sample (macrofracture) will be observed. This change in the microvolume should be manifested in the properties independent of the boundary conditions because the process of the fracture of material is investigated (the internal fracture corresponds to the infinite material) and not the effect of the captures of the test machines, cross-sectional shapes, etc. It is obvious that the changes in the macro-volume are determined by the disturbances of displacements, which are described by the system of static equations TLTDBS (Chap. 2).

Thus, it can be considered that the beginning of fracture corresponds to the appearance of solutions of the system of static equations TLTDBS (Chap. 2) for the compressed material, which are not dependent on the boundary conditions (the case of infinite material is considered) and are not of a local nature. At that, the solutions of the type of the homogeneous stress–strain states should be certainly discarded.

The above condition for the system of the static equation of TLTDBS (Chap. 2) can only be fulfilled when this system becomes a hyperbolic type one.

Taking into account the stated above, the basic concept of constructing the continuum theory of internal fracture of the composite materials under compression can be formulated as follows.

The Basic Concept. The beginning of the process *of* fracture can be identified with the moment in the history of loading when the system of static equations TLTDBS (for *the compressed* materials) goes from *the* elliptical system into hyperbolic; i.e., this system loses the property of ellipticity. *The* theoretical ultimate strengths are determined from the same condition. The fracture of composite materials under compression occurs along *the* characteristic planes and surfaces.

The additional information on this subject is presented in the monograph [4], vol. 1, Chap. 2. Below, only some of the results are considered.

Following the basic concept, the notion of the surface Π_T—the surface of theoretical ultimate strengths under compression, which is introduced in the three-dimensional space of the compressing main stresses of the stress tensor $\left((-\sigma_{11}^0),\ (-\sigma_{22}^0),\ (-\sigma_{33}^0)\right)$. Schematically, the Π_T surfaces of theoretical ultimate strengths under compression are presented in Figs. 3.18, 3.19, and 3.20. Figure 3.18 corresponds to the general case, Fig. 3.19 corresponds to the axisymmetrical load, and Fig. 3.20 corresponds to the plane problem. In Figs. 3.18, 3.19, and 3.20 and below, the designation $(\Pi_j^-)_T$ is introduced—the theoretical ultimate strength under

Fig. 3.18

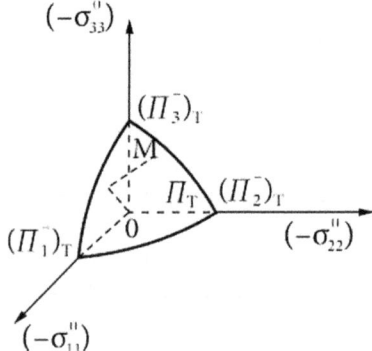

Fig. 3.19

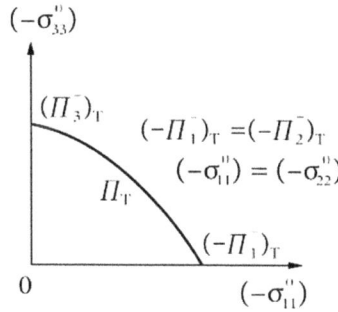

Fig. 3.20

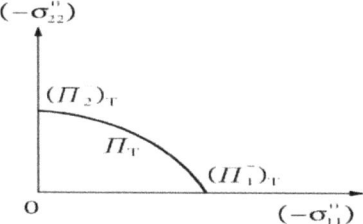

uniaxial compression along the corresponding axes. Figure 3.19 is related to axisymmetrical load and represents a meridional section of the surface of theoretical ultimate strengths for compression since in this case the surface Π_T is the surface of revolution. In the three-dimensional space $\left(\left(-\sigma_{11}^0\right), \left(-\sigma_{22}^0\right), \left(-\sigma_{33}^0\right)\right)$ in Fig. 3.18, the loading trajectory is shown by the bar line OM.

If in the history of loading *the* point M (Fig. 3.18) first gets to the surface Π_T, then a theoretical ultimate strength is reached under corresponding three-axial compression.

In the case of *the* brittle fracture (the composites with a polymeric matrix), the surface Π_T is built in an explicit form for axisymmetrical spatial (Fig. 3.21) and plane (Fig. 3.22) problems.

In Fig. 3.21 for the brittle fracture, as in Fig. 3.19 for the general case, a meridional section of the surface of theoretical ultimate strengths is presented in the form of a surface of revolution (in the case of axisymmetrical load). In Figs. 3.21 and 3.22,

Fig. 3.21

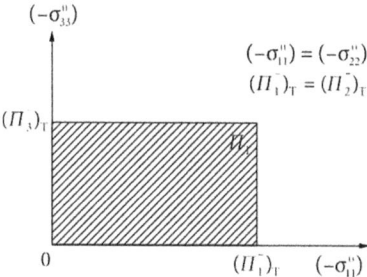

Fig. 3.22

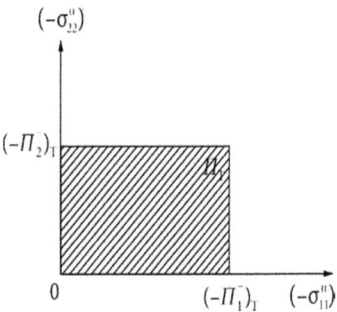

the shaded areas are the areas of ellipticity of the static system of equations of the TLTDBS (Chap. 2).

When the loading parameters are changed inside the area of ellipticity, the fracture at the corresponding values of loading parameters does not occur. The additional information is presented in the monograph [4] (vol. 1, pp. 182–185). In the case of the plastic fracture, more research with using the computer methods is needed to build the appropriate surfaces of theoretical ultimate strength.

Also, it is strictly proven in the case of *the* brittle fracture that the fracture spreads over planes perpendicular to the action of compressing loads. The experimental confirmation of the above theoretical regularity is the results in Figs. 3.7, 3.8, 3.9, 3.10, 3.11, and 3.12 under the uniaxial compression.

Below, the information is shown on a comparison of the theoretical ultimate strengths on compression and theoretical values of ultimate shortenings calculated within the framework of the continuum theory of internal fracture (Sect. 3.3.2.1), with the corresponding values determined in the experimental studies. At that, the focus is made on the information from the monograph [4], vol. 1, Chap. 2 presented separately for the brittle and plastic fracture. These studies are carried out for uniaxial compression along fibers (conditionally along the axis $0x_3$) of a unidirectional fibrous composite (Fig. 3.1), which in the transverse cross section ($x_3 = $ const) does not have a clearly defined structure in the arrangement of fibers. Therefore, in the continuum approximation, this composite is modeled by a transversal–isotropic material, the axis of isotropy of which is directed along the fibers (along the $0x_3$-axis).

Indices "a" and "m" below and in the entire monograph note all the values related to the reinforcing (arming) elements (filler) and matrix (binding).

Brittle fracture. The above comparison will be made for unidirectional fibrous composites with polymer matrix in the form of epoxy resin in the case of very rigid fibers, for the Young modulus of which takes place

$$E_a \gg E_m. \tag{3.5}$$

Introduce also the following designations: $(\Pi_3^-)_T$ is the theoretical ultimate strength under compression along one axis (in this case, along the $0x_3$-axis); $(\Pi_3^-)_{эк}$

is the experimental ultimate strength for the uniaxial compression along the same axis.

In the case (3.5) for a composite with a 50% volume fraction of unidirectional fibers ($S_a = S_m = 0, 5$), the theoretical ultimate strength on compression (within the framework of the continuum fracture theory) is given in the monograph [4], vol. 1, p. 192.

$$(\Pi_3^-)_T = 2.09 - 3.00 \, \text{GPa} \tag{3.6}$$

taking into account the spreading of the epoxy resin properties shown in Table 0.1 [4] (vol. 1, p. 67).

The values of the experimental ultimate strength for different composites (different fibers at $S_a = S_m$) are presented in Russian in the handbook [193] of 1981 (this edition was published in English as [194] in 1978) on page 656. These results are also presented in the [4], vol. 1, p. 192 and are as follows

$$(\Pi_3^-) \begin{cases} 3.10 \, \text{GPa} & - \; in \, the \, case \, of \, boron \, fibers; \\ 1.38 \, \text{GPa} & - \; in \, the \, case \, of \, high - strength \, carbon \, fibers; \\ 1.03 \, \text{GPa} & - \; in \, the \, case \, of \, high - module \, carbon \, fibers. \end{cases}_{exp} \tag{3.7}$$

It follows from the comparison of the results (3.6) and (3.7) that a good match between the theoretical ultimate strength for compression and the experimental ultimate strength at uniaxial compression is observed in these polymer matrix composites.

Plastic fracture. Here, the comparison will be made for a unidirectional fibrous material with a metal matrix, using the experimental results of the article [195]. In this article, the experimental results for a metal composite (filler–stainless steel wire and matrix–pure aluminum) are given. The theoretical ultimate strength and the theoretical value of the ultimate shortening under uniaxial compression (within the framework of the continuum theory of internal fracture, which is considered in Sect. 3.3.2.1) are determined in the first approximation in the monograph [4], vol.1, Chap. 2, pp. 193–202, where the filler was modeled by the linear elastic isotropic compressible body and the matrix—by the elasto-plastic isotropic incompressible body with the power dependence between the stress and strain intensities in the form:

$$\sigma_u^m = A_m \varepsilon_u^{m k_m}; \; A_m, k_m - \text{const.} \tag{3.8}$$

In the article [195], the experimental results for different concentrations of fillers in percent ($S_a = 4.1; 11; 15.3; 21.2; 24.8; 32.8$) are given. To shorten the discussions presented in the monograph [4], vol. 1, Chap. 2, pp. 204–205, the results are shown in Fig. 3.23, unlike Fig. 2.9 from [4], vol.1, Chap. 2, p. 206, only for values S_a (%): 15.3; 21.2; 24.8; and 32.8.

Fig. 3.23

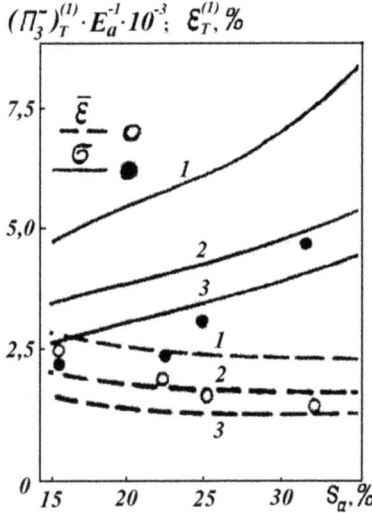

$$(\Pi_3^-)_T^{(1)} \cdot E_a^{-1} \cdot 10^{-3};\ \ \varepsilon_T^{(1)}, \%$$

Note that in Fig. 3.23, the inaccuracy is eliminated in the designations made in Fig. 2.9 of monographs [4], vol. 1, Chap. 2, p. 206, where a multiplier E_a^{-1} (E_a is Young modulus of reinforcing elements according to the designations (3.5) m in this case, the stainless steel wire) is missed.

In the above theoretical studies, the following three approximations for values A_m and k_m in (3.8) were used in the description of the plastic deformation of pure aluminum with the utilizing the relation (3.8):

$$1 \sim A_m = 100\,\text{Mpa},\ k_m = 0.1;$$
$$2 \sim A_m = 100\,\text{MPa},\ k_m = 0.25; \tag{3.9}$$
$$3 \sim A_m = 68\,\text{MPa},\ k_m = 0.25.$$

In the monograph [4], vol. 1, Chap. 2, p. 207, the authors are listed in publications of which the approximations (3.9) were used.

With allowance for the stated above, the dependences on S_a (filler concentrations; in this case stainless steel wires) are represented in Fig. 3.23 for the following values: the solid lines for quantity $(\Pi_3^-)_T^{(1)} \cdot E_a^{-1} \cdot 10^{-3}$—the immeasurable normalized value of the theoretical ultimate strength when compressed along a single axis, calculated in the first approximation; the dashed lines for a quantity $\varepsilon_T^{(1)}$ in %—the theoretical value of the ultimate shortening calculated in the first approximation.

At that, the curves in Fig. 3.23 corresponding to approximations (3.9) are marked with subscripts 1, 2, and 3. The experimental results from [195] are presented in Fig. 3.23: dark circles—for the experimental ultimate strength and light circles—for the experimental value of the ultimate shortening. Note that approximation 2 was used in [195].

It follows from an analysis of the results presented in Fig. 3.23 that satisfactory correspondence between the theoretical and experimental results is observed for the ultimate strengths when the approximation 3 of (3.9) being used and for the ultimate shortenings when the approximation 2 of (3.9) being used.

The above information will be limited to discussing the mechanics of internal fracture within the framework of the continuum theory of fracture under consideration: As was repeatedly noted, enough detailed information on this subject is given in the monograph [4], vol. 1, Chap. 2.

3.3.2.2 Near-the-Surface Fracture

The continuum theory of near-the-surface fracture is based on the study of a near-the-surface form of loss of stability, displacements, and stresses which attenuate when moving away from the boundary surface. This study uses the basic relations of TLTDBS (Chap. 2 of this monograph). In accordance with the basic approach outlined in Sect. 3.3.1, in this case (Sect. 3.3.2.2) the static problems of TLTDBS (Chap. 2) should be solved for the semi-limited areas (to the boundary value problems, in which the eigenfunctions attenuate while moving away from the boundary surface).

In this subsubsection, the continuum theory of the near-the-surface fracture is considered, corresponding to the near-the-surface loss of stability under compression along the reinforcing elements (filler). The continuum theory of the near-the-surface fracture, which refers to this type, is detailed in the monograph [4], vol. 1, Chap. 2, Sect. 2, pp. 209–224.

For example, the design scheme for a plane problem is presented in Fig. 3.24, where the surface $x_2 = 0$ is free and reinforcement elements (filler) are directed along the axis $0x_1$. It can be considered that the monographs [3, 4] sufficiently particularize the continuum theory of near-the-surface fracture of the composite under compression, corresponding to the near-the-surface loss of stability (near the free surface) under compression along the reinforcing elements (filler). Therefore, this monograph will not particularize this problem.

These results from monographs [3, 4] relate to the composites with the polymeric matrix (brittle fracture) and a metal matrix (plastic fracture). At that, Note 3.7 is taken into account in this study of plastic fracture. Besides, note that the two-level continuum fracture mechanics of the composites is also presented in the monograph [4] (vol. 1, Chap. 2, Sect. 2) when the cracks come out of the contour of the hole.

Fig. 3.24

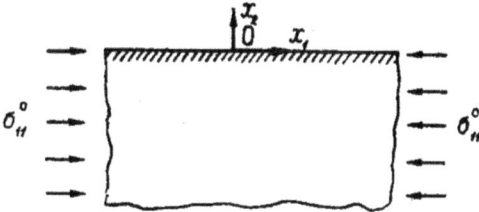

In conclusion in Sect. 3.3.2, note only that the fulfillment of the following condition

$$\left(\Pi_3^-\right)_T^\Pi < \left(\Pi_3^-\right)_T. \tag{3.10}$$

is strictly proven in monographs [3, 4] for composites with polymeric and metallic matrix within the framework of the continuum fracture theory under compression.

In (3.10), in addition to the designation $(\Pi_3^-)_T$ (theoretical ultimate strength for uniaxial compression in *the* case of internal fracture), the designation $(\Pi_3^-)_T^\Pi$ is introduced (theoretical ultimate strength for the uniaxial compression along the free surface in the case of *the* near-the-surface fracture). The condition (3.10) corresponds to the usually accepted position—the fracture of the material begins with the surface of the material.

The above information will be limited to discussing the near-the-surface fracture within the framework of the continuum fracture theory of composites under compression.

Note 3.9. In Sect. 3.3.2.1, as applied to the internal fracture, the fairly good coincidence is shown for the ultimate strengths for the uniaxial compression obtained in the experimental studies, with the corresponding theoretical ultimate strengths of the continuum fracture theory under consideration, both for the brittle fracture (composites with a polymeric matrix) and for the plastic fracture (composites with a metallic matrix). The good coincidence *appears* to be due to *the* fact that the analyzed examples considered the composites in which filler and matrix materials differ significantly in rigidity (conditions (3.5) are fulfilled). For other composites, it seems, one should not expect such a good coincidence. In general, it is worth noting that *the*, considered in this book, continuum *fracture* theory is the most *non*-cumbersome and easy to research compared to any theories built within the model of a piecewise homogeneous medium, both in the first and in the second directions. At that, the continuum theory in some cases produces results corresponding to experiments. In the historical aspect, apparently, the first publication on the construction of this continuum fracture theory was an article [10] of 1969.

The stated information in this subsection when considering the fracture theory of the composite materials under compression along the axes of symmetry of the material properties seems to be sufficient. This theory is based on a model of homogeneous material (medium, body) with the averaged properties (parameters) and the TLTDBS, the mathematical apparatus of which is summarized in Chap. 2 of this monograph.

3.3.3 Layered Composites: Model of a Piece-Wise Homogeneous Medium

This subsection provides very brief information on the fracture mechanics under compression of the layered composites with polymer and metallic matrix, which is constructed based on a piece-wise homogeneous model.

In this case, separately for the materials of each layer of filler and binder the basic relations of TLTDBS are accepted and at the interfaces, certain conditions of continuity of stresses and displacements are accepted. The research is carried out for the brittle and plastic fracture. In the latter case, a generalized concept of continued loading is tentatively adopted.

Following the outlined main approach in Sect. 3.3.1, in the case of the studies in question the analysis of static equations and the corresponding boundary conditions of the TLTDBS for a piece-wise homogeneous (boundary values problem) is used.

The main results are obtained for the loss of stability in the internal structure (internal instability and internal fracture), the results of which are outlined in the monograph [4], vol. 1, Chap. 3, and for the near-the-surface loss of stability (near-the-surface fracture), the results of which are outlined in the monograph [4], vol. 1, Chap. 5. The main results are also in part presented in monographs [1–3] as well as in a very brief form in the review article [162]. Preliminarily, these results for the layered composites were published in articles, the main of which were included in the literature list for this review article and the main articles relating to the entire second direction and given in the introductory part of Chap. 3 of this monograph. The additional information on the published articles can be obtained from the lists of literature to monographs [1–4].

The main results for the layered composites in the form of characteristic determinants for the internal and near-the-surface fracture are stated in the common general form for theories 1, 2, and 3 in accordance with the terminology of Sect. 2.2 and the introductory part of Sect. 2.3.

Below, only the main results on the fracture mechanics of composites under compression are discussed in a very brief form, obtained within the piece-wise homogeneous medium model and the TLTDBS, whose basics of the mathematical apparatus are summarized in Chap. 2. At that, the brief statement of this problem in the review article [162] is taken into account.

3.3.3.1 Internal Fracture

In the study of internal fracture (internal instability), a layered composite is considered that occupies the infinite space (Fig. 3.2), the general solutions of the static equations of the TLTDBS are applied, and analysis is carried out in accordance with the procedure which is stated in Sect. 3.1.1 taking into account the approach corresponding to Fig. 3.3. Note that this paragraph provides only information on the results

Fig. 3.25

Fig. 3.26

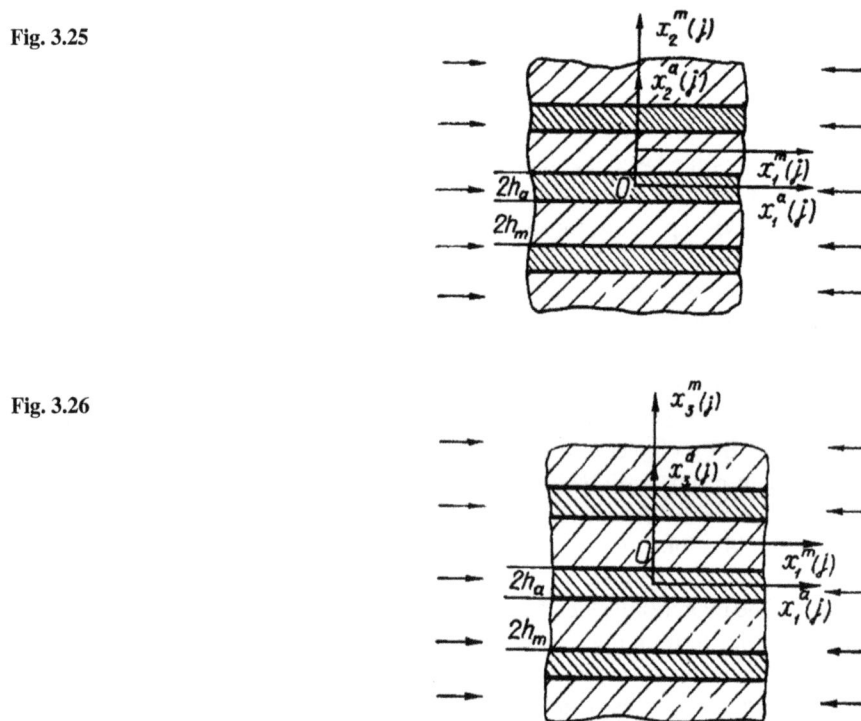

obtained for the composites, on the interfaces of which the defects are absent (on the interfaces, the conditions of continuity of stresses and displacements are met).

The plane and spatial problems for the layered composites with polymeric and metallic matrix consisting of layers of the filler (identical thickness) and layers of the matrix (identical thickness) are studied. The layers alternate periodically along the axis $0x_2$ (Fig. 3.25) in the case of plane problems and along the axis $0x_3$ (Fig. 3.26) in the case of spatial problems. Figure 3.26 corresponds to the plane in the cross section $x_2 = 0$. In Figs. 3.25 and 3.26, index "*a*" marked all values related to the reinforcing elements (filler, layers), and the index "*m*" marked all values related to the matrix (binding, layers). In the case of plane problems, the layers of orthotropic materials (in the particular case, the isotropic materials) are considered. In the case of spatial problems, the layers of transversal isotropic materials with a plane of isotropy $x_3 = \text{const}$ (Fig. 3.26) (in the particular case, the isotropic materials) are considered. **At that,** in all cases, the characteristic determinants are obtained for the materials with constitutive equations of a sufficiently general structure.

The modes of the loss of stability with a period along the vertical axis that is multiple to the period of structure analyzed, i.e., with the period T_k

$$T_k = 2k(h_a + h_m); k = 1, 2, \ldots. \tag{3.11}$$

are analyzed by taking into account the periodicity of the structure along the vertical axis in Figs. 3.25 and 3.26 with the period $2(h_a + h_m)$.

The general method of solving for the plane (Fig. 3.25) and spatial (Fig. 3.26) problems on the internal fracture of layered composites is the following. The static equations of the TLTDBS are applied, which is strictly justified (Sect. 3.3.1), and the general solutions of the TLTDBS (Sects. 2.6.1 and 2.6.2) are used.

The following modes of the loss of stability are considered: along the lines of action of compressing loads (along $0x_1$ for plane problems, Fig. 3.25; along $0x_1$ and $0x_2$ for the spatial problems, Fig. 3.26) and are chosen in the form of trigonometric functions with an unknown wavelength mode of the loss of stability; perpendicular to layers based on the general solutions of the TLTDBS, the solutions by x_2 for the plane problems (Fig. 3.25) and by x_3 for the spatial problems (Fig. 3.26) are built for different types of symmetry, which is shown below on the example of Figs. 3.27, 3.28, 3.29, and 3.30 and the different periodicities according to (3.11).

When the above solutions are applied to the fulfillment of the conditions of continuity of stresses and displacements along the interfaces of the filler and binder, then

Fig. 3.27

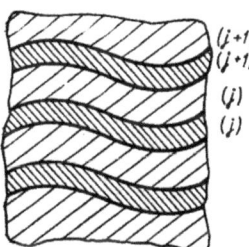

Fig. 3.28

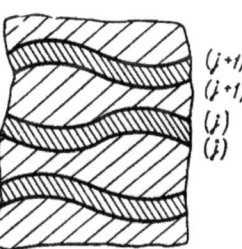

Fig. 3.29

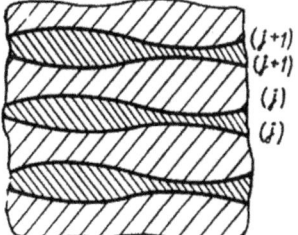

Fig. 3.30

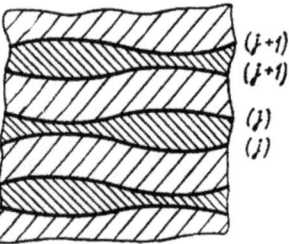

the characteristic determinants of the fourth order for plane problems and the sixth order for the spatial problems are obtained. As a result of the solution of the above characteristic equations, the dependence of the parameters of the loading on the parameters of wave formation is obtained. The minimizing of these relationships is carried out following the results of Sect. 3.1.1 and Fig. 3.5.

The first four modes, which are called the modes of the first, second, third, and fourth kind, were studied. These modes are schematically shown in Figs. 3.27, 3.28, 3.29, and 3.30. The mode of loss of stability of the first kind has a period equal to the period of structure (in (3.11) $k = 1$) and is presented in Fig. 3.27. This mode corresponds to the shear mode in the accepted terminology in a number of publications.

The mode of loss of stability of the second kind has along the vertical axis the period equal to the double period of the structure (in (3.11) $k = 2$) and is represented in Fig. 3.28. This mode corresponds to the stretching mode by the accepted terminology in a number of publications. The form of loss of stability of the third kind has along the vertical axis the period equal to the period of structure (in (3.11) $k = 1$) and is represented in Fig. 3.29. The form of loss of stability of the fourth kind has along the vertical axis the period equal to the double period of the structure (in (3.11) $k = 2$) and is represented in Fig. 3.30.

For the layered composites with polymeric and metallic matrixes in the case of plane and spatial problems, the characteristic determinants are obtained in a closed form for the filler and binding materials where the constitutive equations are quite general form. The critical values of loading parameters are obtained as a result of minimizing the roots of the characteristic determinants, with using the numerical methods.

In this approach, the numerous results for the specific layered composites with the polymeric and metallic matrix are obtained.

As an example of the above results for the specific layered composites below in a very brief form some results are given for a layered composite which is composed of the isotropic layers, when the material of each layer is modeled by a linear elastic body. These results are detailed in the monograph [4], vol. 1, pp. 297–299. Figure 3.31 corresponds to Fig. 3.9 of this monograph. Figure 3.31 represents the dependence of the loading parameter on the wave-forming parameter α_a (of the parameter type (3.2)) for a layered composite with the following parameters: $E_a \cdot E_m^{-1} = 500$; $h_m \cdot h_a^{-1} = 1; \ 5; \ 10; \ 20; \ 30; \ 40; \ 50$ for a plane problem where the notations

Fig. 3.31

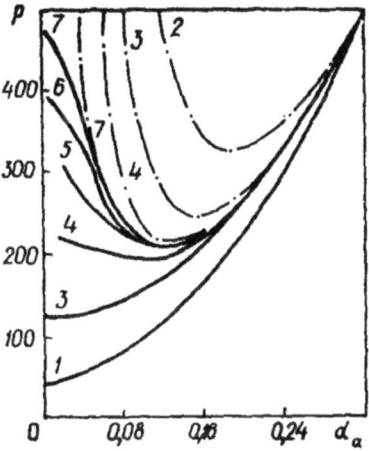

correspond to Fig. 3.25. The numbers 1, 2, 3, 4, 5, 6, and 7 in Fig. 3.31 feature the curves that correspond to the above values of parameter $h_m \cdot h_a^{-1}$. The solid lines in Fig. 3.31 show the results related to the bending mode (the first kind of the loss of stability mode, Fig. 3.27). The dashed-dot lines in Fig. 3.31 show the results related to the stretching mode (the second kind of the loss of stability mode, Fig. 3.28).

Let us take a quick look at the analysis of the results presented in Fig. 3.31 relative to the shear node (the first kind of the loss of stability mode, Fig. 3.27), which are depicted by the solid lines. It follows from Fig. 3.31 that the solid curves with numbers 1 and 3 ($h_m \cdot h_a^{-1} = 1$ and 10) are the curves of type B in Fig. 3.3. Thus, for this composite with the arrangement of the layers $h_m \cdot h_a^{-1} = 1$ *and 10*, the internal loss of stability (loss of stability in the internal structure) *is* not *affected* in a bending mode (the first kind of the loss of stability mode). It follows from Fig. 3.31 that the solid curves with numbers 4, 5, 6, and 7 $\left(h_m \cdot h_a^{-1} = 20, \ 30, \ 40 \text{ and } 50\right)$ are the curves of type *A* in Fig. 3.3.

Thus, according to the approach of Sect. 1.2.1 for this composite with the arrangement of the layers $h_m \cdot h_a^{-1} = 20, \ 30, \ 40,$ and 50 the internal loss of stability (loss of stability in the internal structure) arises by the bending *mode* (the first kind of the loss of stability mode).

Thus, the existence of the next phenomenon is strictly proven—depending on the structure *of the* layered composite, *the* loss of stability in the internal structure (internal instability) in *the* bending fashion (a form of loss *of* stability of *the* first kind, Fig. 3.27) may or may not occur.

The more information on the results of the internal fracture of the layered composites discussed in this subsubsection can be obtained from the monograph [4], vol. 1, Chap. 3.

3.3.3.2   Near-the-Surface Fracture

Consider in a very brief form the results of the near-the-surface fracture of the layered composites obtained within the model of piece-wise homogeneous medium. The sufficient detailed information on this subject is contained in the monograph [4], vol. 1, chap. 5.

In the study of the near-the-surface instability, a layered composite occupying the lower semi-space const $\geq x_2 > -\infty$ in the case of the plane problems according to Fig. 3.25 and lower semi-space const $\geq x_3 > -\infty$ in the case of the spatial problems according to Fig. 3.26 is considered. All designations and considerations stated in the beginning of Sect. 3.3.3.1 up to expression (3.11) remain valid for the near-the-surface fracture (surface instability) analyzed in Sect. 3.3.3.2.

The specific results are obtained for the layered composites with the polymeric and metallic matrix (brittle and plastic fracture). Two methods are developed to study the plane and spatial problems of the near-the-surface fracture (near-the-surface loss of stability).

The first method, which is practically an exact method, is based on the reduction of the problems to the infinite systems of algebraic equations with the following careful analysis of these problems.

The second method, which is a purely approximate method, is based on the application of the variational principles of the TLTDBS relative to the near-the-surface instability.

In the next part of this monograph, the examples of analysis of the results for the specific layered composites with polymeric and metallic matrix relative to the near-the-surface instability will not be shown.

Further, the information on a qualitatively new phenomenon which is described in the monograph [4], vol. 1, Chap. 5, p. 513 will be discussed. It consists in the following:

A phenomenon of *the* surface instability is observed (exists) not for all concentrations *of* filler and not for all relative parameters of stiffness.

So, with a continuous increase in external compressible load, the internal instability may occur at first, or the critical loads for the internal and surface instability may coincide.

The above information seems to be sufficient for a very brief discussion of the near-the-surface fracture (surface instability) in the layered composites with the polymeric and metallic matrix, obtained within the framework of the model of a piece-wise homogeneous medium. A sufficiently detailed presentation of the results on this problem is presented in the monograph [4], vol. 1, Chap. 5.

3.3.3.3   Additional Information About the Mechanics of Layered Composite Materials

This subsubsection provides the additional information in a very concise form pertaining to the following questions on the fracture mechanics of layered composites under compression:

1. An analysis of the continuum mechanics of composite destruction in compression outlined in Sect. 3.3;
2. Constructing a research method for more complex modes of the loss of stability for the layered composites;
3. The conclusions from a consistent analysis of the Dou–Grundfest–Rosen–Schuertz theory;

Analysis of the Continuum Fracture Mechanics of Composites

As was repeatedly noted in this monograph, that, when researching the various problems of stability in the mechanics of composites, the model of a piece-wise homogeneous medium, when the TLTDBS is used separately for the filler and matrix and at interfaces, the conditions of continuity of stresses and displacement vectors are provided, is the most rigorous and exact in the mechanics of deformable bodies.

   Naturally, the concrete results obtained in the most rigorous and exact model of the above are the basis for assessing the exactness and substantiating any more approximate theories, and the concrete results are obtained based on them. All the results for the fracture theory under compression of the layered composites outlined in Sects. 3.3.3.1 and 3.3.3.2 and the corresponding chapters of monographs [1–4, 196] are obtained based on this model. In this regard, the above-mentioned results on the fracture mechanics of the layered composites under compression are the basis for assessing the accuracy and substantiating the more approximate theories in this part of mechanics.

   The continuum theory of the fracture of composites under compression along the axes of the symmetry of the material properties, which is outlined in Sect. 3.3.2, is based on a much more approximate model (compared to the above strong and exact model)—a model of a homogeneous medium with averaged parameters and the TLTDBS. In this regard, the results on the fracture theory of the layered composites under compression obtained based on the most strong and exact model and set out in Sects. 3.3.3.1 and 3.3.3.2 can be used to assess the exactness and justification of the continuum fracture theory of composite under compression along the axes of the symmetry of material properties set out in Sect. 3.3.2 applied to layered composites.

   One of the main results of the continuum fracture theory of the composites under compression along the axes of symmetry of the material properties, which is detailed in the monographs [3, Chap. 2] of 1990 and [4], vol. 1, Chap. 2 of 2008 and in reduced form in Sect. 3.3.2, is the determination of the theoretical ultimate strength under uniaxial compression.

In the case of composites with a polymeric matrix (as the materials with the lower shear stiffness) for a plane problem (Fig. 3.25) in the plane $x_1 0 x_2$ with one-axial compression along the axis $0x_1$ for $(\Pi_1^-)_T$—the theoretical ultimate strength—the estimate is obtained (57, vol. 1, Chap. 2, formulas (2.60) and (2.61))

$$(\Pi_1^-)_T \leq G_{12}, \ (\Pi_1^-)_T \approx G_{12}, \tag{3.12}$$

where G_{12} is the modulus in a continuum approximation of the composite as an orthotropic material.

Note that the expressions (3.12) refer to a composite of an arbitrary structure, in which the axis $0x_1$ (Fig. 3.25) is the axis of symmetry of the material properties.

Since in the present subsubsection the analysis of the accuracy and justification of the continuum theory (Sect. 3.3.2) are carried out in terms of a strict and exact theory for the layered composites (Sects. 3.3.3.1 and 3.3.3.2), then the average value G_{12} of the shear modulus for the layered composites (3.12) is determined by the following expression

$$G_{12} = G_a G_m (S_a G_m + S_m G_a)^{-1}, \tag{3.13}$$

where G_a and G_m are the shear moduli of the reinforcing elements (filler) and matrix (binding), and S_a and S_m are the volume fractions of reinforcing elements (filler) and matrix (binding) in the case of isotropic layers.

Note that the expression (3.13) is obtained when the situation in Fig. 3.25 was strictly examined for a plane problem in the case of statics and with the method of "long waves" in the theory of wave propagation as applied to the classical linear theory of elasticity of the isotropic body. This expression is now well known and generally accepted.

Thus, for layered composites for a plane problem (Fig. 3.25) as part of the continuum fracture theory (Sect. 3.3.2), the theoretical ultimate strength under one-axial compression has the next form

$$(\Pi_1^-)_T = G_a G_m (S_a G_m + S_m G_a)^{-1}. \tag{3.14}$$

In the Foreword to the book [4], vol. 1, pp. 20, 21, the conditions for composites are formulated when the continuum model or theory (a model of a homogeneous anisotropic body with averaged properties or parameters) can be applied. Two geometric parameters are introduced for the formulation of these conditions:

The parameter L characterizes the mechanical processes (the character of the variability of the mechanical fields by the spatial variables).

The parameter h^* characterizes the averaged distance between the centers of neighboring particles in the internal structure of the composite.

The geometric parameter L corresponds: in the static problems—to the minimum distances, on which the fields of stresses and deformations are significantly changed; in the problems of wave propagation—to the wavelength; and in the problems of stability—the wavelength of the mode of loss of stability.

Taking into account the introduced designations, the condition of applicability of the model of a homogeneous anisotropic body with averaged parameters (continuum model and continuum theory) can be presented in the form

$$L \gg h^*. \tag{3.15}$$

In the layered composite with alternating layered composite (see Sect. 3.3.3) for a plane problem (Fig. 3.25) when the stability is lost on the periodic modes along $0x_1$ (Fig. 3.25) according to [4], vol. 1, p. 305, the following values as parameters L and h^* in (3.15) are chosen:
   $L \sim l$, where l is the length of half-wave (along the layers) mode of the loss of stability.
   $h^* \sim 2(h_a + h_m)$, where $2(h_a + h_m)$ is the total thickness of the filler layer and the binder layer.
   Taking into account the selected parameters and rations (3.15), the condition of applicability of the continuum model for the plane problem (Fig. 3.25) can be presented as a form

$$2(h_a + h_m) \ll l. \tag{3.16}$$

The parameters of wave formation (in the mode of the loss of stability) α_a and α_m can be obtained from (3.16)

$$\alpha_a = \pi \frac{h_a}{l} \ll 1; \quad \alpha_m = \pi \frac{h_m}{l} \ll 1. \tag{3.17}$$

Note that within the model of piece-wise homogeneous medium for the layered composites (Sect. 3.3.3) for all four forms of loss of stability, which are represented in Figs. 3.27, 3.28, 3.29, and 3.30, the characteristic determinants are built, elements of which also depend on the parameters of wave formation α_a and α_m (3.17). These results are presented in the monograph [4], vol. 1, Chap. 3.
   In this monograph on page 314, the concept of asymptotically exact continuum theories of internal instability of the layered composites is introduced.
   The continuum theory *of* internal fracture, constructed on a model of*a* homogeneous body with averaged parameters (properties), is asymptotically exact if its results follow from the results of the theory constructed on *the* model of a piece-wise homogeneous body, *when the parameters of wave formation* α_a *and* α_m *tend to zero with allowance for (3.16) and (3.17), that is, when*

$$\alpha_a = \pi \frac{h_a}{l} \to 0; \alpha_m = \pi \frac{h_m}{l} \to 0. \tag{3.18}$$

Of course, the proof of the asymptotic exactness of this theory for a layered composite in the case of a plane problem (Fig. 3.25) can only be used for the four modes of the loss of stability presented in Figs. 3.27, 3.28, 3.29, and 3.30, for which the characteristic determinants are obtained within the model of a piece-wise homogeneous body (Sect. 3.3.2.1).

To prove, it is necessary to present all elements of the characteristic determinants in the form of series by parameters α_a and α_m (3.18) and to calculate the first member of such expansion of the entire characteristic determinant.

In the monograph [4], vol. 1, p. 316, it *is* strongly proven for a plane problem (Fig. 3.25) in the case of layered composites with a polymeric matrix that the continuum theory of internal fracture (Sect. 3.3.2.1) is asymptotically exact and corresponds to the mode of loss of stability of the first kind (the bending mode, Fig. 3.27). In this case, for the theoretical ultimate strength under a one-axial compression, the expression (3.122) of the monograph [4], vol. 1, p. 316 is obtained, which fully coincides with the expression (3.14) obtained within the framework of the continuum theory of internal fracture (Sect. 3.3.2.1).

It also follows from the above-mentioned proof that the continuum theory of internal fracture does not describe the fracture associated with the modes of the loss of stability of the second, third, and fourth kinds which are represented in Figs. 3.28, 3.29, and 3.30 due to the following considerations.

In the case of modes of the loss of stability of the second, third, and fourth kinds in the analysis corresponding to the conditions (3.18), the results that have direct physical meaning cannot be reached.

This issue is discussed in more detail in the monographs [3, Chap. 3] and [4], vol. 1, p. 304–312.

Also, it was noted as long ago as in the monographs [1] of 1971 and [2] of 1973, as well as in the articles [47] of 1969, [197] of 1982, and [198] of 2001 that the numerical studies showed that consideration of modes of the loss of stability of the third (Fig. 3.29) and fourth (Fig. 3.30) kinds did not result in the critical loads lower than in the modes of loss of stability of the first (Fig. 3.27) and second (Fig. 3.28) kinds.

In the following years, the above information was consistently presented in monographs [3, Chap. 3] and [4], vol. 1, Chap. 3.

The conclusion follows from the above information that in the analysis of the asymptotic exactness *of* the continuum theory of internal fracture (internal instability) set out in Sect. 3.3.2.1, it is sufficient to consider the modes *of the* loss *of* stability *of* the first (Fig. 3.27) and the second (Fig. 3.28) kinds which are studied in Sect. 3.3.3.1 within the model of the piece-wise homogeneous body (medium and material).

Note that similar studies were carried out within the model of piece-wise homogeneous medium and the TLTDBS for the layered composites (within the approach and method of Sect. 3.3.3.1) in the case of spatial problem in the case of composites with polymeric matrix (brittle fracture), as well as for the plane and spatial problems in the case of composites with a metallic matrix (plastic fracture).

In all of the above cases, the study produced the same results as for the layered composites with a polymeric matrix for a plane problem.

These results are presented in great detail in monographs [3, Chap. 3] and [4], vol. 1, Chap. 3.

Thus, it can be considered that above in the sufficiently general form the following statement is proven.

Statement. For **the** layered composite materials with polymeric and metallic matrix, the continuum theory of internal fracture, based on the model of **a** homogeneous material with averaged parameters and **the** TLTDBS, has the following properties:

1. **It** is asymptotically exact and corresponds to the mode of loss of stability of **the** first kind (bending mode, Fig. 3.27).
2. **It** does not describe the internal fracture corresponding to the mode of loss of stability of **the** second kind (stretching mode, Fig. 3.28).

Since the loss of stability in the internal structure of the layered composite and the resulting start of fracture occurs with the fewer values of the critical load (of the two critical loads corresponding to the modes of the loss of stability of the first (Fig. 3.27) and the second (Fig. 3.28) kinds), the analysis under discussion can be continued and it is advisable to identify three characteristic cases.

The first case. The mode of the loss of stability of the first kind *is* realized, and the form of loss of stability of the second kind is not realized in the fracture. In this case, the continuum theory under discussion describes the fracture.

The second case. Both the mode of loss of stability of the first kind and the mode of loss of stability of the second kind can be realized under fracture. In this case, as it follows from Tables 3.1 and 3.2 of monographs [3, pp. 182, 183] and [4], vol. 1, pp. 300, 301, the values of the critical loads calculated within the model of the piecewise homogeneous medium (Sect. 3.3.3.1) and the TLTDBS for the modes of loss of stability of the first kind (Fig. 3.27) and the second (Fig. 3.28, stretching mode differ slightly in the examples considered in [3] and , vol. 1. Thus, in the second case also, the continuum theory provides reliable information on the value of the theoretical ultimate strength. It seemed that in the present case it was appropriate to increase the number of the analyzed numerical examples.

The third case. The mode of loss of stability of the first kind is not realized, and the mode of loss of stability of the second kind is realized under fracture. In this case, according to the above-formulated statement, the continuum theory of internal fracture (Sect. 3.3.2.1) under compression does not describe the internal fracture of layered composites under compression. To clear up the suitability of expression (3.14) to determine the theoretical ultimate strength of the layered composites in the third case, it may be appropriate to carry out a numerical study for a number of layered composites, involving the approach of Sect. 3.3.3.1 within the model of piece-wise homogeneous material in relation to the mode of loss of stability of the second kind (Fig. 3.28, stretching mode).

The above information in this section seems to be sufficient in the analysis of the continuum theory of internal fracture of composites under compression along the axes of the symmetry of the material properties outlined in Sect. 3.3.2.1, for the layered composites under compression along layers (Figs. 3.25 and 3.26).

On Constructing a Research Method for the Complex Modes of Loss
of Stability of Layered Composites

In the study of the stability of layered composites within the model of piece-
wise homogeneous medium (material and body) and the TLTDBS, the multipliers
$\sin \pi l_1^{-1} x_1$ (in the case of plane problems, Fig. 3.25) and $\left(\sin \pi l_1^{-1} x_1\right)\left(\cos \pi l_2^{-1} x_2\right)$
(in the case of spatial problems, Fig. 3.26) where l_1 and l_2 are the lengths of the half-
wave modes of loss of stability along the layers were singled out in all the analyzed
modes of loss of fracture.

Thus, it was accepted within the "infinitely long layers" model that each layer loses
stability in the same periodic (in the phase or anti-phase) mode of loss of stability
along layers. In these modes of loss of stability, *the* plane with *the* same phase in
the mode of loss of stability (along the coordinate x_1 in the case of plane problems
(Fig. 3.25) and along the coordinates x_1 and x_2 in the case of spatial problems
(Fig. 3.27), in the general case along the layers) are located perpendicular to layers.
However, the modes of loss of stability are more complex, which are periodic along
the layers, but *the* planes with *the* same phase in the mode of loss of stability are
arranged arbitrarily relative to the plane of layers.

For a clearer characteristic of the forms of loss of stability in this case, the notion
of a plane Π is introduced, which consists of the points of composite that have the
same phase (by coordinates along the layers) in the form of the loss of stability.
At that, the plane Π occupies an arbitrary position, defined by the ort of normal $\boldsymbol{n}$
(Fig. 3.32), and the mode of loss of stability is periodic by the coordinates along the
layers (by x_1 in the case of a plane problem, Fig. 3.25; by x_1 and x_2 in the case of
spatial problem, Fig. 3.26). In the first octant in Fig. 3.32, the plane Π is shaded and
determined by the ort $\boldsymbol{n}$ of normal to the constituents

$$n_1, n_2, n_3; n_1^2 + n_2^2 + n_3^2 = 1. \tag{3.19}$$

In the article [83], the solutions of the system of static equation of the TLTDBS
for the layered composites are explicitly constructed in relation to the above modes
of loss of stability, and it is shown that the proposed approach provides a finite order
of the characteristic determinant, the elements of which are represented in a closed
form.

Note that for these modes of loss of stability the information of Sect. 3.3.1 is
related to Fig. 3.32, the generality of the approach of the second direction, and
its full correspondence to the generally accepted and rigorous method of research
of the loss of stability—the analysis of the behavior of small perturbations within
the linearized three-dimensional dynamical problems. Only because the sufficient
conditions of the applicability of the Euler method are met, authors of the [83] were
limited to the construction of solutions of the statical equations of the TLTDBS.

Fig. 3.32

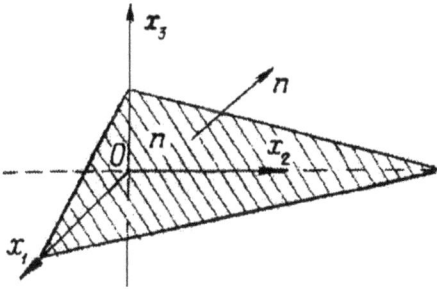

Conclusions from a Consistent Analysis
of the Dou–Grundfest–Rosen–Schuertz Theory

A short information on the formation of the name "The Theory of Dou–Grundfest–Rosen–Schuertz" is presented in the final part of Sect. 3.1.2. The discussed here results of this theory (by the classification of this monograph) refer to the first direction (very approximate approaches), to a very brief characteristic of which Sect. 3.1.2 is devoted. It is worth noting that the five groups of the main approximate assumptions are formulated in Sect. 3.1.2 that are characteristic of the first direction of research.

A characteristic representative of the first direction, as already noted in Sect. 3.1.2, is the article [164] published in English in 1965 and in Russian in 1967 in the form of an article [165], which appears to have been *the* first publication (in the first direction) with the results of theoretical character and which was included into the multivolume editions of an encyclopedic character on the fracture [166] and composites [168].

In connection with the foregoing, the analysis of the Dou–Grundfest–Rosen–Schuertz theory is carried out following the results of articles [164, 165]. Note that all, shown in Sect. 3.1.2, five groups of approximate assumptions are included in the publications under discussion.

The analysis of results of this theory should be carried out on the basis of the corresponding results of the second direction, which, as noted at the beginning of Sect. 3.1.3.1, do not accept the above five groups of approximate assumptions. Such analysis of the Dou–Grundfest–Rosen–Schuertz theory was carried out in the monograph [3, pp. 206–214] of 1990, in the monograph [4], vol. 1, pp. 328–335 of 2008, and in reduced form in the review article [162, pp. 22–26] of 2016. This analysis was carried out based on results for the layered composites obtained within the second direction and, in part, presented in Sect. 3.3.3. In this regard, Sect. 3.3.3.3.3, devoted to set out of the presentation of the conclusions from the analysis under discussion, is included in Sect. 3.3.3.

Note that the main results of the articles [164, 165] are shown in Fig. 3.33, which includes Figs. 3.22, 3.23, and 3.24 of the article [165].

For the convenience of analyzing these results, it is worth noting that all the quantities related to the filler (reinforcing elements, fibers, layers, etc.) and the matrix (binding) are marked with indices: "a" and "m" in this monograph, and "B" and "M" in the article [165] and in Figs. 3.22, 3.23, and 3.24 of the article [165].

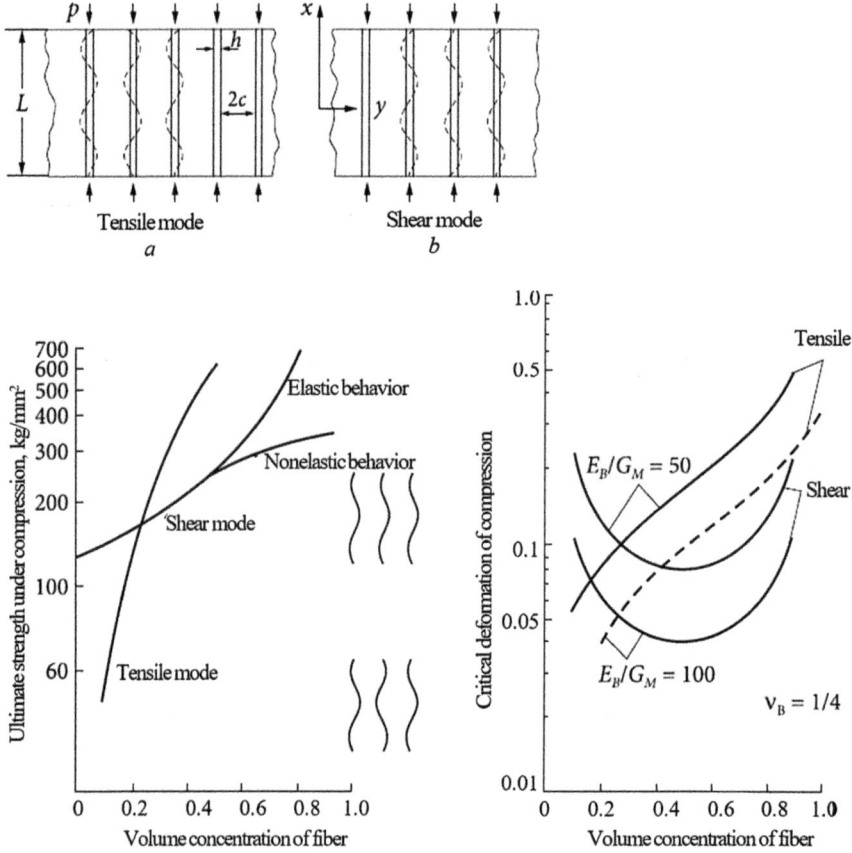

Fig. 3.33

The following brief conclusions can be drawn from the discussion and analysis of the results of the articles [164, 165], presented in the monographs of [3, 4].

1. In the analysis of regularities in the fibrous composite, the model of the layered composite is applied in fact, as the studies are carried out within the plane problem in the plane xOy (Figs. 3.33 and 3.22), although the terminology corresponding to the fibrous composite is used. At that, the above modeling is justified in any way.

2. The theory under discussion has **the qualitative contradictions**, and therefore below the formulation of two positions is only shown.

    2.1. *The theory from* [164, 165]*does not allow (does not describe)* the loss of stability *(internal instability) by the shear mode of the composite* (at any ratio between stiffness and geometric parameters of the composite).

Note that "shear" mode (Figs. 3.22b in Fig. 3.33) corresponds to the mode of loss of the first kind of stability (Fig. 3.27).

This conclusion follows from the strict analysis (in accordance with the approaches set out in Sect. 3.1.1 near Fig. 3.3) of the expression (B.26) on page 96 of [165], which indicates that the type $p = p(\alpha)$ (3.2) dependence for the expression (B.26) has a monotonous character—of the type of the curve B in Fig. 3.2. In the monograph [3], the formulated conclusion is actually presented in more detail on pages 207–209.

It is strongly proven (e.g., in the final part of Sect. 3.3.3.1) within the second direction (more strong and exact as compared to [164, 165] that the internal instability in the "shear" or "bending" modes may arise or not arise depending on *the* ratio between the stiffness and geometric parameters of the composite. In particular, the instability in *the* "shear" or "bending" modes (the mode of loss of stability of the first kind, Fig. 3.27) occurs at *the* low concentrations *of* filler, which is noted in the monograph [3] on page 209.

In the foregoing is the first qualitative contradiction of the theory [164, 165].

2.2. At the high concentrations of filler (reinforcing elements, fibers, etc.), the theory of [164, 165] leads to a physically incorrect result for the "shear" or "bending" modes (loss of stability of the first kind, Fig. 3.27). Thus, it follows from the expression (3.29) on page 82 of [165] that the theoretical ultimate strength under compression tends to "infinity" when $v_B \rightarrow 1$ (v_B indicates in [165] the volume fraction of filler).

It *is* strongly proven within the second **direction** (more strong and exact as compared to [164, 165] that in the above situation the theoretical ultimate strength under compression tends to the finite value [4], vol. 1, p. 189, the expression (2.71). The noted result also follows from the expression (3.14) at $S_a \rightarrow 1$ and $S_m = 1 - S_a \rightarrow 0$, where $S_a \sim v_B$.

Thus, the above is the second qualitative contradiction of the theory [164, 165].

3. This theory has **the quantitative errors**. Below, the formulation of two positions is presented.

3.1. The results of the theory [164, 165] and the corresponding results of the second direction for the layered composites differ significantly from each other at small and large volume fractions of *the* filler (reinforcing elements, fibers, layers, etc.). The noted conclusion is presented in the monograph [3] on page 210.

3.2. The results of the theory [164, 165] and the corresponding results of the second direction for the layered composites at sufficiently low volume fractions of the filler (reinforcing elements, fibers, layers, etc.) can differ from each other three times or more. The noted conclusion is laid out at the top of page 211 of the monograph [3].

Thus, the theory of Dou–Grundfest–Rosen–Schuertz (in the form of publications [164, 165]) and its quantitative results have *the* significant qualitative contradictions and quantitative errors in relation to the theory and results, built with the usual accepted accuracy in the solid mechanics (the second direction, built on the basis of the TLTDBS). The further research is therefore needed to determine *the* validity of some specific results of this theory.

Note that presented in the reduced form the above-formulated conclusions from the analysis of the articles [164, 165] correspond to the review article [162, pp. 24–26] of 2016. In the most general form, the corresponding analysis and conclusions are presented in the monograph [3] of 1990 at the pages 206–214.

It is also worth noting that the above-formulated conclusions from the analysis of the articles [164, 165] together with the five groups of the main approximate assumptions, which are formulated in Sect. 3.1.2 and which are characteristic of the first direction of research (Sect. 3.1.2), give a definite understanding about all the studies of the first direction.

However, as has been repeatedly emphasized, the present monograph did not aim to provide a review of the research in the first direction for the layered composites.

The above-formulated information seems to be sufficient when considering the results related to Sect. 3.3.3 (Layered Composites: Model of a Piece-Wise Medium).

3.3.4 Fibrous Unidirectional Composites: Model of Piece-Wise Homogeneous Medium

In this subsection, very brief information is given on the fracture mechanics under compression (along fibers) of the fibrous unidirectional composites with the polymeric and metallic matrixes, based on a model of a piece-wise homogeneous medium (body and material). In this case, the basic relations of the TLTDBS are used separately for the materials of the matrix (binding) and each fiber (reinforcing element), and conditions of continuity of stresses and displacements are accepted at the interfaces in the form of cylindrical surfaces.

The research is carried out for the brittle and plastic fracture. In the latter case, a generalized concept of the continued loading is tentatively adopted, stated briefly in Chap. 2 of this monograph.

In accordance with the basic approach outlined in Sect. 3.3.1, in the case of the studies performed here the analysis of static equations and corresponding boundary conditions of the TLTDBS for the piece-wise homogeneous medium should be carried out and in the final stage of this research, the boundary value problem with the involvement of the TLTDBS apparatus arises. It is appropriate to note that this subsection provides only information on the results obtained for the fibrous unidirectional composites, in the surfaces of which the defects are absent. Therefore, the results presented here are based on conditions of continuity of stress and displacement vectors on all cylindrical interfaces.

The main results are obtained for the internal fracture (loss of stability in the internal structure and internal instability), the results of which are outlined in the monograph [4], vol. 1, Chap. 4, and for the near-the-surface fracture (loss of stability in the near-the-surface layers of the internal structure and near-the-surface instability), the results of which are outlined in the monographs [4], vol. 1, Chap. 6.

The main results are also partially presented in monographs [1–3] and a very brief form in the review article [162]. The results for the fibrous unidirectional composites were published preliminary in the articles, the main of which are included in the reference of this review article and the list of the main articles relating to the second direction and presented in the introductory part of Chap. 3 of this monograph.

The additional information on the published articles can be obtained from the references of the monographs [1–4].

The main results for the fibrous unidirectional composites in the form of the characteristic determinants for the internal and near-the-surface fracture are laid out in a common general form for **the** theories 1, 2, and 3 in accordance with the terminology of Sect. 2.2 and the introductory part of Sect. 2.4.

Below, in this subsection, only the main results on the fracture mechanics are discussed briefly for the fibrous unidirectional composites under compression along the fibers, obtained within the model of piece-wise medium and the TLTDBS, the basis of the mathematical apparatus of which is summarized in a brief form in Chap. 2 of this monograph. At that, the short presentation of this issue in the review article [162] is taken into account.

3.3.4.1 Internal Fracture

In the study of internal fracture (internal instability), the fibrous one-directional composite is considered, which occupies an infinite space (Fig. 3.1), and the static equations of the TLTDBS are applied.

This is justified in Sect. 3.3.1. The analysis is carried out in accordance with the procedure outlined in Sect. 3.1.1, taking into account the approach outlined in Fig. 3.3. It is also accepted that the shortenings along the fibers in the matrix and fibers (in the subcritical state under compression along the fibers) coincide, which corresponds to Note 3.8.

It is worth noting the following situation. In the fibrous unidirectional composites under compression along fibers (Fig. 3.1), in contrary to the layered composites under compression along layers (Fig. 3.2), *the* heterogeneous stress state can occur in the precritical state, depending on the spatial variables in the plane of cross section. A *homogeneous* precritical state can only occur in two cases: The first is that the fiber and matrix materials are incompressible, and the second is that the fiber and matrix materials have the same Poisson ratio ($v_a = v_m$). This condition is not met when the Poisson ratios of fiber materials are not equal ($v_a \neq v_m$). In this case, the heterogeneous precritical state arises.

Fig. 3.34

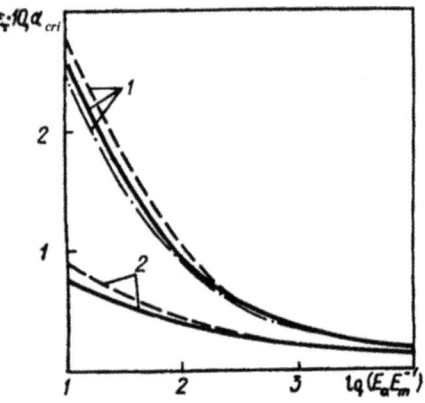

Therefore, the question of the need to take into account (in the problems of the stability of fibrous unidirectional composites) the heterogeneity of the precritical state for the case $v_a \neq v_m$ is raised.

In the monograph [4], vol. 1, Chap. 4, Sect. 1, pp. 391–396, this issue is explored on the example of one fiber. At that, for a heterogeneous precritical state, quite accurate results are obtained numerically. So, Fig. 3.34 represents the dependence of quantities $\varepsilon_T \cdot 10$ (ε_T is the theoretical value of the ultimate shortening) and $\alpha_{cri} = \pi R / l_{cri}$ (the critical value of the wave formation parameter $\alpha = \pi R / l$, (R is the radius of the cross section of the fiber, and l is the length of the semi-wave (along the fiber axis) of the form of loss of stability) on the parameter $\lg\left(E_a \cdot E_m^{-1}\right)$.

The curves for the quantity α_{cri} are marked with numbers 1, the curves for the quantity $\varepsilon_T \cdot 10$ are marked with numbers 2, the solid curves refer to the case (the precritical state is heterogeneous), the solid curves refer to the cases $v_a = 0.2$ and $v_m = 0.4$ (the precritical state is homogeneous one), the dashed curves refer to the case $v_a = v_m = 0.2$ (the precritical state is homogeneous one), and the dash-dot curves refer to the case $v_a = v_m = 0.4$ (the precritical state is homogeneous one).

Note that for the quantity $\varepsilon_T \cdot 10$ the solid and dash-dot lines are practically identical in Fig. 3.34.

Also note that according to Tables 0.1 and 0.2 [4], vol. 1, pp. 67, 68 the cases $v_a = 0.2$ and $v_m = 0.4$ correspond to the greatest difference in the Poisson ratios for the typical fillers and binders.

The following conclusion is made from the analysis of the results in Fig. 3.34 and Table 4.1 [4], vol. 1, p. 395:

In the case of $E_a \cdot E_m^{-1} \geq 20$ while *researching* with *an accuracy of 5%, it is possible not to take into account the heterogeneity of the precritical state caused by the difference in the values of Poisson ratios for the filler and binder, and to carry out the studies at equal values of the Poisson ratios $v_a = v_m = 0.3$.*

The formulated conclusion makes it possible to reasonably apply the general solutions of the static equations of the TLTDBS (Chap. 2, Sects. 2.6.1 and 2.6.2),

Fig. 3.35

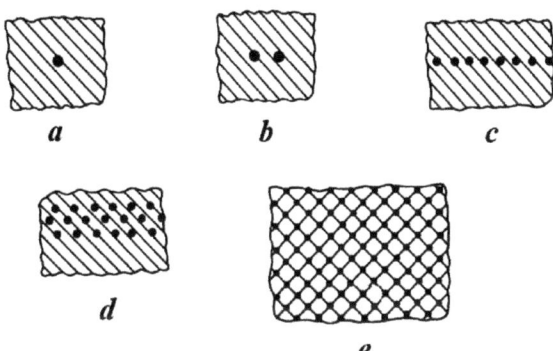

which are obtained for the homogeneous precritical stress states. All of the results outlined below for the fibrous unidirectional composites are obtained in this way.

For the fibrous unidirectional composites under compression along fibers (Fig. 3.1), the different problems arise depending on the structure of the composite in the cross-sectional plane and the design schemes under consideration. Figure 3.35 (in the cross-sectional plane) shows the following design schemes.

1.   One fiber (Fig. 3.35a) for the fibrous composites with a small volume fraction of the filler when the adjacent fibers do not interact with each other;
2.   Two fibers (Fig. 3.35b) for the fibrous composites with a small volume fraction of the filler when, due to irregular structure with the loss of stability, two adjacent fibers can interact;
3.   One periodic series of fibers (Fig. 3.35c) for the fibrous composites of the periodic structure, when in the loss of stability of the fiber within one row the fibers interact with each other, and the adjacent rows of fibers do not interact with each other (very small distances between fibers in one row, very long distances between neighboring rows);
4.   Several periodic rows of fibers (Fig. 3.35d) for the fibrous composites of the periodic structure, when in the loss of stability within each row the fibers interact with each other, and different groups of rows do not interact with each other;
5.   The double-periodic structure of fibers (Fig. 3.35e) is for the fibrous composites of the double-periodic structure at very short distances between neighboring fibers when the interaction within the double-periodic structure model must be taken into account when stability is lost.

The number of variants to research for the fibrous unidirectional composites is not limited to the number of design schemes given in Fig. 3.35 (in total, 5 design schemes), but also the fact that each design scheme should investigate different modes (or types of modes) of the loss of stability, which are determined by the properties of symmetry in the cross-sectional plane. Of course, after the research of different modes of loss of stability, it is necessary to minimize all the eigenvalues obtained (usually the first eigenvalues for each type of the mode of loss of stability) to determine *the* critical value of shortening along the axis *of* fibers.

Fig. 3.36

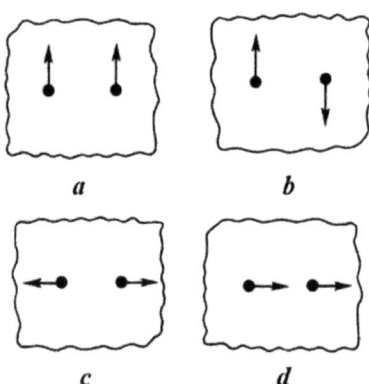

For example, different types of modes of loss of stability are shown in Fig. 3.36 (in the cross-sectional plane) for two fibers (Fig. 3.35b). So, Fig. 3.36a corresponds to the modes of loss of stability in the phase from the fiber plane. Figure 3.36b corresponds to the modes of loss of stability in the anti-phase from the plane of fibers. Figure 3.36c corresponds to the modes of loss of stability in the anti-phase in the fiber plane. Figure 3.36, g corresponds to the modes of loss of stability in the phase in the fiber plane.

The general method of solving the above-mentioned problems, which was used in the study of all problems on the mechanics of internal fracture in the axial compression of fibrous unidirectional composites within the second direction, includes the following components.

Application for the fibers and matrix of the general solutions of static equations of the TLTDBS (Chap. 2, Sects. 2.6.1 and 2.6.2) in the circular cylindrical coordinates;

Representation of the solution in the form of a sum of solutions in the local cylindrical coordinates in the form of Fourier series with uncertain coefficients, which include the special functions of a circular cylinder;

Obtaining the characteristic equations in the form of the infinite characteristic determinants with the calculation of elements in an explicit form;

Proof that these infinite characteristic determinants for non-touching fibers are the determinants of the normal type. This proof substantiates the replacement in the approximate determination of roots of the infinite determinants with the finite; i.e., it substantiates the method of truncation;

To justify the practical convergence of this method by comparing the values of the roots obtained by increasing the order of truncated determinants.

The numerous results on the internal fracture under compression of the fibrous unidirectional composites with the polymeric and metallic matrix are obtained using the above method.

The sufficiently detailed and consistent presentation of these results is given in the monograph [4], vol. 1, Chap. 4. These results appear at present to be the most exact and rigorous. Besides, this method allows refining the results for the non-touching fibers.

Example. Let us consider very briefly the determination of experimental and theoretical ultimate strengths at unidirectional compression along the fibers for the metallic composite (a unidirectional fibrous boro-aluminum composite with 50% boron fiber volume fraction, $S_a = S_m = 0, 5$) VKA-1 with the boron fibers of a diameter of 140 microns. A general view of samples made of boro-aluminum is presented in Figs. 3.37 and 3.38 where the internal structure of the boro-aluminum composite is represented in the cross section with a significant increase.

Fig. 3.37

Fig. 3.38

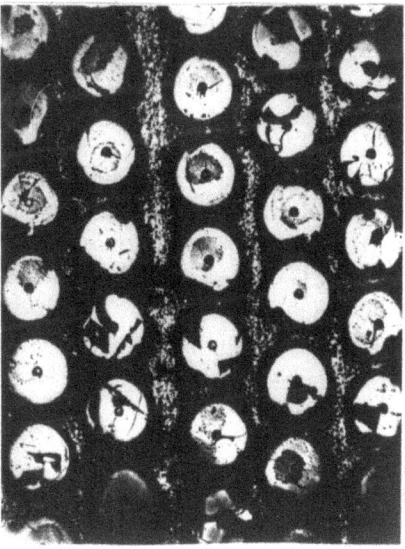

These results are published in the article [26] and relatively detailed in the monograph [4], vol. 1, Chap. 4, pp. 486–488, where more information can be obtained.

The theoretical studies used theory 3 (the second version of the theory of *the* small precritical deformations) following the terminology of Sect. 2.2. The plastic fracture was also considered, which, according to Note 3.7, refers to the loading stage, when the entire matrix is in *a* state of plastic deformation. In this regard, the relation (3.8) was used to describe the plastic deformation of the matrix as part of the model of incompressible isotropic plastic material.

In the fiber of the unidirectional boro-aluminum composite, technical aluminum AD—1 was used as a matrix, for which (concerning the annealed and non-annealed samples) the parameters are included in the relationship (3.8); more detailed information on this issue is outlined in the monograph (i.e., 1, Chap. 4, p. 486–488). This is how it was defined:

In this fibrous unidirectional boro-aluminum composite, the technical aluminum AD—1 was used as a matrix, for which (concerning the annealed and non-annealed samples) the parameters A_m and k_m are included in the relationship (3.8). The more detailed information on this issue is outlined in the monograph [4], vol. 1, Chap. 4, pp. 486–488. So, the following values were determined [26]:

$$A_m = 130 \text{MPa}; \ k_m = 0.43 - \text{for annealed matrix (aluminum)},$$
$$A_m = 70 \text{MPa}; \ k_m = 0.25 - \text{for non} - \text{annealed matrix (aluminum).} \qquad (3.20)$$

It is interesting to note that the values A_m and k_m for non-annealed aluminum (3.20) are exactly the same as the numerical values of the corresponding values at the third approximation (3.9).

It should be noted that in essence two metallic composites (annealed and non-annealed) were considered.

In the experimental studies, the 32 annealed and 14 non-annealed samples were destroyed for which the experimental ultimate strengths were determined.

The theoretical ultimate strength was defined within the framework of the continuum fracture theory (Sect. 3.3.2.1) and within the model of piece-wise homogeneous medium for the fibrous unidirectional composites (Sect. 3.3.4.1) for the design scheme of the composite of the double-periodic structure (Fig. 3.35d).

The above results are presented in Table 3.1, which corresponds to Table 4.10 [4], vol. 1, Chap. 4, p. 487. It follows from an analysis of the results presented in Table 3.1 that the continuum fracture theory gives the results very close to the average experimental results, and the fracture mechanics within the piece-wise homogeneous medium model gives the results very close to the maximal experimental results.

Table 3.1

Material	$(\Pi_3^-)_{exp}$, MPa			$(\Pi_3^-)_T$, MPa	
	Max	Min	Average	Continuum theory	Piece-wise homogeneous medium model
Annealed	965	501	665	736	958
Non-annealed	1716	1049	1282	1467	1972

Additional information about the results on the internal fracture of the fibrous unidirectional composites discussed in the present subsubsection can be obtained from the monograph [4], vol. 1, Chap. 4.

It should be noted that in determining the theoretical ultimate strengths under uniaxial compression along the fibers of the boro-aluminum composite during the plastic fracture within the continuum fracture theory (Sect. 3.3.2.1), the calculations were carried out in the first approximation, as in the example of Sect. 3.3.2.1, the results of which are given in Fig. 3.23.

A comparison of the theoretical and experimental results (Table 3.1) in determining the ultimate strength under uniaxial compression along the fibers of the considered boro-aluminum unidirectional composites allows drawing the following conclusions.

1. For the continual fracture theory (Sect. 3.3.2) the close values of the theoretical ultimate strength under compression compared to the average values of the shown value, which are obtained from the experiment. So, in the case of the annealed material, these results differ from each other by 11%, and in the case of the non-annealed material—by 14.4%;
2. The fracture mechanics using the model of piece-wise homogeneous medium (Sect. 3.3.4.1) gives the values of the theoretical ultimate strength under uniaxial compression close to the maximal values of the ultimate strength under compression, obtained experimentally. So, in the case of the annealed material, the difference between these results is 1%, and in the case of the non-annealed—15%.

The above information will be limited when considering the results of internal fracture of the fibrous unidirectional composites under compression along the fibers. These results are obtained within the framework of the model of piece-wise homogeneous medium and the TLTDBS, the main clauses of which and the corresponding mathematical apparatus are outlined in of this monograph.

3.3.4.2 Near-the-Surface Fracture

Consider in a very brief form the results on the near-the-surface fracture of the fibrous unidirectional composites obtained within the model of the piece-wise homogeneous

Fig. 3.39

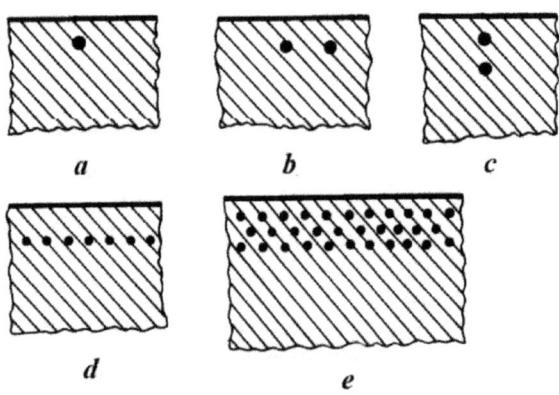

medium. The sufficiently detailed information on this subject is contained in the monograph [57], vol. 1, chap. 6.

When the surface instability being studied, the fibrous unidirectional composite (Fig. 3.1) is considered, which occupies a semi-space. At that, a boundary surface is a plane that passes parallel to the unidirectional fibers (Fig. 3.1).

In connection with the stated before, in the study of surface instability, the half-plane (Fig. 3.39) in the plane of the cross-section of the fibrous unidirectional composite (Fig. 3.1) with different structures corresponding to the characteristic design schemes is considered.

Figure 3.39 shows five characteristic design schemes, which can be described in a similar way to Sect. 3.3.4.1, where this description is given concerning the internal fracture. Note that the dark circles depict the cross-sections of fibers in Fig. 3.39, as in Figs. 3.35 and 3.36.

In the study of the surface instability of the fibrous unidirectional composites, the modes of loss of stability are analyzed, which attenuate away from the boundary of the lower semi-space in Fig. 3.39, which is determined by an additional condition.

In the formation of the structure of the solution, in addition to the structure of the solution of Sect. 3.3.4.1, an additional component in the form of the Fourier integral transform is introduced, ensuring fulfillment of the boundary conditions on the plane boundary of the lower semi-space. Taking into account the above structure of the presentation of the solution, the method of solution was later applied, the information about which is presented before the example of Sect. 3.3.4.1.

The specific results were obtained on the near-the-surface fracture of the unidirectional fibrous composites with the polymeric and metallic matrixes. A rather detailed presentation of these results is presented in the monograph [4], vol. 1, Chap. 6.

3.3.4.3   On Constructing a Research Method for Complex Modes of Loss of Stability of Fibrous Unidirectional Composites

In the study of the stability of fibrous unidirectional composites (Sects. 3.3.4.1 and 3.3.4.2) within the model of piece-wise homogeneous medium (material and body) and the TLTDBS in all analyzed forms of loss of stability, a multiplier $\sin \pi l^{-1} x_3$ was singled out, where the coordinate x_3 is counted along the fibers and l is the length of the half-wave of the mode of loss of stability also along the fibers.

Thus, within the "infinitely long fibers" model, it was accepted that each of the reinforcing element (fibers) stabilities loses in the same periodic mode of loss of stability along the fibers. In some sense, the results of experimental studies for the composites with the polymeric (epoxy resin) matrix are proof of the implementation of the above modes of loss of stability. The results of such experiments are presented in Fig. 3.5 for a composite with the glass fibers and in Fig. 3.6 for a composite with the carbon fibers.

For these modes of loss of stability, *the* planes with *the* same phase in the form of loss of stability (along coordinates x_3, along the fibers) in the composite are located perpendicular to the axis $0x_3$ (perpendicular to the fibers). Therefore, it seems that the fracture propagates across these planes. In some sense, the evidence of the above procedure relative to the modes of loss of fracture is the rigorous proof (Sect. 3.3.2.1) within the continuum fracture theory (Sect. 3.3.2) that (under brittle fracture) the fracture propagates across the planes that are perpendicular to the action of compressing loads.

Note that the continuum fracture theory (Sect. 3.3.2) considers compression along the direction of preferred reinforcement (in this case—along the fibers and layers).

However, the given reasoning, involving the results of a continuum theory to substantiate one of the constituent points of the theory within the model of a piece-wise homogeneous medium, is not logical enough and consistent, since the continuum fracture theory is approximate and less rigorous compared to the theory based on the model of a piece-wise homogeneous medium.

In connection with the foregoing, it seems quite appropriate to develop a method of research in the theory of the stability of fibrous unidirectional composites, allowing to analyze the more general modes of loss of stability compared to the modes of loss of stability, which are considered in Sects. 3.3.4.1 and 3.3.4.2. Of course, along with the development of the research method, it is also essential to analyze the classes of specific problems to form general conclusions.

In the article [68], a method of research for the more complex modes of loss of stability is proposed for the fibrous unidirectional composites (Fig. 3.1) under uniaxial compression along the fibers for the most complex double-periodic structure, which is shown in the plane of cross section (Fig. 3.35d). The used in article [68] method is intended for research within the model of "infinitely long waves" (along the axis $0x_3$) when compression is carried out along the axis $0x_3$ and modes of loss of stability are also periodic along the axis $0x_3$, but the planes Π in Fig. 3.32 with the same phase along $0x_3$ (in the form of loss of stability) but the planes in Fig. 3.32 with the same phase along (in the form of loss of stability) occupy an arbitrary position, determined by the ort **n** of normal to the plane Π.

For the fibrous unidirectional composites, the solutions of the static equations of the TLTDBS corresponding to the arbitrary position of the plane Π, concerning the most complex double-periodic structure, which is shown in Fig. 3.35d in the cross-sectional plane, are built in [68] in an explicit form in the circular cylindrical (local for each fiber) coordinates. It is noted in the case under discussion that in the implementation of this method the [68] characteristic equation in the form of an infinite determinant is obtained.

At that, it is proven that the infinite determinant is a normal-type determinant for non-touching fibers, which in turn justifies the possibility of determining the roots by the method of truncating the infinite determinant in the numerical study. The practical convergence is achieved by comparing the results with an increase in the order of the truncated determinant.

It is worth noting that in the case of these modes of loss of stability under compression of the fibrous unidirectional composite, the information of Sect. 3.3.1 is valid on the generality of the second direction approach and its full conformity to the generally accepted and rigorous method of research into the phenomenon of loss of stability—the analysis of the behavior of small perturbations within the linearized dynamic three-dimensional problems. Only because of the sufficient conditions of the applicability of the Euler method, the construction of solutions of the static equations of the TLTDBS was limited in [68].

Of course, the approach in [68] allows (with appropriate changes) the transition to the simpler periodic structures for the fibrous unidirectional composites, which are shown in the cross-sectional plane in Fig. 3.35c, d.

At present, the concrete results are currently obtained only for the simplest periodic structure for which the cross section of the one-directional fibrous composite is represented in Fig. 3.35c. This is one periodic series of fibers in an infinite matrix.

In the case of fibrous single-directional composites of the periodic structure, in Fig. 3.35b design scheme is used when under the loss of stability the fibers within one row interact with each other and the neighboring rows of fibers do not interact with each other (very small distances between neighboring fibers in one row, very long distances between neighboring rows).

These results are presented in the articles [90] of 2002, [91] of 2005, and [94] (original character) of 1991 within theory 3 (the second variant of the theory of small precritical deformations) following the terminology of Sect. 2.2 of this monograph for the brittle fracture (when modeling the materials of fibers and the matrix by the linearly-elastic isotropic body). A summary of these results is presented in the review article [162] on pp. 43–46.

The following conclusion is made in the above-mentioned publications. In this situation, **the** least critical stress corresponds to the mode of loss of stability, when the plane Π (Fig. 3.32) is perpendicular to **the** fibers under compression along the fibers.

This conclusion corresponds to the experimental results outlined in Sect. 3.2.1 and the methods of internal fracture research used in Sect. 3.3.3.1 for the layered composites and in Sect. 3.3.4.1 for the fibrous one-directional composites, as part of the model of piece-wise homogeneous material (medium and body) and the TLTDBS.

In conclusion in Sect. 3.3.4, it is appropriate to note the following situation.

This subsection discusses in a very brief form the main results on the mechanics of the brittle and plastic internal and near-the-surface fracture of the fibrous unidirectional composites under compression along the fibers which are obtained within the model of the piece-wise homogeneous medium and the TLTDBS *concerning the* fibers of *the* circular cross section.

The corresponding results for the fibrous unidirectional composites with *the* fibers of *the* non-circular cross section were also obtained within the above statement. These results are presented in articles [21, 92, 93, 125, 126] as well as in other articles that are not included in the references of this monograph.

The above-formulated information seems to be sufficient when considering the results related to Problem 1 (fracture in composite materials under compression along the reinforcing elements).

3.4 Conclusion

It should be noted that dedicated to Problem 1 of this chapter is a significant part of this monograph. This situation is explained by the following three considerations.

1. The author of this monograph and his pupils have been doing research on Problem 1 for the last 50 years. The first articles were published in 1969 ([9, 10]).
2. Problem 1 has produced significant results published in monographs, articles, and reports at international conferences. The main publications listed in the references of this monograph are noted in the introductory part of this chapter.
3. The results obtained in the S.P. Timoshenko Institute of Mechanics of NAS of Ukraine on Problem 1 have already received some recognition from the world scientific community. As an example, the publication of a special issue of the well-known journal "Applied Mechanics Reviews" (USA) back in 1992 can be pointed.

    Micromechanics of composite materials: Focus on Ukrainian research // Appl. Mech. Rev. (Special Issue, A.N.Guz—Guest Editor).—1992.—45, N2. pp. 13–101.

This special issue is included in the references of this monograph under number 534, where the titles of articles in this issue and their authors are listed. In connection with the above situation (a large volume of this chapter), the description of the results and their analysis for Problems 2–8 will be presented in a more reduced form.

References

1. Guz, A.N.: Ustoichivost trekhmernykh deformiruemykh tel (Stability of Three-Dimensional Deformable Bodies). Naukova Dumka, Kyiv (1971)
2. Guz, A.N.: Ustoichivost uprugikh tel pri konechnykh deformatsiiakh (Stability of Elastic Bodies Under Finite Deformations). Naukova Dumka, Kyiv (1973)
3. Guz, A.N.: Mekhanika razrusheniia kompozitnykh materialov pri szhatii (Fracture Mechanics of Composite Materials Under Compression). Naukova Dumka, Kyiv (1990)
4. Guz, A.N.: Osnovy mekhaniki razrusheniia kompozitov pri szhatii: V 2-kh tomakh (Fundamentals of the fracture mechanics of composites under compression: In 2 volumes). (Litera, Kyiv, 2008) , T. 1. Razrushenie v strukture materiala. (Fracture in structure of materials), T. 2. Rodstvennye mekhanizmy razrusheniia. (Related mechanisms of fracture)
5. Babich, I.Yu.: O neustoichivosti deformirovaniia kompozitnykh materialov pri malykh deformatsiiakh (On deformation instability of composite materials at small deformations). Doklady Akademii nauk USSR, Ser.A. **10**, 909–913 (1973)
6. Babich, IYu., Guz, A.N.: K teorii uprugoi ustoichivosti szhimaemykh i neszhimaemykh kompozitnykh sred (Towards theory of elastic stability of compressible and incompressible composite media). Mekhanika Polimerov **2**, 267–275 (1972)
7. Babich, IYu., Guz, A.N.: Trekhmernaia zadacha ob ustoichivosti volokna v matritse pri vysokoelasticheskikh deformatsiiakh (Three-dimensional problem of stability of a fiber in a matrix under highly elastic deformations). Izvestiya Akademii Nauk SSSR **3**, 44–48 (1973)
8. Garashchuk, I.N.: Ob ustoichivosti volokna v matritse pri neodnorodnykh dokriticheskikh deformatsiiakh (On stability of a fiber in a matrix under inhomogeneous subcritical deformations). Doklady Akademii nauk USSR, Ser. A. **8**, 24–27 (1983)
9. Guz, A.N.: O postroenii teorii ustoichivosti odnonapravlennykh voloknistykh materialov (On stability theory construction for unidirectional fibrous materials). Prikladnaya Mekhanika **5**(2), 62–70 (1969)
10. Guz, O.M.: Pro vyznachennia teoretychnoi hranytsi mitsnosti na stysk armovanykh materialiv (Determining the theoretical compressive strength of reinforced materials). Dopovidi Akademii Nauk URSR, Ser. A. **3**, 236–238 (1969)
11. Guz, A.N.: O postroenii teorii prochnosti odnonapravlennykh armirovannykh materialov na szhatie (On construction of the strength theory of unidirectional reinforced materials on compression). Probl. Prochn. **3**(3), 37–40 (1971)
12. Guz, A.N.: O kontinualnoi teorii razrusheniia pri szhatii kompozitnogo materiala s uprugo-plasticheskoi matritsei (A continuum theory of fracture upon compression of a composite material with an elastic-plastic matrix). Dokl. Akad. Nauk SSSR **262**(3), 556–560 (1982)
13. Guz, A.N.: O kontinualnoi teorii razrusheniia kompozitnykh materialov pri dvukhosnom szhatii (Continuum fracture theory of composite materials under biaxial compression). Dokl. Akad. Nauk SSSR **293**(4), 805–809 (1987)
14. Guz, A.N.: O lokalnoi ustoichivosti voloknistykh kompozitov (On local stability of fiber composites). Dokl. Akad. Nauk SSSR **314**(4), 806–809 (1990)
15. Guz, A.N.: O neklassicheskikh problemakh mekhaniki razrusheniia (On non-classical problems of fracture mechanics). Fiziko-himicheskaya mekhanika materialov **29**(3), 86–97 (1993)
16. Guz, A.N., Guz, I.A.: O kontinualnom priblizhenii v teorii ustoichivosti sloistykh kompozitnykh materialov (Continual approximation in the theory of stability of layered composite materials). Dokl. Akad. Nauk SSSR **305**(5), 1073–1076 (1989)
17. Guz, A.N., Guz, I.A.: O lokalnoi neustoichivosti sloistykh kompozitnykh materialov (Local instabilities of layered composite materials). Dokl. Akad. Nauk SSSR **311**(4), 812–814 (1990)
18. Guz, A.N., Korzh, V.P., Chekhov, V.N.: Ustoichivost sloistoi poluploskosti pod deistviem poverkhnostnykh raspredelennykh nagruzok (Stability of a layered half-plane under the influence of surface distributed loads). Dokl. Akad. Nauk SSSR **313**(6), 1381–1385 (1990)

19. Guz, A.N., Lapusta, Yu.N.: O metode issledovaniia ustoichivosti volokna v uprugoi polubeskonechnoi matritse vblizi svobodnoi poverkhnosti (Method of studying the stability of a fiber in a semi-infinite elastic matrix near a free surface). Prikladnaya Matematika i Mekhanika 53(4), 693–697 (1989)
20. Guz, A.N., Lapusta, Yu.N.: Pripoverkhnostnaia neustoichivost riada volokon v kompozite (Near-surface instability of a number of fibers in a composite). DAN 325(4), 679–683 (1992)
21. Guz, A.N., Musayev, D.A.: O razrushenii lentochnykh kompozitnykh materialov pri szhatii (Fracture of tape composite materials under compression). Dokl. Akad. Nauk SSSR 301(3), 565–568 (1988)
22. Guz, A.N., Tkachenko, E.A., Chekhov, V.N.: Raschety na ustoichivost sloistykh kompozitnykh pokrytii v tribotekhnike (Calculations of stability of layered composite coatings in tribotechnics). Mekhanika kompozitnykh materialov 36(2), 229–236 (2000)
23. Guz, A.N., Tkachenko, E.A., Chekhov, V.N.: Pripoverkhnostnaia neustoichivost sloistykh pokrytii pri neuprugom deformirovanii (Near-surface instability of layered coatings under inelastic deformation). Mekhanika kompozitnykh materialov 36(6), 791–800 (2000)
24. Guz, A.N., Cherevko, M.A.: K mekhanike razrusheniia voloknistogo kompozitnogo materiala pri szhatii (On fracture mechanics of a fibrous composite material under compression). Dokl. Akad. Nauk SSSR 268(4), 806–808 (1981)
25. Guz, A.N., Cherevko, M.A.: O razrushenii odnonapravlennogo voloknistogo kompozita s uprugo-plasticheskoi matritsei pri szhatii (Fracture of a unidirectional fibrous composite with an elastic-plastic matrix under compression). Mekhanika kompozitnykh materialov 6, 987–994 (1982)
26. Guz, A.N., Cherevko, M.A., Margolin, G.G., Romashko, I.M.: O razrushenii odnonapravlennykh boroaliuminievykh kompozitov pri szhatii (Fracture of unidirectional boron-aluminum composites under compression). Mekhanika kompozitnykh materialov 2, 226–230 (1986)
27. Guz, A.N., Chekhov, V.N.: Poverkhnostnaia poteria ustoichivosti sloistoi poluploskosti pri uprugo-plasticheskom deformirovanii sloev (Surface stability loss of a layered half-plane under elastic-plastic deformation of layers). Dokl. Akad. Nauk SSSR 272(3), 546–550 (1983)
28. Guz, A.N., Chekhov, V.N.: Poverkhnostnaia neustoichivost sloistykh kompozitov pri malykh i konechnykh dokriticheskikh deformatsiiakh (Surface instability of layered composites at small and finite subcritical deformations). Mekhanika kompozitnykh materialov 5, 838–843 (1984)
29. Guz, A.N., Chekhov, V.N.: Issledovanie ustoichivosti polubeskonechnykh sloistykh sred s uchetom ikh uprugikh i uprugo-plasticheskikh svoistv (Stability investigation of semi-infinite layered media allowing their elastic and elastic-plastic properties). Izvestiya Akademii Nauk SSSR 1, 87–96 (1985)
30. Guz, A.N., Chekhov, V.N.: Primenenie variatsionnykh metodov v zadachakh ustoichivosti sloistykh poluogranichennykh sred (Application of variational methods in stability problems for layered semi-bounded media). Dokl. Akad. Nauk SSSR 283(5), 1123–1126 (1985)
31. Guz, A.N., Chekhov, V.N., Shulga, N.A.: Poverkhnostnaia neustoichivost poluprostranstva periodicheskoi struktury (Surface instability of a half-space with a periodic structure). Dokl. Akad. Nauk SSSR 266(6), 1306–1310 (1982)
32. Lapusta, Yu.N.: Metod issledovaniya ustoychivosti dvuh volokon v uprugoy polubeskonechnoy matritse (Research method for stability of two fibers in an elastic semi-infinite matrix). Doklady Akademii nauk USSR, Ser. A. 1, 42–45 (1989)
33. Lapusta, Yu.N.: Uchet vliyaniya svobodnoy granicy na ustoychivost periodicheskogo ryada volokon v uprugoy polubeskonechnoy matritse (Concideration of a free boundary effect on stability of a periodic row of fibers in an elastic semi-infinite matrix). Doklady Akademii nauk USSR, Ser. A. 5, 34–37 (1989)
34. Lapusta, Yu.N.: K resheniu zadachi pripoverhnostnogo vypuchivaniya periodicheskoy sistemy volokon v uprugoy matritse (Solution of the problem of near-surface buckling of a periodic system of fibers in an elastic matrix). Doklady Akademii nauk USSR, Ser. A. 7, 48–52 (1989)
35. Lapusta, Yu.N.: O vozmozhnykh formah poteri ustoychivosti volokna v polubeskonechnoy matritse (On possible forms of fiber stability loss in a semi-infinite matrix). Doklady Akademii nauk USSR, Ser. A. 11, 42–45 (1989)

36. Lapusta, Yu.N.: Ustoychivost ryada volokon vblizi svobodnogo ploskogo kraya svyazu-uschego pri osevom szhatii (Stability of a row of fibers near the free flat edge of binder under axial compression). Mekhanika kompozitnykh materialov **4**, 739–742 (1990)

37. Lapusta, Yu.N.: Ob ustoychivosti volokna vblizi polosti v uprugo-plasticheskoy matritse (On stability of a fiber near a cavity in an elastic-plastic matrix). Doklady Akademii nauk USSR, Ser. A. **9**, 80–84 (1991)

38. Lapusta, Yu.N.: Pripoverhnostnaya neustoychivost periodicheskoy sistemy volokon v uprugoy matritse (Near-surface instability of a periodic system of fibers in an elastic matrix). Doklady Akademii nauk USSR, Ser. A. **8**, 70–75 (1992)

39. Cherevko, M.A.: Ustoychivost volokna v uprugo-plasticheskoy matritse (Fiber stability in an elastic-plastic matrix). Doklady Akademii nauk USSR, Ser.A. **9**, 43–46 (1982)

40. Cherevko, M.A.: Ustoychivost pologo volokna v uprugo-plasticheskoy matritse (Stability of a hollow fiber in an elastic-plastic matrix). Doklady Akademii nauk USSR, Ser.A. **11**, 35–38 (1982)

41. Cherevko, M.A.: Ustoychivost volokna v uprugo-plasticheskoy matritse pri nepolnom kontakte (Stability of a fiber in an elastic-plastic matrix with incomplete contact). Prikladnaya Mekhanika **20**(9), 122–123 (1984)

42. Cherevko, M.A.: Ustoychivost volokna v uprugo-plasticheskoy matritse pri nepolnom kontakte (Stability of a row of circular fibers in an elastic-plastic matrix). Prikladnaya Mekhanika **21**(12), 35–40 (1985)

43. Chekhov, V.N.: Poverkhnostnaya neustojckhivost sloistykh sred pri konechnykh deforma-ciyakh (Surface instability of layered media under finite deformations). Doklady Akademii nauk USSR, Ser.A. **5**, 48–50 (1983)

44. Babich, IYu.: On the stability of a fiber in a matrix under small deformations. Sov. Appl. Mech. **9**(4), 370–375 (1973)

45. Babich, IYu., Chekhov, V.N.: Surface and internal instability in laminated composites. Sov. Appl. Mech. **25**(1), 21–28 (1989)

46. Babich, IYu., Garashchuk, I.N., Guz, A.N.: Stability of a fiber in an elastic matrix with nonuniform subcritical state. Sov. Appl. Mech. **19**(11), 941–947 (1983)

47. Babich, IYu., Guz, A.N.: Deformation instability of laminated materials. Sov. Appl. Mech. **5**(5), 488–491 (1969)

48. Babich, IYu., Guz, A.N.: Methods of studing three-dimensional problems of stability in highly-elastic deformations. Sov. Appl. Mech. **8**(6), 596–599 (1972)

49. Babich, IYu., Guz, A.N., Chekhov, V.N.: The three-dimensional theory of stability of fibrous and laminated materials. Int. Appl. Mech. **37**(9), 1103–1141 (2001)

50. Babich, IYu., Guz, A.N., Kilin, V.I.: Aspects of the fracture and stability of laminated structures with elastic strains. Sov. Appl. Mech. **22**(7), 601–605 (1986)

51. Babich, IYu., Guz, A.N., Shulga, N.A.: Investigation of the dynamics and stability of composite materials in a three-dimensional formulation (survey). Sov. Appl. Mech. **18**(1), 1–21 (1982)

52. Chekhov, V.N.: Folding of rocks with periodic structure. Sov. Appl. Mech. **20**(3), 216–221 (1984)

53. Chekhov, V.N.: Effect of the hereditary properties of a medium on the surface instability of a layered half-space. Sov. Appl. Mech. **20**(7), 613–618 (1984)

54. Chekhov, V.N.: Surface instability of a layered medium connected to a uniform half-space. Sov. Appl. Mech. **20**(11), 1018–1024 (1984)

55. Chekhov, V.N.: Influence of a surface load on stability of laminar bodies. Sov. Appl. Mech. **24**(9), 839–845 (1988)

56. Chekhov, V.N.: On the formation of linear folds in regularly layered rock masses under biaxial loading. Int. Appl. Mech. **41**(12), 1350–1356 (2005)

57. Cherevko, M.A.: Stability of a biperiodic system of circular fibers in an elastoplastic matrix. Sov. Appl. Mech. **22**(4), 316–321 (1986)

58. Guz, A.N.: The stability of orthotropic bodies. Sov. Appl. Mech. **3**(5), 17–22 (1967)

59. Guz, A.N.: Mechanics of composite material failure under axial compression (brittle failure). Sov. Appl. Mech. **18**(10), 863–872 (1982)
60. Guz, A.N.: Mechanics of composite material failure under axial compression (plastic failure). Sov. Appl. Mech. **18**(11), 970–976 (1982)
61. Guz, A.N.: Continuum theory of fracture in the compression of composite materials with metallic matrix. Sov. Appl. Mech. **18**(12), 1045–1052 (1982)
62. Guz, A.N.: Fracture of Unidirectional Composite Materials Under the Axial Compression. In: Sih G.C. (ed.)
63. Tamuzs, V.P.: Fracture of composite materials, pp. 173–182. Nijhoff (1982)
64. Guz, A.N.: Continuous theory of failure of composite materials under compression in the case of a complex stresses state. Sov. Appl. Mech. **22**(4), 301–315 (1986)
65. Guz, A.N.: Construction of a theory of failure of composites in triaxial and biaxial compression. Sov. Appl. Mech. **25**(1), 29–33 (1989)
66. Guz, A.N.: Principles of the continual theory of plastic fracture of unidirectional fiber composite materials with metallic matrix under compression. Sov. Appl. Mech. **26**(1), 1–8 (1990)
67. Guz, A.N.: Plastic failure of unidirectional fibrous composite material with metal matrix in compression. In: Vautrin, A., Sol. H. (eds.) Mechanical Identification of Composite, pp. 278–286. Elsevier, London New-York (1991)
68. Guz, A.N.: Construction of a theory of the local instability of unidirectional fiber composites. Int. Appl. Mech. **28**(1), 18–24 (1992)
69. Guz, A.N.: Stability theory for unidirectional fiber reinforced composites. Int. Appl. Mech. **32**(8), 577–586 (1996)
70. Guz, A.N.: Non-classical problems of composite failure. In: Proceedings of ICCST/1, pp.161–166. Durban, South. Africa, 18–20 June (1996)
71. Guz, A.N.: Conditions of hyperbolicity and mechanics of failure of composites in compression. ZAMM **78**(1), 427–428 (1998)
72. Guz, A.N.: In On study of nonclassical problems of fracture and failure mechanics and related mechanisms. ANNALS of the European Academy of Sciences, pp. 35–68. Liège, Belgium (2006–2007)
73. Guz, A.N.: On study of nonclassical problems of fracture and failure mechanics and related mechanisms. Int. Appl. Mech. **45**(1), 1–31 (2009)
74. Guz, A.N., Babich, IYu.: Three-dimensional stability problems of composite materials and composite construction components. Rozpr. Inz. **27**(4), 613–631 (1979)
75. Guz, A.N., Chekhov, V.N.: Linearized theory of folding in the interior of the earth's crust. Sov. Appl. Mech. **11**(1), 1–10 (1975)
76. Guz, A.N., Chekhov, V.N.: Variational method of investigating the stability of laminar semiinfinite media. Sov. Appl. Mech. **21**(7), 639–646 (1985)
77. Guz, A.N., Chekhov, V.N.: Investigation of surface instability of stratified bodies in three-dimensional formulation. Sov. Appl. Mech. **26**(2), 107–125 (1990)
78. Guz, A.N., Chekhov, V.N.: Problems of folding in the earth's stratified crust. Int. Appl. Mech. **43**(2), 127–159 (2007)
79. Guz, A.N., Chekhov, V.N., Stukotilov, V.S.: Effect of anisotropy in the physicomechanical properties of a material on the surface instability of layered semiinfinite media. Int. Appl. Mech. **33**(2), 87–92 (1997)
80. Guz, A.N., Cherevko, M.A.: Fracture mechanics of unidirectional fibrous composites with metal matrix under compression. Theor. Appl. Frac. Mech. **3**(2), 151–155 (1985)
81. Guz, A.N., Cherevko, M.A.: Stability of a biperiodic system of fibers in a matrix with finite deformations. Sov. Appl. Mech. **22**(6), 514–518 (1986)
82. Guz, A.N., Guz, I.A.: Substantiation of a continuum theory of the fracture of laminated composite in compression. Sov. Appl. Mech. **24**(7), 648–657 (1988)
83. Guz, A.N., Guz, I.A.: Foundation for the continual theory of fracture during compression of laminar composites with a metal matrix. Sov. Appl. Mech. **24**(11), 1041–1047 (1988)

84. Guz, A.N., Guz, I.A.: On the theory of stability of laminated composites. Int. Appl. Mech. **35**(4), 323–329 (1999)
85. Guz, A.N., Kritsuk, A.A., Emel'yanov, R.F.: Character of the failure of unidirectional glass-reinforced plastic in compression. Sov. Appl. Mech. **5**(9), 997–999 (1969)
86. Guz, A.N., Lapusta, Yu.N.: Stability of a fiber near a free surface. Sov. Appl. Mech. **22**(8), 711–718 (1986)
87. Guz, A.N., Lapusta, Yu.N.: Stability of a fiber near a free cylindrical surface. Sov. Appl. Mech. **24**(10), 939–944 (1988)
88. Guz, A.N., Lapusta, Yu.N.: Three-dimensional problem on the stability of a row of fibers perpendicular to the free boundary of a matrix. Int. Appl. Mech. **30**(12), 919–926 (1994)
89. Guz, A.N., Lapusta, Yu.N.: Three-dimensional problems of the near-surface instability of fiber composites in compression (Model of a piecewise-uniform medium) (Survey). Int. Appl. Mech. **35**(7), 641–670 (1999)
90. Guz, A.N., Lapusta, Yu.N., Mamzenko, Yu.A.: Stability of a two fibers in an elasto-plastic matrix under compression. Int. Appl. Mech. **34**(5), 405–413 (1998)
91. Guz, A.N., Lapusta, Yu.N., Samborskaya, A.N.: A micromechanics solution of a 3D internal instability problem for a fiber series on an infinite matrix. Int. J. Fract. **116**(3), L55–L60 (2002)
92. Guz, A.N., Lapusta, Yu.N., Samborskaya, A.N.: 3D model and estimation of fiber interaction effects during internal instability in non-linear composites. Int. J. Fract. **134**(44289), L45–L51 (2005)
93. Guz, A.N., Musaev, D.A.: Fracture of a unidirectional ribbon composite with elasto-plastic matrix in compression. Sov. Appl. Mech. **26**(5), 425–429 (1990)
94. Guz, A.N., Musaev, D.A., Yusubov, Ch.A.: Stability of two noncircular cylinder in an elastic matrix with small subcritical strains. Sov. Appl. Mech. **25**(11), 1059–1064 (1989)
95. Guz, A.N., Samborskaya, A.N.: General stability problem of a series of fibers in an elastic matrix. Sov. Appl. Mech. **27**(3), 223–230 (1991)
96. Guz, A.N., Sporykhin, A.N.: Three-dimensional theory of inelastic stability (General questions). Sov. Appl. Mech. **18**(7), 581–596 (1982)
97. Guz, A.N., Sporykhin, A.N.: Three-dimensional theory of inelastic stability. Specific results. Sov. Appl. Mech. **18**(8), 671–692 (1982)
98. Guz, A.N., Tkachenko, E.A., Chekhov, V.N.: Stability of layered antifriction coating. Int. Appl. Mech. **32**(9), 669–676 (1996)
99. Guz, A.N., Tkachenko, E.A., Chekhov, V.N., Stukotilov, V.S.: Stability of multilayer antifriction coating for small subcritical strains. Int. Appl. Mech. **32**(10), 772–779 (1996)
100. Guz, I.A.: Spatial nonaxisymmetric problems of the theory of stability of laminar highly elastic composite materials. Sov. Appl. Mech. **25**(11), 1080–1085 (1989)
101. Guz, I.A.: Three-dimensional nonaxisymmetric problems of the theory of stability of composite materials with metallic matrix. Sov. Appl. Mech. **25**(12), 1196–1200 (1989)
102. Guz, I.A.: Continuum approximation in three-dimensional nonaxisymmetyric problems of the stability theory of laminar compressible materials. Sov. Appl. Mech. **26**(3), 233–236 (1990)
103. Guz, I.A.: Asymptotic accuracy of the continual theory of the internal instability of laminar composites with an incompressible matrix. Sov. Appl. Mech. **27**(7), 680–684 (1991)
104. Guz, I.A.: The strength of a composite formed by longitudinal–transverse stacking of orthotropic layers with a crack at the boundary. Int. Appl. Mech. **29**(11), 921–924 (1993)
105. Guz, I.A., Soutis, C.: Continuum fracture theory for layered materials: investigation of accuracy. ZAMM **35**(5), 462–468 (1999)
106. Guz, I.A., Soutis, C.: A 3-D stability theory applied to layered rocks undergoing finite deformations in biaxial compression. Eur. J. of Mech. A/Solids. **20**(1), 139–153 (2001)
107. Guz, I.A., Soutis, C.: Accuracy of a continuum fracture theory for non-linear composite materials under large deformations in biaxial compression. ZAMM **81**(4), S849–S850 (2001)
108. Guz, I.A., Soutis, C.: Compressive fracture of non-linear composites undergoing large deformations. Int. J. Solids Struct. **38**(21), 3759–3770 (2001)
109. Guz, I.A., Soutis, C.: Predicting fracture of composites. In: Soutis, C., Beaumont. P.W.R. (eds.) Multi-scale Modelling of Composite Material Systems. The Art of Predictive Damage Modelling, pp. 278–302. Woodhead Publ. Ltd., Cambridge England (2005)

110. Guz, I.A., Soutis, C.: Compressive strength of laminated composites: on application of the continuum fracture theory. In: M.H. Aliabadi, M. Guagliano. Fracture and Damage of Composites (WIT Press, Cambridge England, 2006), 1–24

111. Kokhanenko, Yu.V.: Numerical solution of problems of the theory of elasticity and the three-dimensional stability of piecewise-homogeneous media. Sov. Appl. Mech. **22**(11), 1052–1058 (1986)

112. Korzh, V.P., Chekhov, V.N.: Surface instability of laminated bodies of regular structure under combination loading. Sov. Appl. Mech. **27**(5), 443–449 (1991)

113. Korzh, V.P., Chekhov, V.N.: Combined analysis of internal and surface instability in laminar bodies of regular structure. Sov. Appl. Mech. **27**(11), 1058–1063 (1991)

114. Korzh, V.P., Chekhov, V.N.: Surface instability of laminated materials of regular structure under triaxial compression. Int. Appl. Mech. **38**(9), 1119–1124 (2002)

115. Lapusta, Yu.N.: Fiber stability in a semiinfinite elastic matrix with highly elastic deformation. Sov. Appl. Mech. **23**(8), 718–721 (1987)

116. Lapusta, Yu.N.: Stability of fibers near the free surface of a cavity during finite precritical strains. Sov. Appl. Mech. **24**(5), 453–457 (1988)

117. Lapusta, Yu.N.: Surface instability of two fibers in a matrix. Sov. Appl. Mech. **26**(8), 739–744 (1990)

118. Lapusta, Yu.N.: Stability of a fiber in semi-infinite elastic matrix with sliding contact at the interface. Sov. Appl. Mech. **26**(10), 924–928 (1990)

119. Lapusta, Yu.N.: Stability of a periodic series of fibers in a semi-infinite matrix. Sov. Appl. Mech. **27**(2), 124–130 (1991)

120. Lapusta, Yu.N.: Stability of a fiber in an elastoplastic matrix near a free cylindrical surface. Int. Appl. Mech. **28**(1), 33–41 (1992)

121. Lapusta, Yu.N.: A 3-D model for possible micro-instability patterns in a boundary layer of a fibre composite under compression. Composite Sciences and Technology. **62**, 805–817 (2002)

122. Lapusta, Yu.N., Wagner, W.: An estimation on the influence of matrix cavity and damaged fibre-matrix interface on stability of composites. ZAMM. **81**, 855–856 (2001)

123. Lapusta, Yu.N., Wagner, W.: On various material and fibre-matrix interface models in the near-surface instability problems for fibrous composites. Compos. A Appl. Sci. Manuf. **32**, 413–423 (2001)

124. Lapusta, Yu.N., Wagner, W.: A numerical estimation of the effects of a cylindrical hole and imperfect bounding on stability of a fiber in an elastic matrix. Int. J. Num. Methods Eng. **51**, 631–646 (2001)

125. Guz, A.N. (guest ed.): Micromechanics of composite materials: Focus on Ukrainian research. Appl. Mech. Reviews. Special Issue **45**(2), 13–101 (1992). A.N. Guz, Introduction, pp. 14–15, About the authors, p.16, S.D. Akbarov, A.N. Guz, Statics of laminated and fibrous composites, pp. 17–34, A.N. Guz, N.A. Shulga, Dynamics of laminated and fibrous composites, pp. 35–60, I.Yu. Babich, A.N. Guz, Stability of fibrous composites, pp. 61–80, A.N. Guz, Vic.N. Chekhov, Stability of laminated composites, pp. 81–101

126. Musaev, D.A.: Stability of a noncircular cylinder in an elastic matrix under finite deformations. Sov. Appl. Mech. **38**(6), 566–569 (1988)

127. Musaev, D.A., Nagiev, F.M.: Stability of a row of noncircular cylinders in an elastic matrix with finite strains. Sov. Appl. Mech. **26**(10), 929–933 (1990)

128. Skachenko, A.V.: Stability of multilayer composite under inelastic deformations. Sov. Appl. Mech. **15**(8), 756–757 (1979)

129. Tkachenko, E.A., Chekhov, V.N.: Combined effect of temperature and compressive surface loads on the stability of elastic multilayer coating with small subcritical strains. Int. Appl. Mech. **34**(8), 729–735 (1998)

130. Tkachenko, E.A., Chekhov, V.N.: The stability of tribotechnical laminated polymeric coatings. Int. Appl. Mech. **36**(9), 1198–1204 (2000)

131. Tkachenko, E.A., Chekhov, V.N.: The stability of laminated elastomer coatings under surface loading. Int. Appl. Mech. **36**(10), 1355–1362 (2000)

132. Tkachenko, E.A., Chekhov, V.N.: Stability of laminated coating under elastoplastic deformations. Int. Appl. Mech. **37**(3), 361–368 (2001)
133. Tkachenko, E.A., Chekhov, V.N.: Stability of an elastic layer stack between two half-space under compressive loads. Int. Appl. Mech. **38**(11), 1381–1387 (2002)
134. Tkachenko, E.A., Chekhov, V.N.: Stability of a lamine between two homogeneous half-space under inelastic deformation. Int. Appl. Mech. **41**(5), 481–489 (2005)
135. Guz, A.N.: The study and analysis of non-classical problems of fracture and failure mechanics. In: Abstracts of IUTAM Symposium of Nonlinear Analysis of Fracture, p. 19. Cambridge, 3–7 Sept. (1995)
136. Guz, A.N.: On failure propagation in composite materials in compression (Three-dimensional continual theory). In: Proceedings of ECF 11 Poitiers–Futuroscope, Vol. III, pp.1769–1774. France, Sept. 3–6 (1996)
137. Guz, A.N.: Non-classical problems of composite failure. In: Proceedings of ICF9 Advance in Fracture Research, Vol. 4, pp.1911–1921 Sydney, Australia (1997)
138. Guz, A.N.: Study and Analysis of Non-classical Problems of Fracture and Failure Mechanics and Corresponding Mechanisms. Lecture presented at Institute of Mechanics HANOI (1998)
139. Guz, A.N.: On the plastic failure of unidirectional fibrous composite materials with metal matrix in compression. Continuum approximation. In: Proceedings of the ICCE/6, pp. 279–280. Orlando, Florida, USA, June 27–July 3 (1999)
140. Guz, I.A.: Investigation of local form of stability loss in laminated composites (three-dimensional problem). In: Proceedings of ICCM/9: Composites. Properties and Applications, vol. VI, pp. 377–383. Madrid 12–16 July (1993)
141. Guz, I.A.: On the local stability loss in laminated composite structures. In: Proceedings of the 6th European on Comp. Materials: Development in the Science and Techn. Comp. Materials, Sept. 20–24, 1993, pp. 263–268. Woodhead Publ. Ltd., Bordeaux, France (1993)
142. Guz, I.A.: Metal matrix composites in compression. Substantiation of the bounds. In: Proceedings of 5th International Conference on Automated Composites, 4–5 Sept. 1997, Glasgow, UK. (Institute of Materials, London, UK, 1997), pp. 387–393
143. Guz, I.A.: On asymptotic accuracy of the theory of plastic fracture in compression for layered materials. In: Proceedings EUROMECH Coll. 378: Nonlocal Aspects in Solid Mechanics, 20–22 April, 1998 (Mulhouse, France, 1998), pp. 118–123
144. Guz, I.A.: On continuum approximation in compressive fracture theory for metal matrix composites: asymptotic accuracy. In: Proceedings of ICCST/2, 9–11 June, 1998. (Durban, South Africa, 1998), pp. 501–506
145. Guz, I.A.: Investigation of accuracy of continuum fracture theory for piecewise-homogeneous medium. In: Proceedings of ICNM-III, 17–20 Aug., 1998. Shanghai University Press, Shanghai, China (1998), pp. 224–227
146. Guz, I.A.: On two approaches to compressive fracture problems. Proc. of 12th Eur. Conf. on Fracture, 14–16 Sept., 1998, Sheffield, UK, Vol. 3. (EMAS Publ., 1998), 1447–1452
147. Guz, I.A.: Asymptotic analysis of fracture theory for layered rocks in compression. In Modelling and Simulation Based Engineering. Vol. 1 (Tech. Science Press, Palmdale, USA, 1998), pp. 375–380
148. Guz, I.A.: On calculation of accuracy for continuum fracture theory of metal matrix composites in compression. Proc. of ICAC 96, 15–18 Dec., 1998. (Hurghada, Egypt, 1998), 757–764
149. Guz, I.A.: On estimation of critical loads for rocks in compression: 3-D approach. In: Proc. of ARCOM'99, 15–17 Dec., 1999, Singapore. Vol. 2. (Elsevier, 1999), 847–852
150. Guz, I.A.: Bounds for critical parameters in the stability theory of piecewise-homogeneous media: Laminated rocks. Proc. of SASAM 2000, 11–13 Jan., 2000. (Durban, South Africa, 2000), 479–484
151. Guz, I.A.: Compressive behaviour of metal matrix composites: Accuracy of homogenization. ZAMM. **80**(2), S473–S474 (2000)
152. Guz, I.A.: The effect of the multi-axiality of compressive loading on the accuracy of a continuum model for layered materials. Int. J. Solids and Struct. **42**, 439–453 (2005)

153. Lapusta, Yu.N.: Stability of a row of fibers near the free plane border in matrix in axial compression. Abstracts of the Ist European Solid Mechanics Conf., (Munich, FRG, Sept. 9–13, 1991), 131–132
154. Lapusta, Yu.N.: Near-the-surface fracture of fibrous materials in compression. Abstracts of ICF-8, Part II, pp. 393–394. Kiev, UKRAINE (1993)
155. Lapusta, Yu.N.: On stability loss of fibers in composite materials near the boundaries. Proc. of ECF 11, 3–6 Sept., 1996, Vol. III, pp. 1633–1638. Poitiers, France, (1996)
156. Lapusta, Yu.N.: Study of near-surface buckling of composite materials in zones of compression (model of a piecewise-uniform medium). In: Proc. of the ICCST/2, 9–11 June, 1998, pp. 145–148 Durban, South Africa (1998)
157. Lapusta, Yu.N.: A general micromechanical approach to the study of the near-surface buckling in fibrous composites. Proc. of ECF 12, 13–18.09.1998, (Sheffield, UK, 1998), 1477–1482
158. Lapusta, Yu.N.: Prediction of compressive strength of fiber-reinforced composite based on 3-D microlevel consideration. In: Proc. of ARQUIMACOM 98, 7–9.10.1998, pp. 165–169. Bordeaux, France (1998)
159. Dow, N.F., Gruntfest, I.J.: Deformation of most needed potentially possible improvements in materials for ballistic and space vehicles. General Electric Co., Space Sci. Lab., TIRS 60 SD 389, June (1960)
160. Biot, M.A.: Interface instability in finite elasticity under initial stress. Proc. Roy. Soc. **273**(1354) (1963)
161. Guz, A.N.: Osnovy trekhmernoi teorii ustoichivosti deformiruemykh tel (Fundamentals of the three-dimensional theory of stability of deformable bodies). (Vyshcha Shkola, Kyiv, 1986)
162. Guz, A.N.: Fundamentals of the Three-Dimensional Theory of Stability of Deformable Bodies. Springer, Berlin Hiedelberg New York (1999)
163. Guz, A.N., Dekret, V.A.: Model volokon konechnykh razmerov v trekhmernoi teorii usto-ichivosti kompozitnykh materialov (obzor) (Finite-fiber model in the three-dimensional theory of stability of composite materials (Review)). Prikladnaya Mekhanika. **52**(1), 3–77 (2016)
164. Guz, A.N., Dekret, V.A.: Model korotkikh volokon v teorii ustoichivosti kompozitov (The model of short fibers in the theory of stability of composites). LAP LAMBERT Academic Publishing, Saarbrücken, Deutschland (2015)
165. Rosen, B.W.: Mechanics of Composite Strengthening, presented at Fiber Composite Materials. American Society for Metals (Metals Park, Ohio, 1965), pp.37–75
166. Rosen, B.U.: In Mekhanika uprochneniya kompozitov (Mechanics of hardening composites), ed.by R.H. Krock, L.J. Broutman. Fiber composites [Russian translation], pp.54–94 Mir, Moscow (1967)
167. Libowitz, G. (ed.): Razrushenie, v 7 tomah (Fracture, in 7 volumes) [Russian translation] (Vol. 1, 2, 3, 7 - Mir, Vol. 4, 5—Mashinostroenie, Vol.6—Metallurgia, Moscow, 1973–1977), T. 1. Mikroskopicheskie i makroskopicheskie osnovy mekhaniki razrusheniii (Vol.1. Microscopic and macroscopic foundations of fracture mechanics) (1973), T. 2. Matematicheskie osnovy mekhaniki razrusheniia (Vol.2. Mathematical foundations of fracture mechanics) (1975), T. 3. Inzhenernye osnovy i vozdeistvie vneshnei sredy (Vol.3. Engineering foundations and the impact of the external environmen) (1976), T. 4. Issledovanie razrusheniia dlia inzhenernykh raschetov (Vol.4. Investigation of fracture for engineering calculations) (1977), T. 5. Raschet konstruktsii na khrupkuiu prochnost (Vol.5. Calculation of structures for brittle strength) (1977), T. 6. Razrushenie metallov (Vol.6. Fracture of metals) (1976), T.7, ch. 1. Razrushenie nemetallov i kompozitnykh materialov. Neorganicheskie materialy (stekla, gornye porody, kompozity, keramiki, led) (Vol.7, Part 1. Fracture of non-metals and composite materials. Inorganic materials (glass, rocks, composites, ceramics, ice) (1976), T.7, ch. 2. Razrushenie nemetallov i kompozitnykh materialov. Organicheskie materialy (stekloobraznye polimery, elastomery, kost) (Vol.7, Part 2. Fracture of non-metals and composite materials. Organic materials (glassy polymers, elastomers, bone)) (1976)
168. Rosen, B.U., Dau, N.F.: In Mekhanika razrusheniya voloknistykh kompozitov (Fracture mechanics of fiber composites), ed.by G. Libowitz. Fracture, Vol. 7, Part 1. Fracture of non-metals and composite materials. Inorganic materials (glass, rocks, composites, ceramics, ice) (Mir, Moscow, 1976), pp.300–366

169. Krock, R.H., Broutman, L.J. (eds.): Kompozitnye materialy (Composite materials), [Russian translation], (Vol. 1, 2, 5, 6—Mir, Vol. 3, 4, 7, 8—Mashinostroenie, Moscow, 1978–1979), T. 1. Poverkhnosti razdela v metallicheskikh kompozitakh (V.1. Interfaces in metal matrix composites) (1978), T. 2. Mekhanika kompozitnykh materialov (V.2. Mechanics of composite materials) (1978), T. 3. Primenenie kompozitnykh materialov v tekhnike (V.3. Engineering applications of composites) (1979), T. 4. Kompozitnye materialy s metallicheskoi matritsei (V.4. Metal matrix composites) (1978), T.5. Razrushenie i ustalost (V.5. Fracture and fatigue) (1978), T. 6. Poverkhnosti razdela v polimernykh kompozitakh (V. 6. Interfaces in polymer matrix composites) (1978), T. 7. Analiz i proektirovanie konstruktsii. Ch. 1 (T. 7. Structural design and analysis. Part 1) (1979), T. 8. Analiz i proektirovanie konstruktsii. Ch. 2 (T. 8. Structural design and analysis. Part 2) (1979)
170. Broutman, L.J., Krock, R.H. (eds.): Sovremennye kompozicionnye materialy (Modern composite materials) [Russian translation]. Mir, Moscow (1970)
171. Kelly, A., Zweden C. (eds.): Comprehensive Composite Materials Vol. 1–6 (Elsevier, 2006), Vol.1. Fiber reinforcements and general theory of composites, ed. by T.-W. Chou, Vol.2. Polymer matrix composites, Talreja, R., Mänson, J.-A.E. (eds.) Vol.3. Metal matrix composites, ed. by T.W. Clyne, Vol.4. Carbon/Carbon, Cement and Ceramic Composites, ed. by R. Warren, Vol.5. Test methods, non-destructive evaluation and smart materials, eds. by L. Carlsson, R.L. Crane, K. Uchino, Vol.6. Design and applications, eds. by W.G. Bader, K. Kedsvard, Y. Sawada
172. Milne, I., Ritchie, R.O., Karihaloo, B. (eds.): Comprehensive Structural Integrity, Vol. 1–10 (Elsevier, 2006), Vol.1. Structural integrity assessment–examples and case studies, eds. by I. Milne, R.O. Ritchie, B. Karihaloo, Vol.2. Fundamental theories and mechanisms of failure, ed. by B. Karihaloo, W.G. Knauss, Vol.3. Numerical and computational methods, ed. by R. de Borst, H.A. Mang, Vol.4. Cyclic loading and fatigue, eds. by R.O. Ritchie, Y. Marakami, Vol.5. Creep and high-temperature failure, ed. by A. Saxena, Vol.6. Environmentally-assisted fracture, ed. by J. Petit, P. Scott, Vol.7. Practical failure assessment methods, eds. by R.A. Ainsworth, K.-H. Schwable, Vol.8. Interfacial and nanoscale failure, ed. by W. Gerberich, W. Yang, Vol.9. Bioengineering, ed. by Y.-W. Mai, S.-H. Tech, Vol.10. Indexes
173. H. Schuerch Prediction of compressive strength in uniaxial boron fiber metal matrix composite material. AIAA Journal. **4**(1), 102–106 (1966)
174. Sadovsky, M.A., Pu, S.L., Hussain, M.A.: Buckling of microfibers. J. Appl. Mech. **34**(4), 295–302 (1967)
175. Hayashi, T.: On the shear instability of structures caused by compressive loads. Proc. of 16th Japan Congr. Appl. Mech. 1966, pp. 149–157
176. Hayashi, T.: On the shear instability of structures caused by compressive loads. AIAA Paper **65**, 770 (1970)
177. Cherepanov, G.P.: Mekhanika razrusheniya kompozicionnykh materialov (Fracture mechanics of composite materials). Nauka, Moscow (1983)
178. Testing and Micromechanical Theories: C.R. Schultheisz, A.M. Waas, Compressive Failure of Composites. Part 1. Prog. Aerospace Sci. **32**, 1–42 (1996)
179. Biot, M.A.: Surface instability in finite anisotropic elasticity under initial stress. Proc. Roy. Soc. **273**(1354) (1963)
180. Biot, M.A.: Mechanics of incremental deformations. Willey, New York (1965)
181. Guz, A.N., Rushchitskyi, Ya.Ya., Guz, I.A.: Vvedenie v mekhaniku nanokompozitov (Introduction to mechanics of nanocomposites). (S.P. Timoshenko Institute of Mechanics, Kyiv, 2010)
182. Guz, A.N., Rushchitskii, J.J.: Short Introduction to Mechanics of Nanocomposites. Scientific & Academic Publishing Co., LTD., USA (2013)
183. Guz, A.N.: Three-dimensional theory of stability of a carbon nanotube in a matrix. Int. Appl. Mech. **42**(1), 19–31 (2006)
184. Guz, A.N., Rushchitskii, J.J., Guz, I.A.: Establishing fundamentals of the mechanics of nanocomposites. Int. Appl. Mech. **43**(3), 247–271 (2007)

185. Studies, E.: A.M. Waas, C.R. Schultheisz, Compressive Failure of Composites. Part 2. Prog. Aerospace Sci. **32**, 43–78 (1996)
186. Jochum, Ch., Grandidier, J.-C.: Microbuckling elastic modeling approach of a single carbon fiber embedded in an epoxy matrix. Compos. Sci. Technol. **64**, 2441–2449 (2004)
187. Goldman, A.Y., Savelieva, N.F., Smirnov, V.I.: Issledovanie mekhanicheskikh svoistv tkanevykh stekloplastikov pri rastiazhenii i szhatii normalno k ploskosti armirovaniia (Research of mechanical properties of fiberglass fabrics under strain and compression normal to the plane of reinforcement). Mekhanika Polymerov. **5**, 803–809 (1968)
188. Shetty, H.R., Chou, T.W.: Mechanical properties and failure characteristic of FP-aluminium and W-aluminium composites. Metall. Trans. A. **16**(5), 833–864 (1985)
189. Guz, A.N., Dekret, V.A.: Interaction of two parallel short fibers in the matrix at loss of stability. Comput. Model. Eng. Sci. **13**(3), 165–170 (2006)
190. Wadee, M.A., Hunt, C.W., Peletier, M.A.: Kink band instability in layered structures. J. Mech. Phys. Solids. **52**, 1071–1091 (2004)
191. Budiansky, H.: Micromechanics. Composites and Structures. **16**(1), 3–13 (1983)
192. Flek, N.A.: Compressive failure of fiber composites, Advances in Applied Mechanics, vol. 33, pp. 43–119. Academic Press, New-York (1997)
193. Guz, A.N., Guz, I.A.: On models in the theory of stability of multi-walled carbon nanotubes. Int. Appl. Mech. **42**(6), 617–628 (2006)
194. Katz, G.S., Milevsky, D.V. (ed.): Napolniteli dlya polimernykh kompozicionnykh materialov. Spravochnoe posobie (Fillers for polymer composite materials. Handbook) [Russian translation] Khimia, Moscow (1981)
195. Katz, H., Milevski, I.V. (eds.): Handbook of Fillers and Reinforcements for Plastics. Van Nostrand Reinhold Company, New York (1978)
196. Pinnel, M.R., Lawley, F.: Correlation on uniaxial yielding and substructe in aluminium-stainless steel composites. Metall. Trans. A. **1**(5), 929–933 (1970)
197. Guz, A.N.: Ustoichivost uprugikh tel pri vsestoronnem szhatii (Stability of elastic bodies under all-round compression). Naukova Dumka, Kyiv (1979)
198. Akbarov, S.D.: Stress distribution in multi-layered composite with small-scale antiphase curvatures in structure. Sov. Appl. Mech. **23**(2), 107–111 (1987)
199. Akbarov, S.D.: Normal stresses in a fiber composite with curved structures having a low concentration of filler. Sov. Appl. Mech. **21**(11), 1065–1069 (1985)

Chapter 4
Problem 2. Model of Short Fibers in Theory of Stability and Fracture Mechanics of Composite Materials Under Compression

In this chapter, in a rather brief form (in comparison with **Problem 1**, Chap. 3 of this monograph), the main results on this problem are presented, which are obtained in the Department of Dynamics and Stability of Continua of the S. P. Timoshenko Institute of Mechanics of the NASU since 1999. The presentation of these results is written in the style announced in the Introduction to this monograph (without involving the aspects of a mathematical nature).

It also provides the information on some of the results of experimental studies corresponding to the formation of this problem.

The main results on **Problem 2**, obtained by employees of the Department of Dynamics and Stability of Continua, are presented in the monograph [1] of 2015, in the review article [2] of 2016, and in the articles [3–15], as well as in a number of others that were not included in the list of references to this monograph. A more detailed list of references is presented in the monograph [1] and in the review article [2].

In this scientific area, the dissertation for the degree of Doctor of Physical and Mathematical Sciences (DSc) was prepared and defended by V. A. Dekret.

It is expedient to note that the composite materials are created both with the relatively long fibers (the subject of consideration of the fracture mechanics for these materials under compression in **Problem 1**) and with the sufficiently short fibers (reinforcing elements).

The experimental studies under compression in the case of the composites with short fibers revealed the phenomenon of loss of stability in the internal structure of the composite with modes of loss of stability that *are not periodic* along the fiber axes and which are characteristic of the short fibers in the matrix.

The theory of stability under compression in this case, which corresponds to the construction of fracture mechanics describing the beginning (start) of fracture, was developed with the involvement of TLTDBS (Chap. 2 of this monograph), built with the usually accepted accuracy in the solid mechanics. The main results were obtained

A. N. Guz, *Eight Non-Classical Problems of Fracture Mechanics*,
Advanced Structured Materials 159,
https://doi.org/10.1007/978-3-030-77501-8_4

by the author and his pupils, including the specific results within the plane problem for the composites of various structures.

4.1   Experimental Results on Loss of Stability in the Internal Structure of Composites Under Compression: Case of Short Fibers

First of all, as in Sect. 3.2.1 for the case of the "long" fibers, and in Sect. 4.1 for the case of the "short" fibers, it should be noted that the analyzed phenomenon (loss of stability in the internal structure) *is not observed for the homogeneous materials (within the framework of model representations).* It is characteristic only for the composites (as for the structurally inhomogeneous materials, in which the presence of an internal structure is taken into account at various levels of their analysis).

The model of "infinitely long fiber" obviously applies to the relatively long fibers (reinforcing elements). A priori, it can be expected that in the case of relatively short reinforcing elements (fibers), the modes of loss of stability in the internal structure of the composite may differ significantly from the modes of loss of stability shown in Figs. 3.4, 3.5, and 3.6 and the corresponding "infinitely long fiber" model.

As an example, consider the results of experimental studies published in the article [16] of 2004 and related to the stability of rather short carbon nanofibers in the polymer matrix. Figure 4.1, corresponding to publication [16], shows the modes of

Fig. 4.1

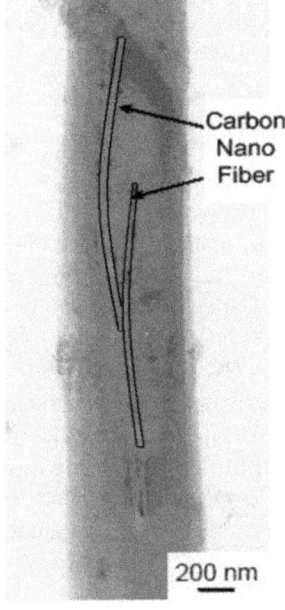

loss of stability of two short carbon nanofibers. The scale of the image in nanometers is shown in the lower right corner of Fig. 4.1.

The modes of loss of stability in Fig. 4.1 have nothing to do with the modes of loss of stability shown in Figs. 3.4, 3.5, and 3.6. So, in Figs. 3.4, 3.5, and 3.6, the modes of loss of stability are periodic sinusoidal (along the fibers) modes with a sufficiently large number of periods. As shown in Fig. 4.1, modes of loss of stability of the short nanofibers can be approximately represented by the sinusoidal modes with *one half-period*. In this case, the values of the critical values of loads and shortenings significantly depend on the boundary conditions at the ends of the reinforcing elements (fibers).

> Thus, it can be considered that the above information related to Fig. 4.1, is the experimental justification of the "fibers of finite dimensions" model.

Taking into account the above information and considerations related to Figs. 3.4, 3.5, and 3.6 and 4.1, it can be considered that the model of "infinitely long fibers" and the model of "fibers of finite dimensions" have the experimental justification, but they are applicable to different types of composites.

> Thus, the model of "infinitely long fiber" is applicable to the composites with relatively long reinforcing elements (fibers).
>
> The model of "fibers of finite dimensions" is applicable to the composites with sufficiently short reinforcing elements (fibers).

Note that these considerations following from the experimental results in Figs. 3.4, 3.5, and 3.6 and 4.1 refer only to the study of the loss of stability phenomenon in the internal structure of the composite.

4.2 Statement of Problems

First of all, note that this chapter deals with the composites formed by the short fibers (as reinforcing elements and filler), which are in the matrix (binder). It is assumed that *these composites have axes of symmetry of material properties.*

It should be noted that it is extremely difficult to create structural elements from materials that do not have axes of symmetry of material properties. In this regard, the above assumption seems to be quite appropriate. The presence of axes of symmetry of the properties of the composite is ensured by the method of stacking the reinforcing elements (in this case, short fibers).

Under compression along the axes of symmetry of the properties of the composite following the **General Concept** of internal fracture (Sect. 3.1.1) and the **General Concept** of near-the-surface fracture (Sect. 3.1.3.2),

> the start (beginning) of fracture in the cases under consideration is determined by the appearance of loss of stability in the internal structure of the composite (internal instability) or in the near-the-surface layers of the internal structure of the composite (near-the-surface instability).

Thus, with this kind of compression, the construction of fracture mechanics is carried out by the construction of the theory of internal and near-the-surface loss of stability of the considered composites.

This chapter summarizes the main results obtained by the author and his pupils on the construction of fracture mechanics (in the above sense) under compression of the composites reinforced with the short fibers (as a filler). These results were obtained within the framework of the model of a piece-wise homogeneous medium (body and material) and TLTDBS, the foundations of which are presented in Chap. 2 of this monograph. As has already been noted many times, this approach (the model of a piece-wise homogeneous medium with the involvement of TLTDBS) is the most consistent and rigorous in the mechanics of deformable bodies.

In Fig. 4.2a–f, the basic simplest design schemes are shown related to **Problem 2** (*model of short fibers in the theory of stability and the mechanics of fracture of composites under compression*), when "at infinity" (at $x_1 = \pm\infty$) the composite is compressed by forces of intensity $P = \text{const}$.

1.   One short fiber (Fig. 4.2a)—for the composites with the low filler concentration, when adjacent fibers do not interact with each other either in the subcritical state or at the loss of stability. In this case, the single fiber in an infinite (in coordinates x_1, x_2, x_3) matrix is considered and the modes of loss of stability that decay with distance from the fiber (at $x_1, x_2, x_3 \to \pm\infty$) are investigated.

2.   Two short fibers located on the same line (Fig. 4.2b)—for the composites with a low filler concentration, when, due to the irregularity of the structure, two adjacent fibers (Fig. 4.2b) can interact in the subcritical state and at the loss of stability. In this case, two fibers are considered (Fig. 4.2b) in an infinite (in coordinates x_1, x_2, x_3) matrix, and the modes of loss of stability that decay with distance from the fiber (at $x_1, x_2, x_3 \to \pm\infty$) are investigated.

3.   Two short fibers located in parallel (Fig. 4.2c)—for the composites with a low filler concentration, when, due to the irregularity of the structure, two adjacent fibers can interact (Fig. 4.2c) in the subcritical state and at the loss of stability. In this case, two fibers are considered (Fig. 4.2c) in an infinite (in coordinates x_1, x_2, x_3) matrix and the modes of loss of stability are investigated, which decay with distance from the fiber (at $x_1, x_2, x_3 \to \pm\infty$).

4.   One periodic row of the short fibers located on the same line (Fig. 4.2d)—for the composite of the periodic structure, when in the subcritical state and at the loss of stability, two adjacent fibers within one periodic row interact with each other, and adjacent rows of fibers do not interact with each other (small distances between adjacent fibers in one row and very large distances between adjacent rows). In this case, the periodic row of fibers (Fig. 4.2d) in an infinite (in coordinates x_2 and x_3) matrix is considered under the conditions of periodicity along the coordinate and the modes of loss of stability are investigated, which decay with distance from the row of fibers (at x_2 and $x_3 \to \pm\infty$) and which are periodic along the coordinate x_1.

5.   One periodic in the direction of the axis $0x_2$ row of the short fibers arranged in parallel (Fig. 4.2e)—for the composite of a periodic structure, when in the

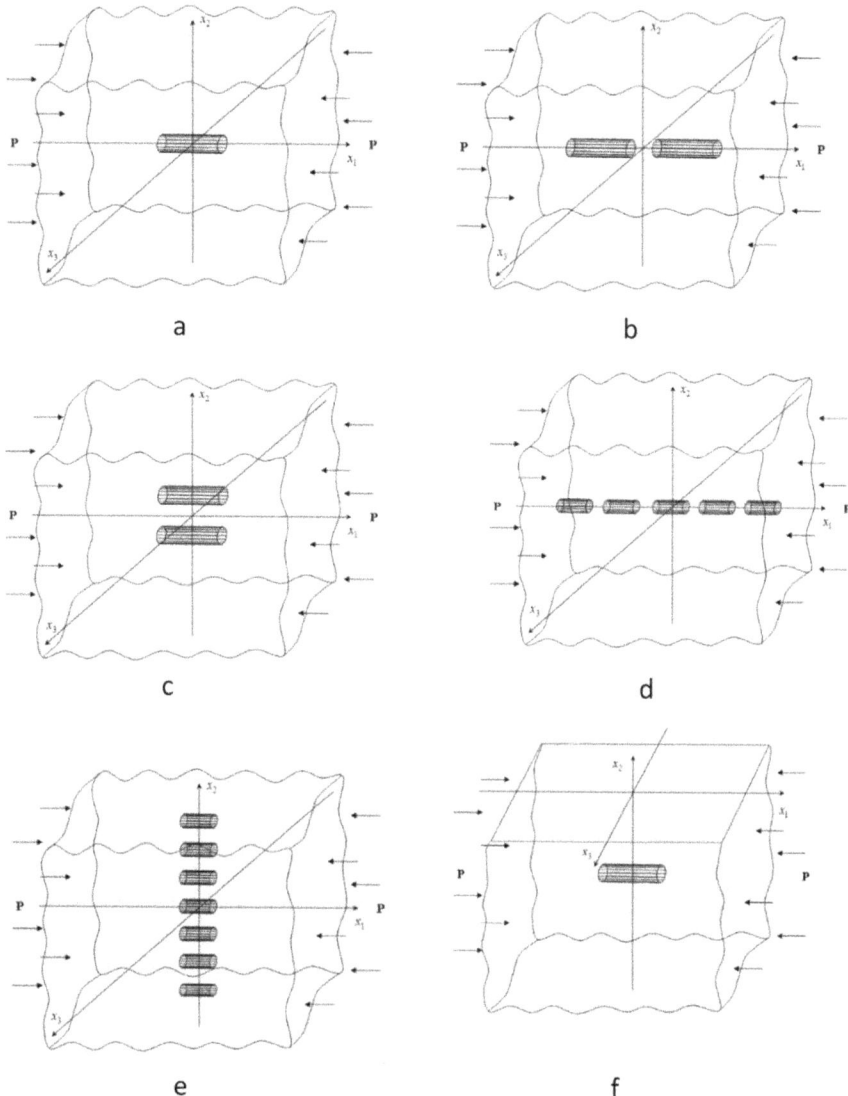

Fig. 4.2

subcritical state and at the loss of stability, two adjacent fibers within one periodic row interact with each other, and the adjacent rows of fibers do not interact with each other (small distances between adjacent fibers within one periodic row and very large distances between adjacent rows). In this case, a periodic (along the coordinate) row of fibers (Fig. 4.2e) in an infinite (along with the coordinates x_1 and x_3) matrix is considered under the conditions of periodicity along the coordinate, and the modes of the loss of stability are investigated, which decay with distance from the row of fibers (for x_1 and $x_3 \rightarrow \pm\infty$) and which are periodic along the coordinate x_2.

6. One short fiber near the free surface (Fig. 4.2f)—for the composites with a low filler concentration, when adjacent fibers do not interact with each other either in the subcritical state or at the loss of stability. Due to the irregularity of the structure, the fibers exist located near the free surface $x_2 = 0$ and interacting with the free surface both in the subcritical state and at the loss of stability. In this case, one fiber is considered in a semi-infinite matrix (in the lower half-space ($x_2 \leq 0$)), which interacts with the boundary of the lower half-space (at $x_2 = 0$ certain boundary conditions are satisfied in the subcritical state and the corresponding homogeneous boundary conditions at the loss of stability). At that, the modes of loss of stability are analyzed, which decay with distance from the boundary of the half-space and the fiber (for $x_2 \rightarrow -\infty$) and with distance only from the fiber (for x_1 and $x_3 \rightarrow \pm\infty$).

It should be noted that the sixth design scheme (Fig. 4.2f) is *the simplest* one for studying near-the-surface fracture. Undoubtedly, the number of the simplest design schemes in the study of near-the-surface fracture can be significantly expanded if the design schemes presented in Fig. 4.2b–d provide for the introduction of the half-space boundary.

The above design schemes (Fig. 4.2a–f), when using the TLTDBS mathematical apparatus described in Chap. 2 of this monograph, are the simplest ones for the fracture mechanics of composites reinforced with short fibers as a filler under compression along the fibers (concerning the internal and near-the-surface fracture) in the case of research in the framework of the model of a piece-wise homogeneous medium (material and body).

The problems arising, in this case, are *purely three-dimensional (spatial) problems of* TLTDBS. As already noted in Sect. 3.3.1, under the action of external loads in the form of "dead" loads, which is generally accepted in the mechanics of composites, *the fulfillment of the sufficient conditions for the applicability of the statical method of studying the stability (Sect. 2.4.2is the first result) is strictly proved.*

Thus, the considered problems are reduced *to the three-dimensional statical eigenvalue problems; i.e., Euler's method is applied.*

Thus, these investigations within the framework of **the second direction or approach** (Sect. 3.1.3) fully correspond to the generally accepted and rigorous method for studying the phenomenon of loss of stability—the analysis of the behavior of small perturbations within the framework of linearized three-dimensional *dynamical* problems.

As noted above, the formulated problems of the fracture mechanics of composites reinforced with the short fibers under compression along the fibers, the design schemes for which are presented in Fig. 4.2a–f, are *the statical three-dimensional (spatial) problems of TLTDBS.* For composites, there are *also two-dimensional static problems of TLTDBS,* for which the design schemes are the two-dimensional analog of the three-dimensional statical problems, and the design schemes shown in

Fig. 4.2a–f are the two-dimensional problems (plane deformation) of composites of the strip structure.

Strip structure composites.
Usually, the composites of the strip structure are understood as the composites in which the thin strips of various shapes are the filler.

In this chapter, the composites are considered in which the filler is rather long plane strips arranged in parallel and having a constant cross-sectional shape along the entire length.

When constructing the design schemes, the reinforcing strips will be considered infinite in the direction of the axis $0x_3$ ($-\infty \leq x_3 \leq +\infty$, the axis $0x_3$ is directed perpendicular to the plane of the figure). In this case, the two-dimensional problems (plane deformation) in the plane $x_1 0 x_2$ can be considered and the cross section of the strip will be called the reinforcing element, which is shown in Fig. 4.3a–e in the dark color.

In Fig. 4.3a–e, the six design schemes are shown in the plane $x_1 0 x_2$, visually corresponding to Fig. 4.2a–e, where the reinforcing elements (cross sections of strips) are represented by the dark rectangles.

Consider briefly the description of six design schemes indicated in Fig. 4.3a–e, taking into account the analogy in the description of the six design schemes, presented three-dimensional (spatial) problems of TLTDVS in Fig. 4.2a–e.

1. One reinforcing element (Fig. 4.3a)—for the composites with a low filler concentration, when the adjacent reinforcing elements do not interact with each other either in the subcritical state or in case of loss of stability. In this case, one reinforcing element is considered in the infinite (in coordinates x_1 and x_2) matrix and the modes of loss of stability are investigated, which decay with the distance from the reinforcing element (at $x_1, x_2 \rightarrow \pm\infty$).
2. Two reinforcing elements located on the same line (Fig. 4.3b)—for the composites with a low filler concentration, when, due to the irregularity of the structure, two adjacent reinforcing elements (Fig. 4.3b), located on the same line, can interact in a subcritical state and with the loss of stability. In this case, two adjacent reinforcing elements located on the same line (Fig. 4.3b) are considered in an infinite (in coordinates x_1 and x_2) matrix, and the modes of loss of stability are investigated, which decay with the distance from two adjacent reinforcing elements (at $x_1, x_2 \rightarrow \pm\infty$).
3. Two reinforcing elements located in parallel (Fig. 4.3c)—for the composites with a low filler concentration, when, due to the irregularity of the structure, two adjacent reinforcing elements can interact (Fig. 4.3c), located in parallel, in a subcritical state and at the loss of stability. In this case, two adjacent reinforcing elements are considered, located in parallel (Fig. 4.3c), in an infinite (in coordinates x_1 and x_2) matrix, and the modes of loss of stability are investigated, which decay with distance from two reinforcing elements (at $x_1, x_2 \rightarrow \pm\infty$).
4. One periodic row of reinforcing elements located on the same line (Fig. 4.3d)—for the composite of a periodic structure, when in the subcritical state and at

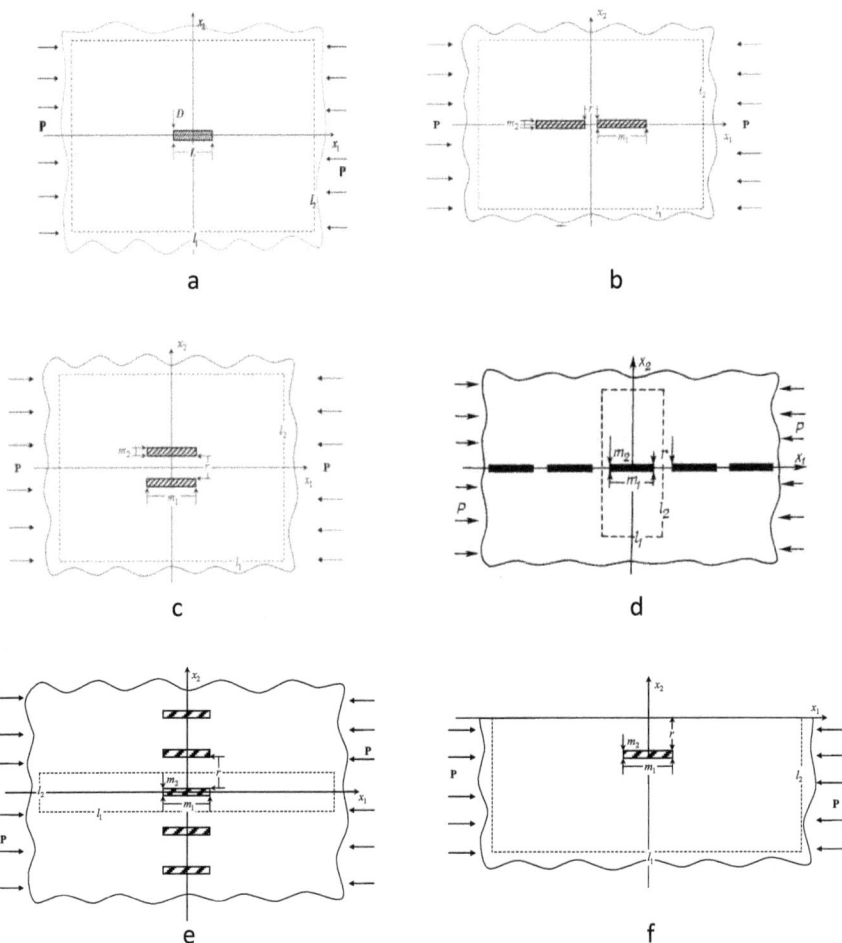

Fig. 4.3

the loss of stability, two adjacent reinforcing elements within one periodic row interact with each other, and the adjacent periodic rows of reinforcing elements do not interact with each other (the small distances between two adjacent reinforcing elements in one periodic row and the very large distances between the adjacent periodic rows of reinforcing elements). In this case, one periodic row of reinforcing elements is considered which is located on one line (Fig. 4.3d), in an infinite (along the coordinate x_2) matrix under the conditions of periodicity along the coordinate x_1, and investigates the modes of loss of stability, which decay with distance from the periodic row of reinforcing elements (at $x_2 \rightarrow \pm\infty$) and which are periodic along the coordinate x_1.

5. One row of reinforcing elements periodic in the direction of the axis $0x_2$, arranged in parallel (Fig. 4.3e)—for the composite of a periodic structure, when in the subcritical state and at the loss of stability, two adjacent reinforcing elements within one periodic row interact with each other, and the adjacent periodic rows of reinforcing elements do not interact with each other (the small distances between two adjacent reinforcing elements in one periodic row and the very large distances between adjacent periodic rows of reinforcing elements). In this case, one periodic (along the coordinate x_2) row of reinforcing elements (Fig. 4.3e), located in parallel, in an infinite (along the coordinate x_1) matrix, is considered under the conditions of periodicity along the coordinate x_2, and the modes of loss of stability are investigated, which decay with the distance from the row of reinforcing elements (at $x_1 \rightarrow \pm\infty$) and which are periodic along the coordinate x_2.

6. One reinforcing element near the free surface (Fig. 4.3f)—for the composites with a low concentration of reinforcing elements, when the adjacent reinforcing elements do not interact with each other either in the subcritical state or in case of loss of stability. In this case, due to the irregularity of the structure, the reinforcing elements exist, which are located near the free surface $x_2 = 0$ and interact with the free surface both in the subcritical state and at the loss of stability. In this case, one reinforcing element is considered in a semi-infinite matrix (in the lower half-plane ($x_2 \leq 0$)), which interacts with the boundary of the lower half-space (certain boundary conditions at the boundary at $x_2 = 0$ are satisfied in the subcritical state and the corresponding homogeneous boundary conditions at the loss of stability). At that, the modes of loss of stability are analyzed, which decay when moving away from the boundary of the half-plane and the reinforcing element (at $x_2 \rightarrow -\infty$) and when moving away only from the reinforcing element.

It should be noted that the sixth design scheme (Fig. 4.3f) *is the simplest* one in the study of near-the-surface fracture in the framework of the plane problem (plane deformation). Undoubtedly, the number of the simplest design schemes in the study of near-the-surface fracture in the framework of the plane problem (plane deformation) can be significantly expanded if the schemes presented in Fig. 4.3b–e provide for the introduction of the half-plane boundary.

The above design schemes (Fig. 4.3a–f) when using the TLTDBS mathematical apparatus, given in Chap. 2 of this monograph, are the simplest design schemes for the fracture mechanics of composites (with a filler in the form of the considered short reinforcing elements) under compression along the reinforcing elements (as applied to the internal and near-the-surface fractures) in the case of research in the framework of the model of a piece-wise homogeneous medium (material and body) for the plane problem (plane deformation).

The problems arising are in this case the *purely two-dimensional* problems of the TLTDBS.

As already noted in Sect. 3.3.1, under the action of external loads in the form of "dead" loads, which is generally accepted in the mechanics of composites, *the fulfillment of the sufficient conditions for the applicability of the statical method of studying the stability* (Sect. 2.4.2 is the first result) *is strictly proved*. The considered problems are reduced to *the two-dimensional statical eigenvalue problems; i.e., Euler's method is applied*.

Thus, these investigations in the framework of **the second direction or approach** (Sect. 3.1.3) fully correspond to the generally accepted and rigorous method of studying the phenomenon of loss of stability—the analysis of the behavior of small perturbations within the framework of linearized *dynamical* problems as applied to the plane problem (plane deformation).

It is advisable to note one more consideration related to the statement of the plane problems (plane deformation) within the framework of the second direction or approach (Sect. 3.1.3), the design schemes for which are shown in Fig. 4.3a–f. Within the framework of the model of the piece-wise homogeneous medium and **the second direction** in Sect. 3.3.3, the results for the layered composites are presented concerning the plane (Fig. 3.25) and spatial (Fig. 3.26) problems in the case of internal and near-the-surface fracture.

In this case, the layers of filler and binder were considered sufficiently long, which made it possible to consider them the infinite ones. If concerning the plane problem (Fig. 3.25), the filler layers are considered to be short (of finite dimensions), then we arrive at the various schemes presented in Fig. 4.3a–f.

Thus, for the plane problems in the plane $x_1 0 x_2$, the name "reinforcing element" (used in the description of the schemes in Fig. 4.3a–f to designate the cross section of the filler of the strip structure) and the name "short layer," which can be used to designate the filler in a layered composite (Fig. 3.35), coincides in meaning.

Consequently, it can assume that the design schemes in Fig. 4.3a–f refer to the plane problems for the layered composites with filler in the form of short layers, which are placed in a matrix (binder) and are oriented along the axis $0 x_1$ (Fig. 4.3a–f).

Thus, it can be assumed that this section sets out the statement of problems for the composites reinforced with the unidirectional short fibers and short layers under compression along the reinforcing elements using a piece-wise homogeneous medium model.

Moreover, Fig. 4.2a–f shows the design schemes for the composite with the short fibers, which correspond to *the three-dimensional (spatial)* problems, and Fig. 4.3a–f shows the design schemes for the composite with the short layers, which correspond to *the two-dimensional (plane)* problems concerning the plane deformation.

It is worth noting that Chaps. 2 and 3 discuss the results that relate to the sufficiently similar problems and that are obtained for the various design schemes, mainly within the framework of the model of piece-wise homogeneous media (materials and bodies). In this regard, it seems appropriate to classify the applied design schemes, which are the subject of the next paragraph.

4.3 Classification of Design Schemes. About Analogies

Consider briefly the classification of the design schemes that are used in Chaps. 2 and 3 of this monograph within the framework of *a piece-wise homogeneous medium model.*

Note that the model of the piece-wise homogeneous medium (material and body) consists in the fact that the relationships of *various* homogeneous deformable bodies are used to describe the deformation of the matrix and separately each element of the filler. At the same time, at the interfaces between the filler and the matrix, certain (different depending on the statement of problems) conditions of continuity of stress vectors and displacements are met.

Besides, we note that in the model of a piece-wise homogeneous medium (in a broad sense), to describe the deformation of the matrix and each element of the filler, *it is not necessary* to use the three-dimensional strict relations of the classical models of the mechanics of deformable bodies. It is possible to use the simplified or refined above relations with the corresponding design schemes, while the conditions at the interface are also formulated within the framework of the accepted design schemes.

Taking into account the above information of an introductory nature, three types of calculation schemes of a general nature are distinguished in this section that are used in the construction of the fracture mechanics of the unidirectional fibrous and layered composites under compression and which are presented below as three separate subsections.

4.3.1 Model of Infinitely Long Fibers and Layers in the First Direction of Research

The sufficiently long fibers and layers are considered (as a filler, while in the case of layered composites, naturally, enough long matrix layers are also considered). In this case, in the design schemes, the fibers and layers are considered to be infinitely long, and periodic (sinusoidal) modes of loss of stability along the fibers and layers are investigated.

In this situation, it is obvious that in the case of the fibrous composites, the studies must be carried out in *the three-dimensional (spatial) statement,* and in the case of the layered composites, one can limit ourselves to the study in *the two-dimensional statement* (within the framework of the plane deformation).

With the application of the approach discussed in this subsection, the studies are carried out in the framework of **the first direction (very approximate approaches)**, a brief description of which, with indicating the characteristic *approximate assumptions,* is presented in Sect. 3.1.2.

It should be noted that the above approximate assumptions are introduced into the basic relationships for the filler and matrix, as well as into the conditions at the interface.

The above-mentioned characteristics of the features of **the first direction** are clearly manifested when considering the well-known and generally accepted publications [17] and [18], the results of which were included in the fundamental collective publications on fracture [19] and composite materials [20].

The brief information on the sequential analysis and conclusions from this analysis from the publications [17] and [18] are presented in Sect. 3.3.3.3.3 of this monograph.

In fact, the studies in [17] and [18] are carried out within the framework of the plane problem in the plane x0y (Fig. 3.33 corresponds to Fig. 3.22 in [17] and [18]) and the reinforcing elements are called the fibers.

Actually, in Fig. 3.33 (Fig. 3.22) the reinforcing elements (in the framework of the plane problem) are the strips corresponding to the cross-section of the layered composites.

Thus, in [17] and [18], the plane problem for the layered composite is actually considered and the obtained concrete results (Figs. 3.33, 3.23, and 3.24) also refer to the plane problem. When discussing the results, the terminology corresponding to *the fibrous composite* is used, although for the fibrous composite *it is necessary* to carry out the studies within the framework of *the three-dimensional* (spatial) statement, corresponding, for example, to Fig. 4.2a–f.

Taking into account the above, it can be assumed that in [17] and [18] **the approximate approach** to the study of the *fibrous* composites was proposed, which consists of the following:

1. The research is carried out within the framework of *the plane problem for layered* composites.
2. The obtained results of a quantitative nature for *the layered* composites are used to analyze the phenomena that occur in *the fibrous* composites.
3. The terminology typical for *the fibrous* composites is used, calling the reinforcing elements *the fibers* in the framework of the plane problem (in the framework of the plane problem, the reinforcing elements are the *stripes*).

It should be noted also that in the above **approximate approach**, the question of the existence of an *analogy* between the *layered* and *fibrous* composites during their study is not stated or commented on. Also, the results [17] and [18] in the framework of the plane problem for the layered composites obtained using **the first approach** (Sect. 3.1.2) are sufficiently approximate. As already noted, the accuracy of the results discussed is considered in Sect. 3.3.3.3.3 of this monograph.

4.3.2 Model of Infinitely Long Fibers and Layers in the Second Direction of Research

As in Sect. 4.3.1, the sufficiently long fibers and layers are considered (as a filler, while in the case of layered composites, enough long matrix layers are naturally also considered). In this case, in the design schemes, the fibers and layers are considered infinitely long and the periodic (sinusoidal) modes of loss of stability along the fibers and layers are investigated, which corresponds to the results of experimental studies (Figs. 3.4, 3.5, and 3.6).

In the above situation, as already noted in Sect. 4.3.1, it is obvious that in the case of the fibrous composites, the studies must be carried out in a *three-dimensional (spatial)* statement, and in the case of the layered composites, one can limit ourselves to the studies in a *two-dimensional* statement (within the framework of the plane deformation).

Using the model discussed in this subsection, the research is carried out in the framework of **the second direction (the strict sequential approaches based on TLTDBS)**, a brief description of which, with indicating the characteristic points, is presented in Sect. 3.1.3.

The information on the results obtained within the framework of the model of a piece-wise homogeneous medium (material and body) and TLTDBS for the elastic and plastic bodies with the constitutive relations of the sufficiently general form is presented for the *fibrous* composites in a *three-dimensional (spatial)* statement in Sect. 3.3.4 of this monograph.

The most complete presentation of the discussed results is placed in this chapter and Chap. 6 of the monograph [21] (vol. 1). The information about the results obtained within the framework of the model of a piece-wise homogeneous medium (material and body) and TLTDBS for the elastic and plastic bodies with the constitutive relations of a sufficiently general form is presented for the *layered* composites in a *three-dimensional (spatial) statement* (Fig. 3.26) and in a particular case (for the plane deformation, Fig. 3.25) in a *two-dimensional statement* in Sect. 3.3.3 of this monograph. The most complete presentation of the discussed results is placed in Chaps. 3 and 5 of the monograph [21] (vol. 1).

It follows from the above information about the discussed results for the fibrous and layered composites obtained using the model of the infinitely long fibers and layers, as well as the model of a piece-wise homogeneous medium in the framework of **the second direction** (Sect. 3.1.3) that in the situation under consideration there *is no need* to set the question of *the existence of analogies between the fibrous and layered composites* in their study.

The basis for the above conclusion is that all the results discussed in Sect. 4.3.2 were obtained in a *three-dimensional (spatial)* statement for the fibrous and layered composites with a polymeric and metallic matrix and with the involvement of TLTDBS, which forms the basis of **the second direction** (Sect. 3.1.3). For only the plane problem (the plane deformation) applied to the layered composites (Fig. 3.25), the studies were carried out in a two-dimensional formulation, which in this case *strictly follows from a three-dimensional (spatial)* statement.

As already noted in Sect. 3.3.1, the studies in the framework of **the second direction** (the second approach) fully correspond to the generally accepted and rigorous method for studying the phenomenon of loss of stability—the analysis of the behavior of small perturbations in the framework of linearized three-dimensional dynamical problems. Besides, the model of a piece-wise homogeneous medium and the involvement of the TLTDBS apparatus, which is characteristic of **the second direction**, constitute the most rigorous approach in the mechanics of fracture of composites under compression, which can be represented within the framework of the mechanics of a deformable body.

The above considerations apply also to all the results discussed in Sect. 4.3.2.

4.3.3 Model of Short Fibers and Layers in the Framework of the Second Direction of Research

In contrast to Sects. 4.3.1 and 4.3.2, this subsection considers the composites with sufficiently short reinforcing elements when the model of infinitely long fibers and layers is not applicable for the situation under discussion. In this case, the modes of loss of stability, as follows from the experimental studies (Fig. 4.2), have a different character—*they are not periodic along the axis of the reinforcing elements* under compression also along the reinforcing elements.

Note 4.1 A characteristic feature of the problems under study in the framework of the model in this section is an *inhomogeneous subcritical state* (*three-dimensional*, in the general case of the fibrous and layered composites, and *two-dimensional*, in the particular case of the plane problems for the layered composites).

Within the framework of the model of this subsection, the problems for the fibrous and layered composites are investigated.

In the case of the fibrous composites, an infinite matrix reinforced with the short fibers is considered. The simplest design schemes, in this case, are shown in Fig. 4.2a–f for studying the internal fracture (the internal instability) and the near-the-surface fracture (the near-the-surface instability).

These problems are reduced to the three-dimensional statical problems of the TLTDBS (due to the study in **the second direction**) for the eigenvalues. The specificity of the discussed problems is the dependence of the coefficients of the corresponding systems of differential equations on *three spatial variables*.

Note 4.2 Because of the above information, the specific results for the fibrous composites reinforced with the short fibers for the design schemes in Fig. 4.2a–f are possible only with *the use of numerical methods and computational mechanics*. Currently, the specific results for the design schemes in Fig. 4.2a–f **have not yet been obtained** either for the first direction, a brief description of which is presented in Sect. 3.1.2, or for **the second direction**, a brief description of which is presented in Sect. 3.1.3.

In the case of the layered composites, it is possible to consider the two-dimensional problems corresponding to the plane deformation in the plane in Fig. 4.3a–f for the composites of a strip structure. In this case, the filler is rather long plane strips, which are parallel along the axis $0x_3$ (the axis $0x_3$ is directed perpendicular to the plane of Fig. 4.3a–f) and have a constant cross-sectional shape along the entire length.

In this situation, the two-dimensional problems in the cross-sectional plane are considered for an infinite two-dimensional matrix reinforced with the short reinforcing elements corresponding to the cross sections of the strips. The simplest design schemes in this case for studying the internal fracture (the internal instability) and near-the-surface fracture (the near-the-surface instability) are shown in Fig. 4.3a–f. These problems are reduced to the two-dimensional statical problems of the TLTDBS (due to the research carried out in the framework of **the second direction**) for the eigenvalues. The specificity of the discussed problems is the dependence of the coefficients of the corresponding systems of differential equations on *two spatial variables*.

Note 4.3 Due to the above information, the specific results for the layered composites with the short reinforcing elements in the framework of the plane problem (the plane deformation) concerning the design schemes presented in Fig. 4.3a–f *can be obtained only with the involvement of the numerical methods and computational mechanics*.

At present, the specific results for the simplest design schemes presented in Fig. 4.3a-f, **have already been obtained** for the plane problem (the plane deformation) **in the framework of the second direction** (the model of a piece-wise homogeneous medium and the mathematical apparatus of TLTDBS). These results are presented in the publications (the monograph, the review article, and selected articles) that are indicated in the introductory part of this chapter.

Note 4.4 The above results were obtained for a plane problem (the design schemes in Fig. 4.3a–f). Nevertheless, taking into account **the approximate approach** to the study of fibrous composites, indicated in the final part of Sect. 4.3.1 and actually proposed in [17] and [18], the results obtained for the plane problem [1, 4–17] within the framework of the statement and model of this subsection can be used to analyze the phenomena that occur in the fibrous composites (within the framework of spatial problems for the design schemes in Fig. 4.2a–f).

In connection with the above situation, the discussed publications related to the plane problem (the plane deformation) use the terminology typical of the fibrous composites. In this case, the short fibers are called the reinforcing elements in the framework of the plane problem, presented in Fig. 4.3a–f, although in reality in Fig. 4.3a–f the reinforcing elements correspond to the cross sections of the filler in the form of strips in the composites of the strip structure.

Note 4.4 can be related to the statement of the plane problems of brittle fracture mechanics for the composites with the short reinforcing elements under compression, which is set out in the next subsection of this chapter.

4.4   Statement of Plane Problems of Brittle Fracture Mechanics of Composites with Short Reinforcing Elements Under Compression

In the previous section of this chapter, quite a lot of attention was paid to the various aspects of the staging nature related to the model of short fibers (in the general sense, the short reinforcing elements) in the theory of stability and in the mechanics of fracture of composites under compression along the above short reinforcing elements, which are the filler in the composite. In the monograph [1] and the review article [2], the discussed model was called the "finite-size fibers" model. Apparently, some equivalent names can be proposed for the discussed model.

4.4.1   On Statement of Problems

Taking into account the above information, below in this subsection the main positions will be briefly summarized related to the statement of the plane problems and methods of their study under the brittle fracture using the simplest design schemes presented in Fig. 4.3a–f. The main results related to the statement of problems, methods of solution, and specific information about the investigated problems are presented in publications [1–15] and some others.

Apparently, it can be considered that these results are *the first results in the world* on a sufficiently rigorous study for the composites with *the short reinforcing elements* as applied to the theory of stability and brittle fracture under compression.

The main assumptions of the discussed approach, which determine the statement of problems and, naturally, the limits of its applicability, can be defined as follows.

1.   The brittle fracture of the considered composites under compression along the reinforcing elements is analyzed. Therefore, the materials of the reinforcing elements (following Note 4.4 hereinafter will be called the fibers) and matrices are modeled by the linear elastic isotropic bodies. It is expedient to note that the above modeling for the brittle fracture of composites can be considered acceptable under the relatively short-term action of external loads and at moderate temperatures.

2.   The research is carried out within the framework of **theory 3** (*the second variant of the theory of small subcritical deformations*) according to the terminology of Sect. 2.2 of this monograph. It is assumed in this theory that the subcritical state is determined by the geometrically linear theory. It is expedient to note that the discussed assumption can be considered acceptable for the relatively rigid fibrous composites, which are fractured mainly at the relatively small deformations.

3.   The loading by the external "dead" loads is considered. In this case, as has been repeatedly noted in Chaps. 3 and 4 of this monograph, the sufficient conditions

for the applicability of the statical TLTDBS method (Euler's method) are satisfied and these problems are reduced to the statical two-dimensional eigenvalue
problems. At that, the coefficients of the corresponding system of differential
equations depend on two variables (on x_1 and x_2 following Fig. 4.3a–f).

4.   At the interfaces between the filler and the matrix (Fig. 4.3a–f), the conditions of
     continuity of stress and displacement vectors are accepted both in determining
     the subcritical state and in studying the corresponding stability problems.

5.   For the design schemes presented in Fig. 4.3a–f, concerning the study of internal
     and near-the-surface fracture, the corresponding conditions of attenuation "at
     infinity" are taken. For the periodic system of fibers, the corresponding condi-
     tions of periodicity are also taken. These conditions are considered in sufficient
     detail in Sect. 4.2 concerning each of the simplest design schemes presented in
     Fig. 4.3a–f.

6.   The study is carried out within the framework of the plane problem (the
     plane deformation) with the involvement of the above version of the TLTDBS
     in Lagrangian coordinates $(x_1, \quad x_2)$, which in the reference state (the first,
     undeformed) coincide with the Cartesian coordinates.

The above main assumptions of the discussed approach are given in accordance
with the monograph [1] and the review article [2].

4.4.2   On the Method of Numerical Study of Problems of Sect. 4.4

As already noted in Note 4.3, the considered two-dimensional statical eigenvalue
problems with the variable coefficients in differential equations depending on two
variables x_1 and x_2 (Fig. 4.3a–f) can be investigated only using the numerical methods
and computational mechanics. In this regard, below the brief information is given
on the numerical study of problems, the design schemes of which are presented in
Fig. 4.3a–f and the statement of which (the main assumptions of the approach) is
given above in Sect. 4.4.1.

At that, it is advisable to note that the above main assumptions of the approach set
forth at the beginning of Sect. 4.4.1 should be considered together with the discussion
of the design schemes in Fig. 4.3a–f, which is stated in Sect. 4.2 after Fig. 4.3a–f.
The discussed brief information following the review article [2] is presented here. A
more detailed information is presented in monograph [1].

The numerical solution of the formulated problems is performed by the finite
difference method using the variational-difference approach and using the basic
schemes. This general method is described in detail in the review article [15]
(concerning the broad classes of problems in the mechanics of composites).

Below, following [2], we will consider the main stages of the implementation
of the numerical method [15] concerning the simplest calculation scheme shown in

Fig. 4.3a. For this, instead of an infinite area (in the design scheme), the finite area is introduced, bounded by an outer rectangle with dimensions $l_1 \times l_2$.

Note that the same finite rectangle with dimensions $l_1 \times l_2$ is indicated in all the simplest design schemes of Fig. 4.3.

When determining the subcritical state and studying the corresponding TLTDBS problem, the attenuation conditions "at infinity" are replaced by the same attenuation conditions at the outer boundary of the rectangle. The dimensions of this rectangle $l_1 \times l_2$ are chosen such that their further increase *does not affect* the final results (the value of the critical load), which is determined as a result of a computational experiment. With the help of straight lines parallel to the axes $0x_1$ and $0x_2$, in the computational area (the rectangle with dimensions $l_1 \times l_2$), a difference grid $\bar\omega = \omega \cup \gamma$, *non-uniform in each direction*, is introduced, where ω is the set of internal nodes and γ is the set of boundary nodes.

Figure 4.4a shows the non-uniform grid and Fig. 4.4b—the cell. In this case, the grid is introduced so that the material (or fibers, or matrix) is homogeneous within each cell. Besides, it is assumed that the grid can be densified in the vicinity of a sharp change in the material properties (e.g., near the line of separation of the matrix and fiber). It is assumed that the densification of the grid near the lines of separation of the matrix and fiber can be carried out to such a level that the further densification of the grid *does not affect* the final results (the value of the critical load), which is determined as a result of a computational experiment. Thus, the grid area, which consists of a plurality of internal and boundary nodes, is a collection of rectangular cells. Each of the cells has the mechanical and geometric characteristics of a composite component (binder or filler) that is contained in this cell.

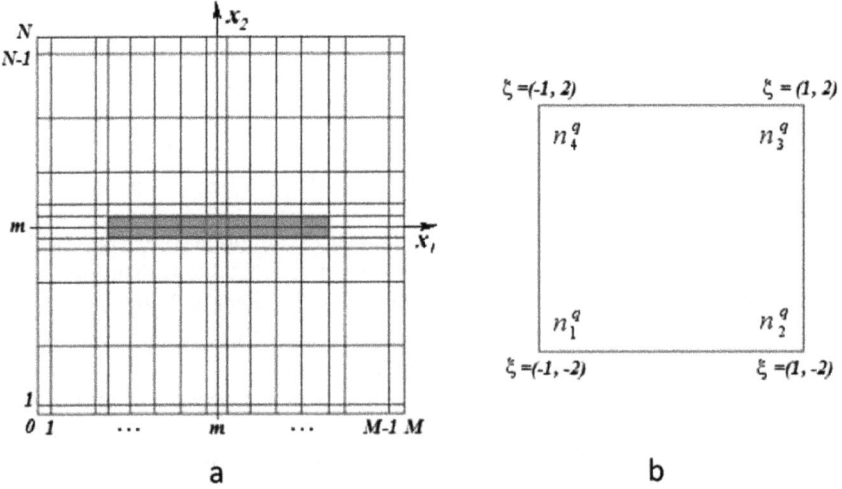

Fig. 4.4

The discrete problems on a grid $\bar{\omega}$ are built by a variational-difference method using the concept of basic schemes. The components of the basic schemes are determined by approximating and minimizing the corresponding functional on the grid cell template.

It should be noted that when implementing this procedure concerning the study of the stability problem, the variational principles of TLTDBS are used, which are summarized in Sect. 2.5 of this monograph.

By summing the values of the basic schemes at each node of the grid area, the difference problems are obtained, which are the discrete analogs of the corresponding continuum problems.

In the above way, the difference problems are formed in the operator form for determining the subcritical state (within the framework of the classical linear theory of elasticity) and for studying the stability problem (within the framework of the TLTDBS in the form of **theory 3** (*the second variant of the theory of small subcritical deformations) in the terminology of*   Sect. *2.2of this monograph*). At that, the difference operators of the corresponding stability problems preserve the properties of self-conjugacy and positive definiteness of the corresponding differential operators. Thus, the stability problem is also reduced to solving the corresponding grid equations, which can be represented as the generalized algebraic eigenvalue problem.

To obtain the numerical results for solving the algebraic problems, the direct and iterative methods known in the theory of difference schemes are used: the Cholesky method, the conjugate gradient method, the subspace iteration method, and the gradient descent method.

We restrict ourselves to the above very brief information with the information on the numerical method for studying the problems, the statement of which is presented in Sects. 4.4.2 and 4.2 after Fig. 4.3a-f.

4.5   Results of Studies of Plane Problems of Brittle Fracture Mechanics of Composites with Short Fibers Under Compression

In this section, the information on specific results on the study of plane problems of the mechanics of brittle fracture of composites reinforced with the short fibers under compression is presented in a very brief form.

The discussed specific results were obtained within the framework of the statement of Sect. 4.4.1 with the involvement of the research method, brief information about which is presented in Sect. 4.4.2, concerning the simplest design schemes in Fig. 4.3a–f, corresponding to the different structures of the considered composites.

Note 4.5. In publications [1–15], the concept was introduced and used—the "mental" modes of stability loss. In this regard, it is advisable to clearly define this concept, since it will also be applied in Sect. 4.5 when analyzing the specific results obtained for the design schemes in Fig. 4.3a–f.

The *imaginary* general patterns of deformation of the fiber and the surrounding matrix during the loss of stability, obtained (in advance before the computational experiments) from the physical considerations, taking into account the relative stiffness and geometric characteristics of the fiber and the matrix, are called the *"mental" modes of loss of stability*.

Thus, the "mental" modes of loss of stability *do not always* correspond to the individual eigenfunctions of the problem under discussion but characterize the general assumed picture of deformation of the fiber and the matrix surrounding it. All the same, the "mental" forms of loss of stability in many cases are convenient when analyzing the specific results.

Example Let us consider the case when the fiber stiffness is much greater than the matrix stiffness. It can be assumed in this case that under the joint deformation (of fibers and matrices with the loss of stability), the nature of fiber deformation approaches the model of a rigid body. In this case, the only possible "mental" mode of loss of stability of the fiber and the surrounding matrix is the rotation of the fiber as a rigid body.

Note 4.5 Should be referred to the introductory part of Sect. 4.5.

4.5.1   Asymptotic Transition to the Model of "Infinitely Long Fibers"

Information on the problem under discussion will be given based on research concerning the design scheme of Fig. 4.3a for one fiber, where the designations are introduced:

L—fiber length; D—size in cross section.

This problem is considered in more detail in [9] and in the review article [2]. It is quite obvious that with a strict and consistent statement of problems in the case $LD^{-1} \to \infty$ the results for the design scheme in Fig. 4.5, corresponding to the "infinitely long fiber" model, have to follow from the results for the design scheme in Fig. 4.3a, corresponding to the "short fiber" model.

Note that Fig. 4.5 shows the upper and lower half-planes (matrix), which are connected through an endless strip (filler) of wide D. The strip is considered which, following Note 4.4, is approximated by the infinite fiber within the plane problem.

The design model in Fig. 4.5 corresponds to Fig. 3.25 (for $h_m \to \infty$ and $2h_a = D$) in Sect. 3.3.3 of this monograph, where the research results for the layered composites are considered using the model of "infinitely long fibers" (in the case under consideration of the layered composites, using the model of "infinitely long layers") in the framework of **the second direction** in the terminology of Sect. 3.1.3 (the model of piece-wise homogeneous medium and the sequential approaches based on TLTDBS).

Fig. 4.5

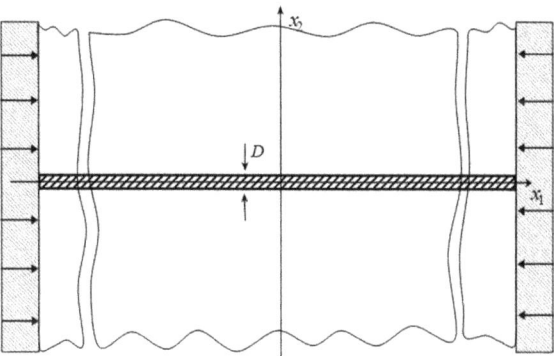

The results of the comparative analysis are presented as a dependence of the value $\left|\varepsilon_{11}^{\text{кра}}\right|$ on the geometric parameter LD^{-1}.

In the case of the "infinitely long fibers" model (Fig. 4.5), this value corresponds to the critical value of the deformation along the axis $0x_1$ for both the reinforcing elements (fiber and layer) and the matrix by **Note 3.3**.

In the case of the model of "fibers of finite size" or "short fibers," the value is introduced by the following expression

$$\varepsilon_{11}^{\text{cri}} = \varepsilon_{11}^{\text{cri}}(x_1, x_2) \text{ for } x_1 = 0 \text{ and } x_2 = 0 \qquad (4.1)$$

concerning Fig. 4.3a. Therefore, the quantity (4.1) corresponds to the critical value of the deformation along the axis $0x_1$ at the midpoint of the reinforcing element (Fig. 4.3a), which characterizes only the critical value of the fiber deformation and does not characterize the critical value of the matrix deformation. In this case, the critical value of the deformation along the axis $0x_1$ (Fig. 4.3a) for the matrix, which is determined "at infinity," can reach significantly different values. This situation should be taken into account when comparing the results obtained using the considered models.

Figures 4.6 and 4.7 present the results, which, respectively, refer to the micro- and nanocomposites with a polymer matrix with the following mechanical properties: $E_m = 2.68$ GPa, $v_m = 0.4$. At that, the distinction between the micro- and nanocomposites is established following the level scale given in monographs [22, 23].

For the microcomposites, the calculations were carried out at the following values of mechanical parameters ($E_a \cdot E_m^{-1} = 10; 30; 50; 100; 150$, E_a and E_m are Young's moduli for fibers and matrix) in the range of variation of the geometric parameter $10 \leq LD^{-1} \leq 1510$. In Fig. 4.6, the corresponding curves are marked with the numbers of the parameter values $E_a \cdot E_m^{-1}$.

For the nanocomposites, the calculations were carried out for the higher values of the parameter $E_a \cdot E_m^{-1} = 285; 373; 448; 500; 1000$ and in the wider range of variation of the parameter LD^{-1} ($10 \leq LD^{-1} \leq 2310$). In Fig. 4.7, the corresponding curves are marked with the numbers of the parameter values $E_a \cdot E_m^{-1}$. In Figs. 4.6

Fig. 4.6

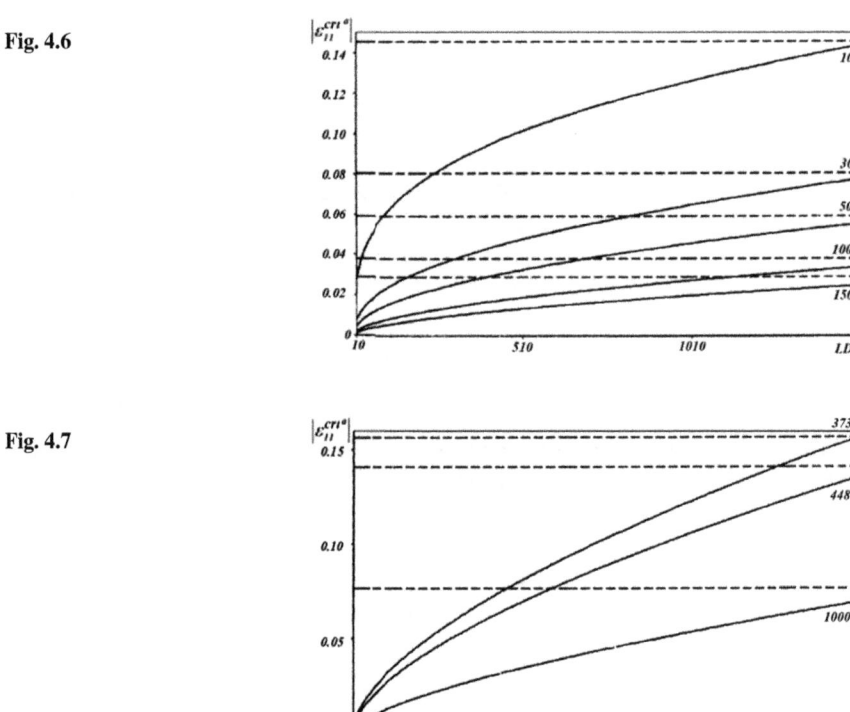

Fig. 4.7

and 4.7, the solid lines show results related to the "short fiber" model, the dashed lines show results related to the "infinitely long fiber" model.

It follows from the analysis of the results presented in Fig. 4.6 and 4.7 that for all considered values of the parameter $E_a \cdot E_m^{-1}$, with an increase in the geometric parameter LD^{-1} in the considered intervals, the critical values of deformation along the axis $0x_1$ (Fig. 4.3a), calculated in the framework of the "short fibers" model, asymptotically approach to the critical values calculated in the framework of the model "Infinitely long fibers".

Besides, for the upper values of the considered intervals of variation of the geometric parameter, the critical values of the above deformations practically coincide, which can be the basis for determining the limits of applicability of the model of "infinitely long fibers" depending on the values of the parameters $E_a \cdot E_m^{-1}$ and LD^{-1}. The results of this type are presented in [9], the review article [2], and the monograph [1].

The above results of this subsection are presented in a form corresponding to the review article [2].

4.5.2 Results for Single Fiber Under Compression Along Fibers

The design scheme is shown in Fig. 4.3a. The statement of the problem includes position **1** after Fig. 4.3a–f in Sects. 4.2, 4.3.3, and 4.4.1. The research was carried out by the method, brief information about which is presented in Sect. 4.4.2.

For the clearer discussion of these results, the specified design scheme (Fig. 4.3a) is shown in Fig. 4.8 together with the "mental" (in the terminology of Note 4.5) modes of loss of stability of one short fiber (Fig. 4.8) under compression along the fiber. The considered "mental" modes (a, b, c, and d) are presented in the lower part of Fig. 4.8.

It should be noted that in Fig. 4.8, as in other similar figures in Sect. 4.5, relating to the "mental" modes of loss of stability, the segments of "dark" lines depict the shapes of the middle lines of the form of fibers (reinforcing elements) after the loss of stability. In this case, it is assumed that the nearest neighborhood of the matrix surrounding the fibers is deformed under the loss of stability correspondingly due to position 4 (Sect. 4.4.1) of the statement of plane problems of the mechanics of brittle fracture of composites with short reinforcing elements under compression.

The aforementioned position 4 (Sect. 4.4.1) is that at the interfaces between the filler and the binder (matrix) (Fig. 4.3a–f), the conditions of continuity of stress and displacement vectors are accepted both in determining the subcritical state and in studying the corresponding problems of stability.

Fig. 4.8

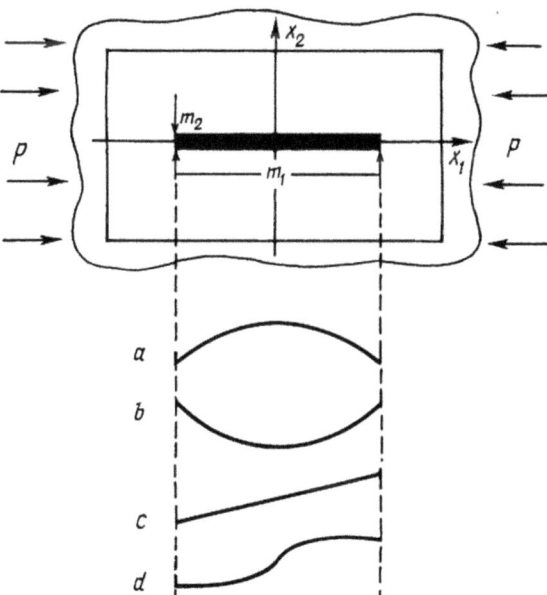

Fig. 4.9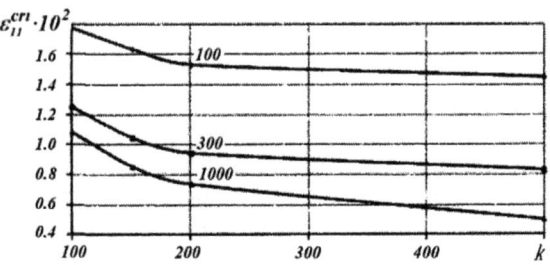

The "mental" modes of loss of stability (a and b in Fig. 4.8) can be called symmetrical about the vertical axis $0x_2$. It should be noted that modes a and b are equivalent (the critical loads coincide) due to the symmetry of the design scheme in Fig. 4.8 (top of the figure) relative to the horizontal axis $0x_1$. The symmetric modes of loss of stability can be called the flexural forms by analogy with the corresponding modes of loss of stability of a strip under axial compression.

The "mental" modes of loss of stability (c and d in Fig. 4.8) can be called antisymmetric modes of loss of stability relative to the vertical axis $0x_2$. The antisymmetric modes of loss of stability c correspond to a kind of "hard" rotation of the fiber (reinforcing element), when the material of the binder (matrix) does not provide a sufficient "supporting" effect, as a result of which, when stability is lost near the ends of the fiber as if the "hinge" is formed, close to the plastic hinge.

Obviously, the discussed mode of loss of stability can occur when the reinforcing elements and the matrix differ significantly in rigidity, which is typical for technological processes. The antisymmetric mode of loss of stability (d) seems to correspond to the rotation of the reinforcing element with a bend.

The obtained results of the numerical solution of the considered stability problem are presented in Fig. 4.9 as a dependence of the quantity $\left|\varepsilon_{11}^{\text{кр}}\right|$ which characterizes *the critical value of deformation along the axis* $0x_1$ (Fig. 4.8) for the matrix "at infinity," on the mode reinforcing element factor k (quantity)

$$k = m_1 \cdot m_2^{-1} = L \cdot D^{-1}. \tag{4.2}$$

The calculations were carried out for the following parameter values: $E_a \cdot E_m^{-1} = 100, 200, 300, 500, 1000$; $E_m = 2.76\,\text{GPa}$, $v_a = v_m = 0.35$; $100 \le k \le 500$.

Figure 4.9 shows the results only for three values of the parameter $E_a \cdot E_m^{-1} = 100; 300; 1000$, which is marked near to each curve.

It should be noted that in the considered calculations for the matrix the mechanical properties (quantities E_m and v_m) are used corresponding to the cast polyamide. For this material, the ultimate shortening corresponding to the ultimate strength is 0.028.

The value of critical deformation for the matrix "at infinity" (value of $\varepsilon_{11}^{\text{cri}}$), corresponding to the loss of stability in all curves of Fig. 4.9 is significantly less than the above value of the limiting shortening for the matrix. This situation indicates the possibility of fracture of the discussed composite under compression due to the loss of stability in the internal structure before reaching the ultimate strength for the matrix.

Fig. 4.10

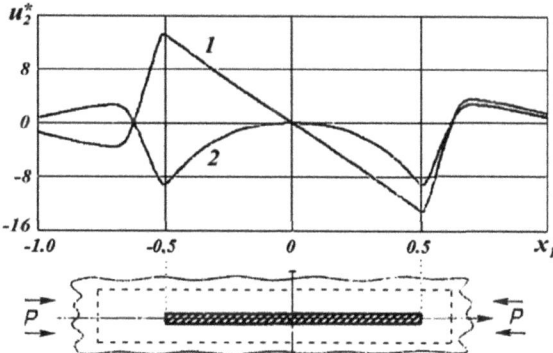

Consider the examples of determining the mode of loss of stability in the numerical study of this problem. The mode of loss of stability will be characterized by the dimensionless displacement u_2^* along the vertical axis $0x_2$ in Fig. 4.8, which is related to the amplitude factor, according to the following expression

$$u_2^*(x_1) = \left[u_2(x_1, x_2)|_{x_2=0} \right] \cdot \left[\left| \max\{ u_2(x_1, x_2)|_{x_2=0} \} \right| \right]^{-1}. \tag{4.3}$$

At the top of Fig. 4.10, the modes of loss of stability are shown, which are characterized by a function $u_2^*(x_1)$ in the form (4.3), for a composite with parameters $E_a \cdot E_m^{-1} = 1000$; $E_m = 2.76\,\text{GPa}$; $\nu_a = \nu_m = 0.35$. Here, *curve 1* refers to the case $k = 10$, and *curve 2* refers to the case $k = 30$, where k is determined by expression (4.2).

At the bottom of Fig. 4.10, the part of the design scheme in Fig. 4.8 is shown. At that, the linear dimensions on the upper and lower parts of Fig. 4.10 along the horizontal axis $0x_1$ coincide, so the fiber ends match at the top $x_1 = \pm 0.5$. Therefore, at the top of Fig. 4.10 the curves illustrating the mode of loss of stability are related to:

at $|x_1| \le 0.5$—to a short fiber; at $0.5 \le |x_1| \le 1.0$—to the matrix.

From the analysis of the results on the numerical determination of modes of loss of stability for this problem, which are presented in the upper part of Fig. 4.10, the following conclusions can be drawn (*concerning the part of curves 1 and 2 within the short line, at $|x_1| \le 0.5$*):

1. With the sufficiently short fibers (in this case at $k = 10$, where k is determined by expression (4.2)), the mode of loss of stability is realized (*the curve 1*), which corresponds to the "mental" antisymmetric form (Fig. 4.8—mode **c**). In this case, as if a "rigid" rotation of the reinforcing element occurs. A more detailed description of mode c is presented above in this subsection.
2. With longer fibers (in this case at $k = 30$, where k is determined by expression (4.2)). The indicated tendency is saved also at ≥ 30, the mode of loss of stability is realized (*the curve 2*), which corresponds to a "mental" symmetric or bending

mode (Fig. 4.8—mode **a**). A more detailed description of mode a is presented above in this subsection.

Thus, the above analysis and conclusions from it indicate that the "mental" modes of loss of stability discussed in Note 4.5 seem to be very useful in analyzing the mode of loss of stability obtained from the numerical study.

Additional information about the results obtained in the study of problems, the design scheme of which is presented in Fig. 4.3a, is presented in the monograph [1], the review article [2], some articles [3, 12, 14], and other publications.

4.5.3   Results for Two Sequentially Located Fibers Under Compression Along the Fibers

The design scheme is shown in Fig. 4.3b. The statement of problems includes position 2 after Fig. 4.3a–f in Sects. 4.2, 4.3.3, and 4.4.1. The research was carried out by the method, brief information about which is presented in Sect. 4.4.2.

For the clearer discussion of these results, the discussed design scheme (Fig. 4.3b) is shown in Fig. 4.11 together with the "mental" (in the terminology of Note 4.5) modes of loss of stability of two short fibers located in series (Fig. 4.11) under compression along the fibers. The considered "mental" modes (**a, b, c**, and **d**) are presented in the lower part of Fig. 4.11.

It should be noted that in Fig. 4.11, as in other similar figures in Sect. 4.5, relating to the "mental" modes of loss of stability, the segments of "dark" lines depict the shapes of the middle lines of the shape of fibers (reinforcing elements) after the loss of stability. In this case, it is meant that the nearest neighborhood of the matrix

Fig. 4.11

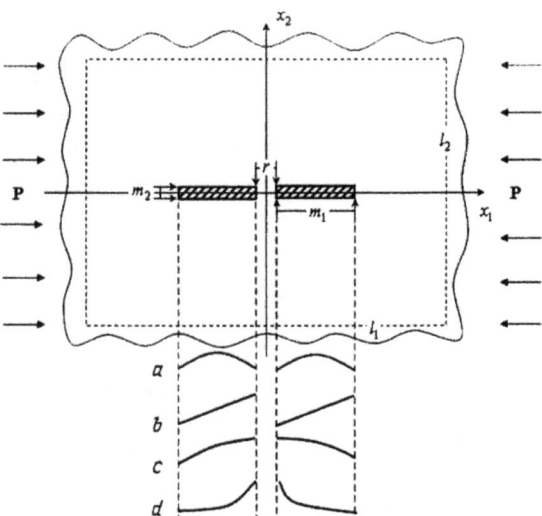

surrounding the fibers is deformed accordingly under the loss of stability due to position 4 (Sect. 4.4.1) of the statement of the plane problems of the brittle fracture mechanics for composites with the short reinforcing elements under compression. The aforementioned position 4 is that at the interfaces between the filler and the binder (matrix) (Fig. 4.3a–f), the conditions for the continuity of the stress and displacement vectors are accepted both in determining the subcritical state and in studying the corresponding problems of stability.

The "mental" modes of loss of stability a, b, c, and d shown in the lower part of Fig. 4.11 can be characterized as follows.

The flexural mode of loss of stability **a** corresponds to the case when each short fiber (reinforcing element) loses its stability as if almost without interaction with another short fiber (reinforcing element).

The mode of loss of stability **b** corresponds to the case when there is, as it were, a rigid rotation of each fiber independently of each other. Such a case can be realized for a sufficiently rigid material of the fibers, when the matrix does not provide the proper supporting action and when stability is lost the "hinge" is formed at the ends, as it were, close to the plastic hinge, and when the fibers interact little with each other.

The mode of loss of stability **c** corresponds to the case when two fibers lose their stability, as it were, in "one bending shape" (the common bending mode of loss of stability). Such a case can be realized, apparently, for the sufficiently flexible fibers with a very small distance between cracks.

The mode of loss of stability **d** corresponds to the case when two adjacent fibers lose stability, as it were, with a relatively rigid mutual rotation of adjacent fibers with some bending. Such a case is realized, apparently, for the sufficiently rigid material of the fibers, when the matrix between the fibers does not provide the proper supporting action and when stability is lost between the fibers, the "hinge" appears, as it were, close to the plastic hinge.

Construction of the "mental" modes of loss of stability for the design scheme presented in the upper part of Fig. 4.11 can be continued by considering the "mental" modes of loss of stability presented in the lower part of Fig. 4.11, as the first "mental" modes of stability loss.

In the numerical study, the dimensionless parameter $r_1^* = r \cdot m_1^{-1}$ is additionally introduced, which characterizes the dimensionless distance between the ends of two adjacent fibers (Fig. 4.3b). The geometric dimensions m_1, m_2 and r are shown in Fig. 4.3a.

The numerical studies were carried out for the composite with the following parameters: $E_a \cdot E_m^{-1} = 1000$; $v_a = v_m = 0.35$; $E_m = 2.76$ GPa; $k = m_1 \cdot m_2^{-1} = 100$; $0.001 \le r_1^* \le 32$.

It should be noted that the above value of the parameter $E_a \cdot E_m^{-1}$ can occur in technological processes, given the significant dependence E_m on temperature.

In Fig. 4.12a, b (a—at $r_1^* = 1$, b—at $r_1^* = 0.001$), the distribution along the axis $0x_1$ of the dimensionless displacement u_2^* along the vertical axis $0x_2$ is shown, which is determined by expression (4.3). These results are presented at the top of Fig. 4.12a, b.

It follows from the results presented in Fig. 4.12a (case $r_1^* = 1$) that the corresponding mode of loss of stability (the upper part of Fig. 4.12a) is close enough to

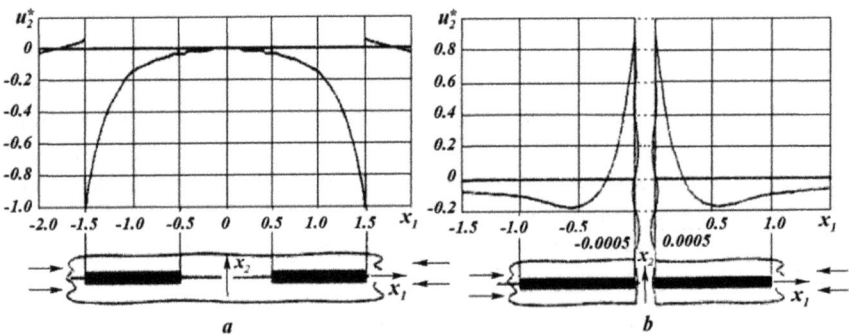

Fig. 4.12

the "mental" mode of loss of stability **c**, which is shown in the lower part of Fig. 4.11 and is called "the common flexural mode of loss of stability."

It follows from the results presented in Fig. 4.12 b (case $r_1^* = 0.001$) that the corresponding mode of loss of stability (upper part of Fig. 4.12b) is close enough to the "mental" mode of loss of stability d, which is shown in the lower part of Fig. 4.11 and is named as "the mutual rotation of adjacent fibers with the bending in the presence of as if the hinge between the ends."

It is advisable to note that at the bottom of Fig. 4.11 the "mental" modes of loss of stability a, b, c, and d, represented as the "dark" segments of curved lines, refer only to the fibers (the "dark" segments of curves in Fig. 4.11 are limited by the vertical dashed lines corresponding to the linear dimensions of the fibers). In this regard, a comparison with the lines representing the calculated modes of loss of stability in Fig. 4.12a, b can be carried out only within the linear dimensions corresponding to the fibers. So, in Fig. 4.12a the above comparison can be made at $0.5 \leq |x_1| \leq 1.5$, and in Fig. 4.12b—at $0.0005 \leq |x_1| \leq 1.0$.

The results of a numerical study of the effect of the interaction of two short fibers in their sequential placement (the design schemes in Figs. 4.3b and 4.11) on the critical deformation of the matrix "at infinity" (at $x_1 \to \pm\infty$), which, as in subsection 4.5.2, is denoted through $\varepsilon_{11}^{\text{kp}}$, are shown in Fig. 4.13 in the form of the dependence $\varepsilon_{11}^{\text{kp}}$

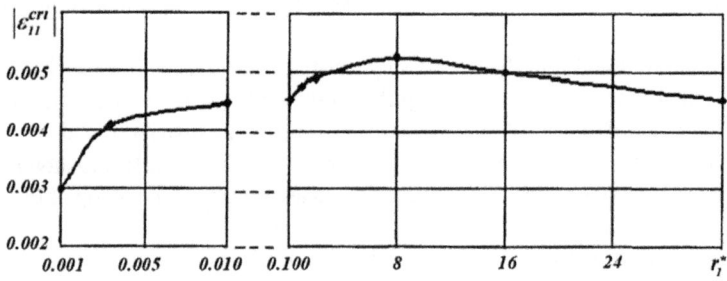

Fig. 4.13

on the quantity r_1^*—the dimensionless distance between the ends in the range of variation of r_1^* in the form of $0.001 \leq r_1^* \leq 32$.

For a more compact presentation of the discussed results in Fig. 4.13, the considered interval $(0.001 \leq r_1^* \leq 32)$ is divided into two parts $(0.001 \leq r_1^* \leq 0.01)$ and $(0.1 \leq r_1^* \leq 32)$, for which different scales on the axis $0r_1^*$ are taken. The convergence of the ends of the two short cylinders in Fig. 4.13 corresponds to the regularities presented in Fig. 4.13 and is observed when moving along the axis r_1^* from right to left from $r_1^* = 32$ to $r_1^* = 0.001$. Thus, in the interval from $r_1^* = 32$ to $r_1^* = 8$, a monotonic increase in the quantity $\left|\varepsilon_{11}^{\text{cri}}\right|$ occurs, and in the interval from $r_1^* = 8$ to $r_1^* = 0.001$, a monotonic decrease in the quantity $\left|\varepsilon_{11}^{\text{cri}}\right|$ occurs. Also, on the interval from $r_1^* = 32$ to $r_1^* = 0.005$, the mode of loss of stability corresponds to the "mental" common bending form c in Fig. 4.11, and on the interval from $r_1^* = 0.005$ to $r_1^* = 0.001$, the mode of loss of stability corresponds to the "mental" form "with mutual rotation and bending in the presence of a hinge between the ends of the fibers" (form d in Fig. 4.11).

Thus, a new mechanical effect is described—the non-monotonic change in the critical deformation when the reinforcing elements come together in the composite.

We restrict ourselves to the above information when discussing the results obtained for the successively arranged two short fibers under compression along the fibers. Additional information can be obtained from the monograph [1], some articles [13, 24], and other publications.

4.5.4 Results for Two Parallel Fibers Under Compression Along the Fibers

The design scheme is shown in Fig. 4.3c. The statement of problems includes position **3** after Fig. 4.3a–f in Sects. 4.2, 4.3.3, and 4.4.1. The researches were carried out by the method, brief information about which is presented in Sect. 4.4.2.

Below, as an example only one result will be given obtained by the indicated numerical method and related to the study of the question of the effect of the interaction of two short fibers with their parallel arrangement (the design scheme in Fig. 4.3c) on the critical deformation of the matrix "at infinity" (at $x_1 \rightarrow \pm\infty$), which, as in Sects. 4.5.2 and 4.5.3, is denoted by $\varepsilon_{11}^{\text{cri}}$.

As in paragraph 4.5.3, the research was carried out for a composite with the following values of the main parameters:

$$E_a \cdot E_m^{-1} = 1000; \; v_a = v_m = 0.35; \quad E_m = 2.76 \text{ GPa}; \, k = m_1 \cdot m_2^{-1} = 100.$$

It should be noted that the above value of the parameter $E_a \cdot E_m^{-1}$ can occur in the technological processes, given the significant dependence E_m on temperature.

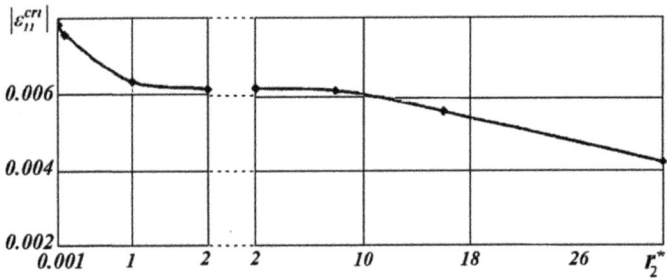

Fig. 4.14

The numerical studies of the dependence of the critical deformation value ε_{11}^{cri} on the distance between two parallel short fibers under compression along the fibers were carried out. In the design scheme in Fig. 4.3c, the distance between parallel fibers is characterized by r. Therefore, the dimensionless parameter $r_2^* = r \cdot m_1^{-1}$ was introduced.

The research results for the considered composite are shown in Fig. 4.14 in the form of the dependence of the quantity $\left|\varepsilon_{11}^{cri}\right|$ on the parameter r_2^* in the interval $0.001 \leq r_2^* \leq 32$.

As in Fig. 4.13, referring to the case of a sequential arrangement of two short fibers (design scheme in Fig. 4.3b), and in Fig. 4.14, referring to the case of parallel arrangement of two short fibers (design scheme in Fig. 4.3c) for a more compact presentation of the discussed results, the considered interval $(0.001 \leq r_2^* \leq 32)$ is divided into two parts $(0.001 \leq r_2^* \leq 0.01)$ and $(0.1 \leq r_2^* \leq 32)$, which are taken on different scales on the axis $0r_2^*$.

It should be noted that the closing in of two parallel short fibers following the design scheme in Fig. 4.3c occurs when the parameter r_2^* decreases, i.e., when moving along the axis $0r_2^*$ from right to left from $r_2^* = 32$ to $r_2^* = 0.001$.

It follows from the results shown in Fig. 4.14 that when the fibers closing in each other (when moving along the axis $0r_2^*$ from right to left), *the monotonic increase* occurs, which corresponds to generally accepted engineering considerations. The noted situation once again emphasizes that in the previous *subsection the new mechanical effect* was discovered:

a non-monotonic change in the value of the critical deformation when the reinforcing elements in the composite come together.

4.5.5 Results for One Periodic Row of Sequentially Located Fibers Under Compression Along the Fibers

The design scheme is shown in Fig. 4.3d. The statement of problems includes position 4 after Fig. 4.3a–f in Sects. 4.2, 4.3.3, and 4.4.1. The research was carried out by the method, brief information about which is presented in Sect. 4.4.2.

It should be emphasized that position 4 after Fig. 4.3a–f provides that for the design model in Fig. 4.3d the attenuation conditions are set "at infinity" only along the coordinate x_2 (at $x_2 \rightarrow \pm\infty$), and the periodicity conditions are set along the coordinate x_1. In this regard, the following situation arises when implementing the numerical research method. As already noted in Sect. 4.4.2, when implementing the numerical method the infinite area is replaced by a finite area in the form of a rectangle $l_1 \times l_2$, which in Fig. 4.3d is indicated by the dotted lines.

Taking into account the above, for the periodic series (Fig. 4.3d) in the indicated "dotted" rectangle the size l_1 is fixed from the conditions of periodicity and only the size l_2 changes to ensure the conditions of attenuation "at infinity" (at $x_2 \rightarrow \pm\infty$); i.e., the parameter l_2 value is determined by a computational experiment.

It should be noted that for a periodic structure (Fig. 4.3d) with a period $T = m_1 + r$, the periodic (along the axis $0x_1$) modes of loss of stability can be considered with a period that is a multiple of the structure period, in the form $N(m_1 + r)$, where N is an integer. In this case, the "dotted" rectangle in Fig. 4.3d when used in a computational experiment has dimensions $l_1 \times l_2$, where $l_1 = N(m_1 + r)$. Besides, the "dotted" rectangle already covers the N short fibers.

Consider the construction of the "mental" modes of loss of stability for a periodic structure (Fig. 4.3d), focusing on the "dotted" rectangle in this figure, which includes one short fiber.

Figure 4.15 shows the "mental" modes (a, b, c) of loss of stability, which are symmetrical concerning vertical lines drawn through the middle by a segment between the ends of adjacent fibers.

Figure 4.16 shows the "mental" modes (a, b, c) of loss of stability, which are antisymmetric concerning the indicated vertical lines.

As shown in Figs. 4.15 and 4.16, "mental" modes of loss of stability can also be characterized by the fact that mode **a** in Fig. 4.15 and modes **b** and **c** in Fig. 4.16 are

Fig. 4.15

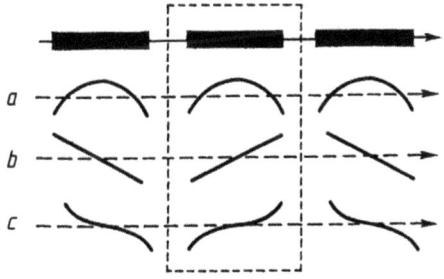

Fig. 4.16

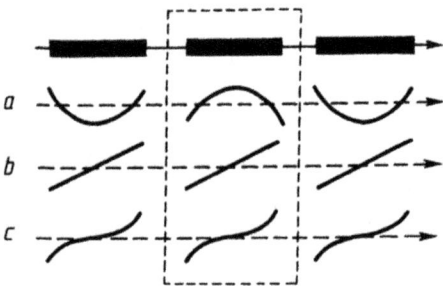

periodic along the axis $0x_1$ with a period equal to the period of the structure T. The modes **b** and **c** in Fig. 4.15 and mode **a** in Fig. 4.16 are periodic along the axis with a period equal to twice the period of structure $2\,T$.

Additionally, a dimensionless parameter $r^* = r \cdot m_1^{-1}$ can be introduced that characterizes the distance between the ends of two adjacent fibers. The geometric parameters m_1, m_2 and r are indicated on the design scheme (Fig. 4.3d).

The "mental" modes of loss of stability presented in Figs. 4.15 and 4.16 can also (by analogy with Sect. 4.5.3, Fig. 4.11) be characterized as follows.

> The "mental" mode of loss of stability **a** in Fig. 4.15 can be called quite close to the bending mode of loss of stability, which is realized in the case of sufficiently distant fibers with practically no mutual influence.

> The "mental" mode of loss of stability **b** in Fig. 4.15 can be called close to a rigid rotation of nearby fibers, which is realized for the sufficiently rigid materials of fibers with their relatively close location, when the matrix does not provide the proper supporting effect and the "plastic hinge" arises between the ends of adjacent fibers.

> The "mental" mode of loss of stability **c** in Fig. 4.15 can be called close to the rotation with bending, which is realized for the relatively stiff fiber materials and corresponds to the mode of loss of stability **b** in Fig. 4.15, supplemented by some bending of the fibers.

It is convenient to use the above considerations about the "mental" mode of loss of stability when interpreting the specific results on describing the mode of loss of stability obtained by numerically solving problems.

So, Fig. 4.17 provides information on the mode of loss of stability of the composite with specific parameters concerning the design scheme in Fig. 4.3d. These results were obtained by a numerical method, brief information about which is presented in Sect. 4.4.2.

The information discussed in Fig. 4.17 is presented within one fiber in dimensionless coordinates x_1 related to the fiber length m_1 in Fig. 4.3d. In this regard, in Fig. 4.17 the interval $-0.5 \leq x_1 \leq +0.5$ is considered. The information in Fig. 4.17 on the mode of loss of stability, which is determined as a result of a numerical solution, is presented as distribution along the axis $0x_1$ of the dimensionless vertical displacement u_2^* calculated by expression (4.3).

The results in Fig. 4.17 are presented for two cases of distances between the ends of adjacent fibers, which is characterized by a dimensionless parameter $r^* = r \cdot m_1^{-1}$ (the geometric parameters are indicated in the design diagram in Fig. 4.3d):

Fig. 4.17

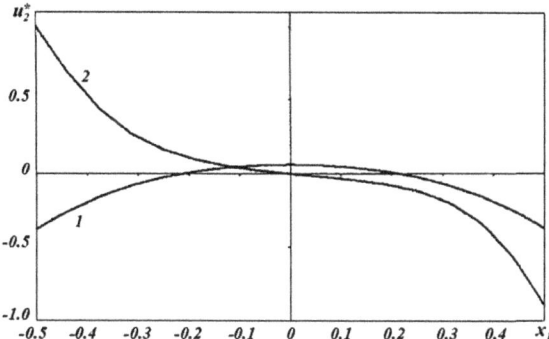

The first case $r^* = 1$, the distance between the ends of two adjacent fibers is equal to the fiber length, corresponds to curve 1;

The second case $r^* = 0.2$ corresponds to curve 2.

It follows from the analysis of the results in Fig. 4.17 that the mode of loss of stability (the curve 1, $r^* = 1$) practically coincides with the "mental" mode of loss of stability **a** in Fig. 4.15 and the mode of loss of stability (curve 2, $r^* = 0, 2$) practically coincides with the "mental" mode **c** of loss of stability in Fig. 4.15.

Besides, it follows from the analysis of the results in Fig. 4.17

that when the successively located short fibers closing in each other in a periodic row, the modes of stability loss change.

a similar situation took place in the case of two sequentially located fibers, as evidenced by the results in Figs. 4.10 and 4.12.

Below, the information is given about the results of the study of the dependence of the quantity $|\varepsilon_{11}^{\text{кр}}|$ (the value of the critical deformation along the axis $0x_1$ in the matrix "at infinity" (at $x_1 \to \pm\infty$) following the design scheme in Fig. 4.3d) on the dimensionless quantity $r^* = r \cdot m_1^{-1}$ (the geometric parameters m_1, m_2 and r are indicated on the design scheme in Fig. 4.3d) characterizing the relative distance between the ends of two adjacent fibers in the periodic row of the successively located short fibers (Fig. 4.3d).

These results were obtained by a numerical method, a summary of which is presented in Sect. 4.4.2. In this case, after obtaining the numerical value of the corresponding quantity $\varepsilon_{11}^{\text{кр}}$, which is the corresponding eigenvalue, the numerical studies on obtaining information about the corresponding eigenfunction, which determines the corresponding form of buckling, were not carried out. The research was carried out for a composite with the following parameters $E_a \cdot E_m^{-1} = 1000$; $v_a = v_m = 0.35$; $E_m = 2.76$ GPa; $k = m_1 \cdot m_2^{-1} = 100; 200; 300; 500$; the dimensionless distance between the ends of two adjacent fibers varied in the interval $0.2 \le r^* \le 4.5$.

These results are shown in Fig. 4.18, where the numbers 1, 2, 3, 4 mark the curves corresponding to the parameter values $k = 100; 200; 300; 500$. Here, the dash-dotted

Fig. 4.18

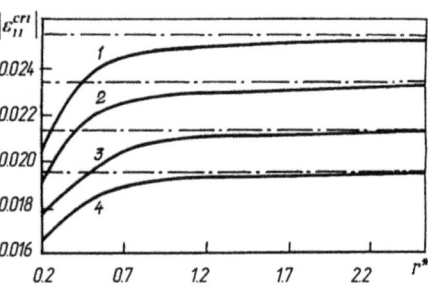

lines correspond to the values of the quantity $\left|\varepsilon_{11}^{\text{cri}}\right|$ for the case of one fiber at the same values of the parameter k.

It can be concluded from the results shown in Fig. 4.18

that when the distances between the ends of two adjacent fibers exceed the fiber length (at $r^* > 1$), the critical deformation $\left|\varepsilon_{11}^{\text{cri}}\right|$ for a periodic row of sequentially located fibers under compression along the fibers practically coincides with the value $\left|\varepsilon_{11}^{\text{cri}}\right|$ for one insulated fiber of the same dimensions under compression along the fiber.

It is possible to draw **the engineering conclusion** from the above **conclusion** related to the mechanics of the interaction of short fibers in a composite. It is related to the creation of composites and can be formulated as follows.

From the point of view of *the criterion of the strength under compression* of composite materials (reinforced with short fibers) *along the fibers*, **the creation of composites at** $S_a \geq$ 50% (S_a is the volume concentration of fibers) **may be ineffective** *for the indicated materials with an inhomogeneous structure.*

The fact is that it follows from Fig. 4.18 that at $r^* < 1$ ($r^* = 1$ corresponds to the distance between the ends of adjacent fibers, which is equal to the fiber length. In this case, $S_a \approx 50\%$), a significant decrease in the value of $\left|\varepsilon_{11}^{\text{cri}}\right|$ occurs, i.e., reduction in tensile strength at $S_a > 50\%$.

In the case of composites with *an ordered structure* (e.g., a doubly periodic structure in the cross-sectional plane of a fiber composite), the supporting influence of the adjacent periodic rows of short fibers can prevent a decrease in the value $\left|\varepsilon_{11}^{\text{cri}}\right|$ with an increase in the value of S_a that occurs (according to the curves in Fig. 4.18) in the interval ($0.2 \leq r^* \leq 1$) when the ends closing in.

It should be noted that the closing on of the ends of two adjacent fibers in a periodic row of the successively arranged short fibers corresponds to the movement along the axis $0r^*$ in Fig. 4.18 from right to left.

In the case of composites *with a disordered (irregular)* structure, the situations may arise in the separate parts of the composite, which correspond to the design scheme in Fig. 4.3d (one periodic row of fibers in an "infinite" space, i.e., without interaction with the fibers of adjacent rows). In this case, for such parts of the composite, **the conclusion** is valid concerning the mechanics of the interaction of short fibers in the composite, which was formulated when analyzing the results in Fig. 4.18.

Consequently, **the conclusion** about the inefficiency of creating these materials at $S_a > 50\%$ also is valid for such parts of the material.

Of course, additional studies are needed for the discussed effect, but the discussed phenomenon exists.

We restrict ourselves to the above information when discussing the results related to the design scheme in Fig. 4.3g. Additional information can be obtained from the monograph [1], the review article [2], some articles [5, 11, 25], and other publications.

4.5.6 Results for One Periodic Row of Parallel Fibers Under Compression Along the Fibers

The design scheme is shown in Fig. 4.3e. The statement of problems includes position **5** after Fig. 4.3a–f in Sects. 4.2, 4.3.3, and 4.4.1. The research was carried out by the method, brief information about which is presented in Sect. 4.4.2. Additionally, a dimensionless parameter $r^* = r \cdot m_1^{-1}$ is introduced. The geometric parameters m_1, m_2 and r are indicated on the design scheme in Fig. 4.3d.

It should be emphasized that position **5** after Fig. 4.3a–f provides that for the design scheme in Fig. 4.3e the attenuation conditions are set "at infinity" only along the coordinate x_1 (at) $x_1 \rightarrow \pm\infty$, and the periodicity conditions are set along coordinate x_2.

In this regard, the following situation arises when implementing the numerical research method.

As already noted in subsection 4.4.2, when implementing the numerical method, the infinite region is replaced by a finite region in the form of a rectangle $l_1 \times l_2$, which in Fig. 4.3e is indicated by the dotted lines. For the periodic series (Fig. 4.3e) in the indicated "dotted" rectangle the size l_2 is fixed from the periodicity condition and only the size l_1 changes to ensure the attenuation conditions "at infinity" (at $x_1 \rightarrow \pm\infty$), i.e. the parameter l_1 value is determined by a computational experiment.

It should be noted that for a periodic structure (Fig. 4.3e) with a period $T = m_2 + r$, it is possible to consider the periodic (along the axis $0x_2$) modes of loss of stability with a period that is a multiple of the period of the structure, in the form $N(m_2 + r)$, where N is an integer. In this case, the "dotted" rectangle already covers the N short fibers.

Consider the construction of the "mental" modes of loss of stability for a periodic structure (Fig. 4.3e), focusing on the "dotted" rectangle in this figure, which includes one short fiber.

This construction is carried out by analogy with the construction in the previous subsection. For example, Fig. 4.19 shows the "mental" modes of loss of stability **a, b, c,** and d, which can also be characterized by analogy with paragraph 4.5.5. We only note that the "mental" modes of loss of stability **a** and **b** in Fig. 4.19 have a period $T = m_2 + r$ along the vertical axis, and the modes **c** and **d** in Fig. 4.19 have a period along the vertical axis equal to $2(m_2 + r)$.

Fig. 4.19

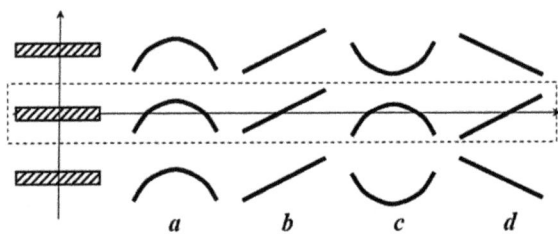

We will restrict ourselves to the above information when discussing the construction of the "mental" modes of loss of stability concerning the design scheme in Fig. 4.3e.

Below, following [6], the information is presented on the results of the study of the dependence of the quantity $\left|\varepsilon_{11}^{\text{кр}}\right|$ (the value of the critical deformation along the axis $0x_1$ in the matrix "at infinity" (at $x_1 \to \pm\infty$) following the design scheme in Fig. 4.3e) on the dimensionless quantity $r^* = r \cdot m_1^{-1}$ (the geometric parameters m_1, m_2 and r are indicated in the design diagram in Fig. 4.3e), which characterizes the relative distance between adjacent parallel fibers in an infinite periodic row of fibers located along the axis $0x_2$. The studies were carried out for composites with the following parameter values [6]: $E_a = 1.2$ TPa; $E_m = 3.5$ GPa; $\nu_a = \nu_m = 0.4$; $k = m_1 \cdot m_2^{-1} = 10; 20; 50; 100; 500$ with a change r^* in the range of $0.2 \le r^* \le 4.5$. The research results are presented in Fig. 4.20, where the numbers near each curve indicate the value of the parameter k.

It can be concluded from the results presented in Fig. 4.20 that with an increase in the distance between adjacent fibers, the value of which exceeds the fiber length (at $r^* > 1$), the value of the quantity $\left|\varepsilon_{11}^{\text{cri}}\right|$ *practically does not change* and corresponds to the results obtained for one fiber in the matrix (the design scheme in Fig. 4.3a).

The value of $\left|\varepsilon_{11}^{\text{cri}}\right|$ for one fiber in the matrix is shown in Fig. 4.20 with a dotted line for each value of the parameter k. With a decrease in the distance between the adjacent fibers in a row of parallel fibers (design scheme in Fig. 4.3e), i.e., at $r^* < 1$, as follows from Fig. 4.20, the value of the critical deformation along the axis $0x_1$

Fig. 4.20

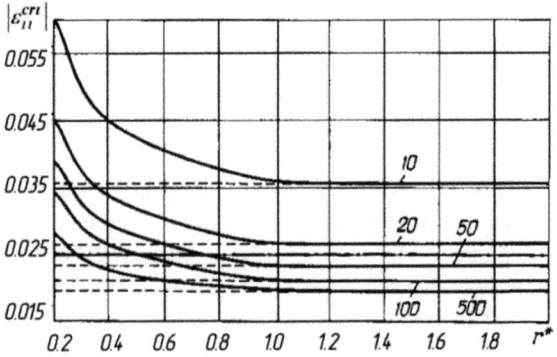

(the value of $\left|\varepsilon_{11}^{\text{cri}}\right|$) *increases* for all cases determined by the value of the parameter k. Thus, in the case of a periodic row of parallel fibers (design scheme in Fig. 4.3e), a situation *does not arise*, which is discussed in detail in the final part of Sect. 4.5.5 concerning the case of the periodic row of sequentially located fibers (design scheme in Fig. 4.3d) and which arose in connection with a decrease in the value of $\left|\varepsilon_{11}^{\text{cri}}\right|$ for $r^* < 1$, which follows from the results in Fig. 4.18 for the design scheme in Fig. 4.3d.

We restrict ourselves to the above information when discussing the results related to the design scheme in Fig. 4.3e. Additional information can be obtained from monograph [1], the review article [2], some articles [6, 26], and other publications.

4.5.7 Results for Single Fiber Located Close to Surface Under Compression Along the Fiber (Analysis of Near-the-Surface Instability)

The design scheme is shown in Fig. 4.3f. The statement of problems includes position **6** after Fig. 4.3a–f in Sects. 4.2, 4.3.3, and 4.4.1. The research was carried out by the method, brief information about which is presented in Sect. 4.4.2. Additionally, a dimensionless parameter $r^* = r \cdot m_1^{-1}$ is introduced, which characterizes the relative distance from the fiber to the boundary of the half-plane (boundary surface of the material). The geometric parameters m_1, m_2 and r are indicated on the design scheme in Fig. 4.3f.

As already noted in Sect. 4.2 after position **6**, the sixth design scheme (Fig. 4.3f), within the framework of which the studies are carried out in this subsection, is *the simplest* design scheme for studying near-the-surface fracture in the framework of **the second direction** (*the strict sequential approaches based on the TLTDBS*), a brief description of which is presented in Sect. 3.1.3, and *the model of the piecewise homogeneous medium* as applied to the plane problem (the plane deformation). Of course, the number of the simplest computational schemes in the study of near-surface fracture in the framework of the plane problem (plane deformation) can be significantly expanded if to provide for the introduction of the half-plane boundary in the computational schemes shown in Fig. 4.3b–e.

When investigating this problem within the framework of the simplest design scheme (Fig. 4.3f), which is the subject of this subsection exclusively, certain boundary conditions are set for the plane problem (plane deformation) on the half-plane boundary at $x_2 = 0$ (Fig. 4.3f). In this subsection, only the boundary conditions in stresses on an unloaded free half-plane are considered when determining the subcritical state. In this case, in the study of the problem of stability for $x_2 = 0$, the homogeneous boundary conditions in stresses are set. Here, in the study of the phenomenon of near-the-surface loss of stability concerning both the determination of the subcritical state and the solution of the corresponding problem of stability, the attenuation conditions "at infinity" are set for $x_1 \to \pm\infty$ and $x_2 \to -\infty$.

Considering the above, when applying the numerical method, the brief information about which is presented in Sect. 4.4.2, the semi-infinite area (the half-plane) is replaced by a finite area in the form of a rectangle with dimensions $l_1 \times l_2$, which is shown in Fig. 4.3f. In this case, one of the sides of the above rectangle passes over the border of the half-plane, and the other three sides of the rectangle are indicated by "dotted" lines in Fig. 4.3f.

On the "dotted" sides of the rectangle, boundary conditions are set that correspond to the attenuation conditions at $x_1 \to \pm\infty$ and $x_2 \to -\infty$. When applying the above numerical method, the dimensions of the discussed rectangle are increased by changing the position of the "dotted" sides. In this case, the dimensions $l_1 \times l_2$ are chosen such that their further increase does not affect the final results (on the value of critical loading or the value of critical deformation), which is determined as a result of a computational experiment.

Note 4.6 It should be noted *a specific situation* that arises in the study within the framework of the mechanics of near-the-surface fracture, the initial stage (start) of which is the near-the-surface loss of stability in the composite near the material surface *when near the material surface the reinforcing element of finite dimensions (e.g., a short fiber, Fig. 4.3f) exists.*

Consider the discussed situation by the example of the simplest computational scheme (Fig. 4.3f) related to the study of near-the-surface fracture. In this case, when loading "at infinity" (at $x_1 \to \pm\infty$) uniformly distributed over a load of constant intensity, an asymmetric (relative to the centerline of the cylinder, with $x_2 = -(r + 0, 5m_2)$ in Fig. 4.3f) distribution of material takes place near the short cylinder, since at $x_2 > 0$ there is no material at all.

Under these conditions, a local bending occurs near *the short fiber*, caused by the noted asymmetry in the distribution of the material. Moreover, the noted local bending increases with an increase in the value of the compressive load. The discussed situation arises in a subcritical state. Consequently, in this case, *a much more complex* mechanism of stability loss takes place in comparison with the loss of stability of a rod under axial compression. It is advisable to note that concerning the design scheme in Fig. 4.3f setting "at infinity" (at $x_1 \to \pm\infty$) the uniformly distributed over a load of constant intensity is equivalent to setting "at infinity" (at $x_1 \to \pm\infty$) constant shortening along the axis $0x_1$ (independent of x_2).

Note 4.7 It should be noted that the situation discussed in Note 4.6 does not apply to studies of near-the-surface fracture of composites in the framework of the model of "infinitely long reinforcing elements" that are parallel to the free surface of composites. Such results in the framework of **the second direction**, a brief description of which is presented in Sect. 3.1.3, and the model of *a piece-wise homogeneous medium*, are presented in a very brief form for the layered composites in Sect. 3.3.3.2 and the fibrous unidirectional composites in Sect. 3.3.4.2.

The fact is that when obtaining the above results "at infinity" (near the boundary surface), *the identical shortenings are set for the reinforcing elements and the matrix,* which is natural within the framework of the model of "infinitely long layers and fibers" and which leads to a homogeneous subcritical state.

In the monograph [1], the review article [2], some articles [7, 27], and in some other publications, the numerous quantitative results are given related to the description of the local bending of short fiber concerning the determination of the subcritical state within the framework of the design scheme in Fig. 4.3f.

Below are just two specific examples. The presence of the discussed bending in the subcritical state for the design scheme in Fig. 4.3f is convenient to characterize by the difference in the vertical displacements (displacements u_2^0) in the following form

$$u_2^{0*} = u_2^0\big|_{x_1=0} - u_2^0\big|_{x_1=0.5m_1}, \tag{4.4}$$

where

$u_2^0\big|_{x_1=0}$ is the vertical displacement (displacement u_2^0) on a vertical line passing through the center of the reinforcing element;

$u_2^0\big|_{x_1=0.5m_1}$ is the vertical displacement (displacement u_2^0) on a vertical line passing through the end face of the reinforcing element.

In this case, putting in (4.4) $x_2 = -(r + 0.5m_2)$, we obtain according to Fig. 4.3f the difference between the vertical displacements of the center of the reinforcing element and the center of the end face of the reinforcing element in the form

$$u_2^{0*}\big|_{x_2=-(r+0.5m_2)} = \left(u_2^0\big|_{x_1=0} - u_2^0\big|_{x_1=0.5m_1}\right)\bigg|_{x_2=-(r+0.5m_2)}. \tag{4.5}$$

Putting in (4.4) $x_2 = 0$, we obtain according to Fig. 4.3f the difference between the vertical displacements of points lying on the free surface and being the projection (of the center of the reinforcing element and the center of the end of the reinforcing element) onto the free surface, in the following form

$$u_2^{0*}\big|_{x_2=0} = \left(u_2^0\big|_{x_1=0} - u_2^0\big|_{x_1=0.5m_1}\right)\bigg|_{x_2=0}. \tag{4.6}$$

Figure 4.21 shows the dependence of the quantity u_2^{0*} (4.4) (the difference of vertical displacements) on the (dimensionless distance from the free surface) quantity $r^* = r \cdot m_1^{-1}$ (the geometric parameters are shown in Fig. 4.3f). Curve **1** in Fig. 4.21 corresponds to the case (4.5), and curve **2** in Fig. 4.21 corresponds to the case (4.6). The results in Fig. 4.21 are obtained for the next interval $(0.0 \le r^* \le 15.0)$ changes in the dimensionless distance to the free surface in Fig. 4.3f. The results presented in Fig. 4.21 are obtained for a specific composite with the following parameters: $E_a \cdot E_m^{-1} = 343$; $E_m = 3.51$ GPa; $v_a = v_m = 0.4$; $k = m_1 \cdot m_2^{-1} = 1000$.

It follows from the results presented in Fig. 4.21 that

in the entire range of variation of $r^*(0 \le r^* \le 15)$ the value of the difference between the vertical displacements of the center of the reinforcing element and the end of the reinforcing

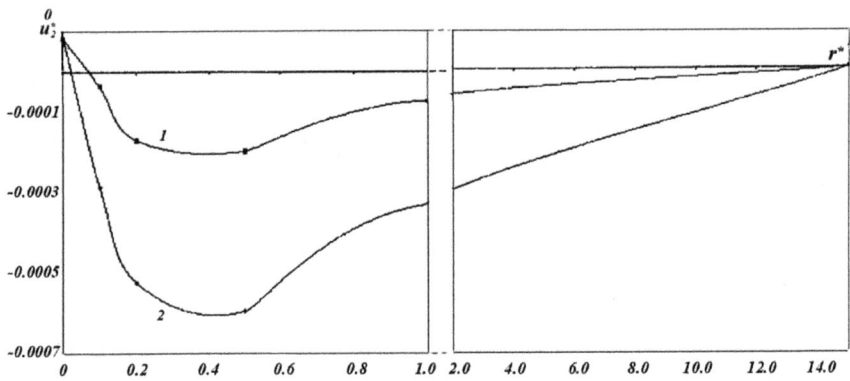

Fig. 4.21

element (curve 1) is not equal to zero, which indicates the existence of a local bending. Only for $r^* \to 15$ the difference under consideration $\to 0$. Thus, at $r^* > 15$, the local bending is absent, and the obtained numerical results correspond to the case of one reinforcing element in an infinite matrix (the design scheme in Fig. 4.3a, results in subsection 4.5.2).

In Fig. 4.22 for the subcritical state in a particular composite, *the dependence of the difference* between the vertical displacements of the center of the reinforcing element and the center of the end of the reinforcing element (value (4.5)) *on the value of the compressive load* "at infinity" is shown following the design scheme in Fig. 4.3f. The dotted line in Fig. 4.22 represents the buckling value.

Figure 4.23 presents the results related to the study of near-the-surface loss of stability for the design model in Fig. 4.3f. These results were obtained for the composites with the following parameter values: $E_a \cdot E_m^{-1} = 343; 1000; E_m = 3.51$ GPa; $v_a = v_m = 0.4; k = m_1 \cdot m_2^{-1} = 200, 1000$ and for the next interval of variation ($0 \le r^* \le 15$) of the dimensionless parameter $r^* = r \cdot m_1^{-1}$. The geometrical parameters m_1, m_2 and r are indicated in Fig. 4.3f.

Fig. 4.22

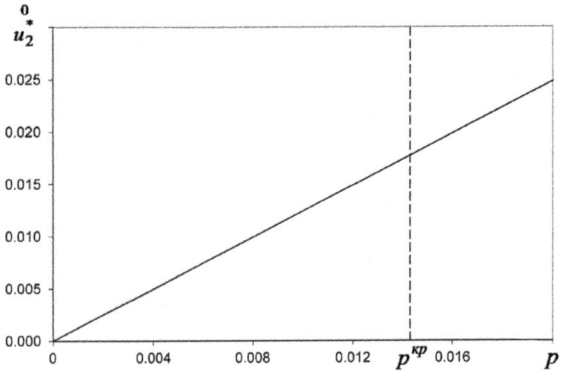

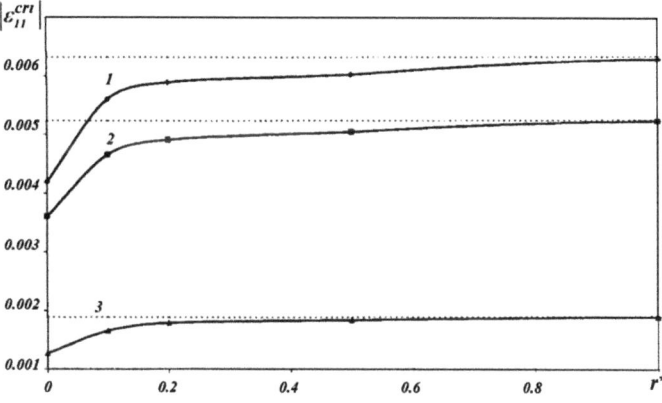

Fig. 4.23

It should be noted that the dashed lines in Fig. 4.23 show the values of $\left|\varepsilon_{11}^{cri}\right|$ for one fiber in the "infinite" matrix (design scheme in Fig. 4.3a, internal loss of stability) at the corresponding values of the geometric and mechanical parameters of the composite. Curves **1**, **2**, and **3** in Fig. 4.23 correspond to the composites with the following parameter values:

curve **1** ~ $E_a \cdot E_m^{-1} = 1000,\ k = 1000$;
curve **2** ~ $E_a \cdot E_m^{-1} = 343,\ k = 200$;
curve **3** ~ $E_a \cdot E_m^{-1} = 1000\ k = 200$.

The curves **1,2**, and **3** in Fig. 4.23 as the parameter r^* grows (as the fiber moves away from the free surface in Fig. 4.3f), asymptotically close into the corresponding dashed lines from below and *at $r^* = 1$,practically coincide with them.* Note that case $r^* = 1$ corresponds to the distance of the fiber from the free surface of the material (Fig. 4.3f), equal to the fiber length.

Thus, it follows from the above analysis that **for the considered composites** two following conclusions can be done.

1. *At $r^* \geq 1$, there is no need to investigate the near-the-surface instability,* since for it the value of $\left|\varepsilon_{11}^{cri}\right|$ practically coincides with the corresponding value $\left|\varepsilon_{11}^{cri}\right|$ calculated for the internal instability (the design scheme in Fig. 4.3a, a fiber in an "infinite" matrix).

2. *At $r^* \leq 1$, the study of the near-the-surface instability is expedient,* since at $r^* \to 0$ it is possible to decrease the value *of* $\left|\varepsilon_{11}^{\kappa p}\right|$ up to 30% in comparison with the value $\left|\varepsilon_{11}^{cri}\right|$ calculated for the internal instability (design scheme in Fig. 4.3a, the fiber in an "infinite" matrix).

We restrict ourselves to the above information when discussing the results related to the design scheme in Fig. 4.3f. Additional information can be obtained from the monograph [1], the review article [2], some articles [7, 14, 27], and other publications.

4.6   Conclusion

First of all, it is advisable to note that **Problem 2** *"model of short fibers in the theory of stability and fracture mechanics of composite materials under compression,"* the analysis of which is devoted to Chapter 4 of this monograph, began to be actively developed only from the beginning of the twenty-first century. In this regard, in this chapter, significant attention is paid to aspects of the statement nature, which in turn led to the increase of the volume of the presented material.

In this chapter, for the model of short fibers, as in the previous chapter for the model of "infinitely long" fibers and layers, the **General Concept** is adopted, which is that when the composites under discussion are compressed along the fibers and layers, *the onset (start) of fracture is determined by the loss of stability in the internal structure of composites* by the type of internal or near-the-surface instability.

It should be noted that the above mechanism of the start of fracture has received a certain experimental confirmation, the corresponding results are presented in the discussed chapters.

The information presented in Chapter 4 on **Problem 2**, related to the statement of problems, approaches, and specific results, is based on the model of *a piece-wise homogeneous medium* with the involvement of *the TLTDBS apparatus*. The noted general approach, apparently, is the most rigorous in the framework of the mechanics of deformable bodies as applied to the problem under study. The concrete results on **Problem 2** with the above general approach can be obtained *only with the use of numerical methods* since in all problems related to **Problem 2**, we arrive at *the inhomogeneous subcritical stress–strain states*.

In this chapter, the spatial (three-dimensional) problems and plane (two-dimensional in the case of plane deformation) problems are analyzed.

The spatial (three-dimensional) problems refer to the composites that are reinforced with filler in the form of short cylinders of circular cross section.

The plane (two-dimensional) problems, when strictly considered, refer to the strip composites. In an approximate consideration, the results of these problems can be used to analyze the phenomena that arise in the composites reinforced with short fibers.

At present, the problems concerning **Problem 2** are investigated only concerning *the brittle fracture*. In the study (within the framework of the above statements) of the corresponding problems concerning *the plastic fracture*, the additional difficulties arise associated with the determination of the subcritical inhomogeneous state, taking into account the unloading zones that change during the loading process.

The actual configuration for **Problem 2** is the study of the influence of the shape of the ends of short fibers on the value of the critical deformation under compression. It should be taken into account that the quantity $\left|\varepsilon_{11}^{\text{кр}}\right|$ (the critical shortening at the loss of stability) is a criterion of fracture, the initial stage (start) of which is determined by the loss of stability. The indicated fracture criterion is an integral fracture criterion since the quantity $\left|\varepsilon_{11}^{\text{кр}}\right|$ is determined through the eigenvalue of the corresponding eigenvalue problem within the TLTDBS.

Due to the integral nature of the applied fracture criterion, it can apparently be assumed that the influence of the shape of the ends of short fibers on the value $\left|\varepsilon_{11}^{\text{кр}}\right|$ will be less perceptible compared to the application of local fracture criteria, which are determined by the values of some invariants of the stress tensor at the point of the material under consideration. Nevertheless, despite the above considerations, *the study of the influence of the shape of the ends of short fibers on the value of* $\left|\varepsilon_{11}^{\text{кр}}\right|$ (the critical shortening corresponding to the loss of stability) *is quite relevant for* **Problem 2** *considered in this chapter.*

References

1. Guz, A.N., Dekret, V.A.: Model korotkikh volokon v teorii ustoichivosti kompozitov (The Model of Short Fibers in the Theory of Stability of Composites). LAP LAMBERT Academic Publishing, Saarbrücken, Deutschland (2015)
2. Guz, A.N., Dekret, V.A.: Model volokon konechnykh razmerov v trekhmernoi teorii ustoichivosti kompozitnykh materialov (obzor) (Finite-Fiber Model in the Three-Dimensional Theory of Stability of Composite Materials (Review)). Prikladnaya Mekhanika **52**(1), 3–77 (2016)
3. Guz, A.N., Dekret, V.A., Dekret, V.A.: Ploskie zadachi ustoichivosti kompozitnykh materialov dlia sluchaia napolnitelia konechnykh razmerov (Plane Problems of Stability of Composite Materials with a Finite-Size Filler). Mekhanika kompozitnykh materialov **36**(1), 77–86 (2000)
4. Guz, A.N., Dekret, V.A., Dekret, V.A.: Vzaimodeistvie korotkikh volokon pri potere ustoichivosti. Ploskaia zadacha (Interaction of short fibers at loss of stability. Plane problem) «Problemy mekhaniki», Sbornik statei k 90-letiiu so dnia rozhdeniia A.Yu. Ishlinskogo, pp. 331–341. Fizmatlit, Moscow (2003)
5. Dekret, V.A.: Two-dimensional buckling problem for a composite reinforced with a periodical row of collinear short fibers. Int. Appl. Mech. **42**(6), 684–691 (2006)
6. Dekret, V.A.: Plane stability problem for a composite reinforced with a periodical row of parallel fibers. Int. Appl. Mech. **44**(5), 498–504 (2008)
7. Dekret, V.A.: Near-surface instability of composites weakly reinforced with short fibers. Int. Appl. Mech. **44**(6), 619–625 (2008)
8. Guz, A.N., Dekret, V.A.: Interaction of two parallel short fibers in the matrix at loss of stability. Comput. Model. Eng. Sci. **13**(3), 165–170 (2006)
9. Guz, A.N., Dekret, V.A.: On two models in the three-dimensional theory of stability of composites. Int. Appl. Mech. **44**(8), 839–854 (2008)
10. Guz, A.N., Dekret, V.A.: Stability loss in nanotube reinforced composites. Comput. Model. Eng. Sci. **49**(1), 69–80 (2009)
11. Guz, A.N., Dekret, V.A.: Stability problem of composite material reinforced by periodic row of short fibers. Comput. Model. Eng. Sci. **42**(3), 179–186 (2009)
12. Guz, A.N., Dekret, V.A., Kokhanenko, Yu.V.: Solution of plane problems of the three-dimensional stability of a ribbon-reinforced composite. Int. Appl. Mech. **36**(10), 1317–1328 (2000)
13. Guz, A.N., Dekret, V.A., Kokhanenko, Yu.V.: Two-dimensional stability problem for interacting short fibers in a composite: in-line arrangement. Int. Appl. Mech. **40**(9), 994–1001 (2004)
14. Guz, A.N., Dekret, V.A., Kokhanenko, Yu.V.: Planar stability problem of composite weakly reinforced by short fibers. Mech. Adv. Mater. Struct. **12**, 313–317 (2005)
15. Kokhanenko, Yu.V.: Numerical study of the three-dimensional stability problems for laminated and ribbon-reinforced composites. Int. Appl. Mech. **37**(3), 317–340 (2001)
16. Kumar, S., Uchida, T., Dang, T., Zhang, X., Park, Y.-B.: Polymer/carbon nano fiber composite fibers. In: Presented at SAMPE, pp. 1–12. Long Beach, CA. May 16–20, 2004

17. Rosen, B.W.: Mechanics of composite strengthening. In: Presented at Fiber Composite Materials, pp. 37–75. American Society for Metals. Metals Park, Ohio (1965)
18. Rosen, B.U.: in Mekhanika uprochneniya kompozitov (Mechanics of hardening composites) [Russian translation]. In: Krock, R.H., Broutman, L.J. (eds.) Fiber Composites, pp.54–94. Mir, Moscow (1967)
19. Libowitz, G. (ed.): Razrushenie, v 7 tomah (Fracture, in 7 volumes) [Russian translation] (Vol. 1, 2, 3, 7 - Mir, Vol. 4, 5 - Mashinostroenie, Vol.6 - Metallurgia, Moscow, 1973–1977). T. 1. Mikroskopicheskie i makroskopicheskie osnovy mekhaniki razrusheniia (Vol.1. Microscopic and macroscopic foundations of fracture mechanics) (1973), T. 2. Matematicheskie osnovy mekhaniki razrusheniia (Vol.2. Mathematical foundations of fracture mechanics) (1975), T. 3. Inzhenernye osnovy i vozdeistvie vneshnei sredy (Vol.3. Engineering foundations and the impact of the external environmen) (1976), T. 4. Issledovanie razrusheniia dlia inzhenernykh raschetov (Vol.4. Investigation of fracture for engineering calculations) (1977), T. 5. Raschet konstruktsii na khrupkuiu prochnost (Vol.5. Calculation of structures for brittle strength) (1977), T. 6. Razrushenie metallov (Vol.6. Fracture of metals) (1976), T.7, ch. 1. Razrushenie nemetallov i kompozitnykh materialov. Neorganicheskie materialy (stekla, gornye porody, kompozity, keramiki, led) (Vol.7, Part 1. Fracture of non-metals and composite materials. Inorganic materials (glass, rocks, composites, ceramics, ice)) (1976), T.7, ch. 2. Razrushenie nemetallov i kompozitnykh materialov. Organicheskie materialy (stekloobraznye polimery, elastomery, kost) (Vol.7, Part 2. Fracture of non-metals and composite materials. Organic materials (glassy polymers, elastomers, bone)) (1976)
20. Krock, R.H., Broutman, L.J. (eds.): Kompozitnye materialy (Composite materials), [Russian translation], (Vol. 1, 2, 5, 6 − Mir, Vol. 3, 4, 7, 8 − Mashinostroenie, Moscow, 1978–1979). T. 1. Poverkhnosti razdela v metallicheskikh kompozitakh (V.1. Interfaces in metal matrix composites) (1978), T. 2. Mekhanika kompozitnykh materialov (V.2. Mechanics of composite materials) (1978), T. 3. Primenenie kompozitnykh materialov v tekhnike (V.3. Engineering applications of composites) (1979), T. 4. Kompozitnye materialy s metallicheskoi matritsei (V.4. Metal matrix composites) (1978), T.5. Razrushenie i ustalost (V.5. Fracture and fatigue) (1978), T. 6. Poverkhnosti razdela v polimernykh kompozitakh (V. 6. Interfaces in polymer matrix composites) (1978), T. 7. Analiz i proektirovanie konstruktsii. Ch. 1 (T. 7. Structural design and analysis. Part 1) (1979), T. 8. Analiz i proektirovanie konstruktsii. Ch. 2 (T. 8. Structural design and analysis. Part 2) (1979)
21. Guz, A.N.: Osnovy mekhaniki razrusheniia kompozitov pri szhatii: V 2-kh tomakh (Fundamentals of the Fracture Mechanics of Composites Under Compression: In 2 volumes). Litera, Kyiv (2008). T. 1. Razrushenie v strukture materiala. (Fracture in structure of materials), T. 2. Rodstvennye mekhanizmy razrusheniia. (Related mechanisms of fracture)
22. Guz, A.N., Rushchitskyi, Ya.Ya., Guz, I.A.: Vvedenie v mekhaniku nanokompozitov (Introduction to mechanics of nanocomposites). S. P. Timoshenko Institute of Mechanics, Kyiv (2010)
23. Guz, A.N., Rushchitskii, J.J.: Short Introduction to Mechanics of Nanocomposites. Scientific & Academic Publishing Co., LTD, USA (2013)
24. Dekret, V.A.: Rozviazannia ploskoi zadachi stiikosti kompozytnoho materialu armovanoho dvoma korotkymy voloknamy (Solving the flat problem of stability of a composite material reinforced with two short fibers). Dopovidi NAN Ukrainy 8, 37–40 (2003)
25. Dekret, V.A.: Pro stiykist kompozitnogo materialu armovanogo periodichnim ryadom korotkikh volokon (Stability of a composite material reinforced with a periodic row of short fibers). Dopovidi NAN Ukrainy 11, 47–50 (2004)
26. Dekret, V.A.: Pro stiykist kompozitnogo materialu armovanogo periodichnim ryadom paralelno rozmischenikh korotkikh volokon (Stability of a composite material reinforced by a periodic row of collinear short fibers). Dopovidi NAN Ukrainy 12, 41–44 (2004)
27. Dekret, V.A.: Pro stiykist kompozitnogo materialu slaboarmovanogo korotkimi voloknami poblizu vilnoï poverhni (Stability of a composite material weakly reinforced by short fibers near the free surface). Dopovidi NAN Ukrainy 10, 49–51 (2006)

Chapter 5
Problem 3. End-Crush Fracture of Composite Materials Under Compression

In this chapter, in a very brief form (compared to Problem 1, Chap. 3, and Problem 2, Chap. 4 of this monograph), the main results are outlined on the problem shown in the title. They are obtained in the Department of Dynamics and Stability of Continua of the S. P. Timoshenko Institute of Mechanics of NASU. The presentation of these results is made in a style announced in the Introduction to this monograph (without involving aspects of a mathematical nature). Some results of experimental studies related to the formation of this problem are also given.

5.1 Introduction

The main results on this problem, which are obtained by the staff of the Department of Dynamics and Stability of Continua, are presented in the monographs [3, Chap. 7, Sect. 5.4, pp. 568–589) and [4, vol. 2, Chap. 11, Sect. 5.4, pp. 529–551), review articles [23–26] and [21] in the list of literature to [24], individual articles [2, 7, 12, 13, 17, 22] and in reports at international conferences [14–16, 18–21] as well as several other publications that were not included in the References of this monograph.

It should be noted that the name of the phenomenon "end-crush fracture under compression of composites," which is explored in the framework of Problem 3, was translated into English as "buckling of the ends" in [12, 13], "bearing strain in end faces"—in [14, 15] and "end-crush fracture of compressed composites" in [25, 26]. In the Russian-language literature, the phenomenon of "crumpling ends in the compression of composites" is also called "brooming."

It should be noted that the Doctor of Sciences dissertation in Physics and Mathematics of Yu.V. Kokhanenko also, in part, refers to this scientific direction. In general, the above dissertation of Yu.V. Kokhanenko is devoted to the development of a numerical method of research which is based on the method of finite differences involving the variation-difference approach and the use of the basic schemes, a brief description

A. N. Guz, *Eight Non-Classical Problems of Fracture Mechanics*,
Advanced Structured Materials 159,
https://doi.org/10.1007/978-3-030-77501-8_5

of which is given in Sect. 4.4.2. This method discussed is intended for the piecewise homogeneous media, which corresponds to the model of the piecewise homogeneous materials in mechanics of composites, including in the studies involving the TLTDBS apparatus. In connection with the foregoing, it should be noted that the References of this monograph includes the publications, the results of which are obtained using this method. For example, [5–7, 9, 10, 27, 28] and a number of others.

The above general information will be limited to the formation of the Introduction to this chapter.

5.2   Experimental Researches

The subject of research in **problem 3** is the analysis of phenomena that occur near the ends of composites in the form of unidirectional fibrous or layered materials under compression along fibers and layers or in the composites of another structure under compression along the axes of the symmetry of the material properties. Therefore, in this section, when considering the results of experimental studies, the main attention is paid to the analysis of the nature of fracture near the ends in samples from composites.

Consider the nature of the fracture near the ends of metal composite samples, the experimental results for which are presented in **the Example** in the final part of Sect. 3.3.4.1. These experimental results relate to the uniaxial compression along the fibers for the metal composite (unidirectional fibrous boro-aluminum composite with 50% boron fiber concentration, $S_a = S_m = 0.5$) VKA-1 with the boron fibers of 140 μm diameter. A general view of the samples from boro-aluminum *before the fracture* is presented in Figs. 3.37. Figure 3.38 shows in the cross section (with a significant increase) the internal structure of boro-aluminum boro-composite.

The general view of *the fractured specimen under the plastic fracture* is presented in Fig. 5.1. It follows from this picture that *the left end of the sample is fractured in*

Fig. 5.1

the form of **a crumpled end** *when the fracture has a local character (near the left end) and does not extend to the entire length of the sample.*

It is worth noting that the general view of the fractured part of the sample on Fig. 5.1 visually has the form of a "broomstick" which causes the term "brooming."

The above results (Fig. 5.1) along with other results on plastic fracture under uniaxial compression of the boro-aluminum composite *in the form of crumpled ends*, as already noted in Sect. 3.3.4.1, published in the article [8] and comparative detail is outlined in the monograph [4, vol.1, Chap. 4, pp. 486–488), where more information can be obtained.

It is worth noting that among the various mechanisms implemented in the fracture of composites, the mechanisms are quite common, which are associated with the fracture of the crumpling of ends od samples and structural elements made of the composites in the case of compression, in the example considered—in the case of uniaxial compression.

In general outline, the phenomenon of fracture *in the crumpling of ends* is that (for the samples and structural elements made of composites) *a local fracture of the material near the ends occurs, while this fracture does not extend far from the ends and decreases at a distance from them.* In this regard, the ultimate strengths of the material corresponding to the fracture with crumpling of ends are slightly less than the ultimate strengths corresponding to the fracture of the entire material (away from the ends, the internal fracture—in the terminology of Sect. 3.3).

With allowance for the foregoing, usually in the experimental determination of ultimate strengths *under compression* (meaning the ultimate strengths corresponding to the fracture of the material away from the ends), *the possibility of crumpling ends* on the samples is excluded with the help of various design and technological techniques. In particular, the appearance of crumpling ends is excluded by strapping the ends of samples over the side surfaces near the ends or by placing the ends of samples in the casings of the more rigid material.

The application of the last technique for the fiberglass samples is implemented in studies, the results of which are presented in Figs. 3.8 and 3.10, where the ends of the fiberglass samples were placed in the specially made metal frames. Non-acceptance of the above techniques can lead to a phenomenon of fracture in the form of crumpling ends, which is shown in Fig. 5.1.

The exclusion of the crumpled ends in the case of experimental studies of samples in the laboratory (through the use of various design or technological techniques) does not exclude the manifestation of this and other similar phenomena in the natural structural elements. The phenomena similar to the fracture of the crumpling ends are observed in various structural elements, for example, in different joints and docks. In the case of wood (as a structural material), these phenomena have been known for a long time and have been repeatedly observed and described by many researchers.

It should also be noted that the phenomenon of fracture by crumpling the ends is also observed in a combined stress state when the applied to the ends external compressive loads reach certain values.

Note that the experimental study of the phenomenon of fracture under crumpling of the ends is quite complex and does not always lead to one-valued results. This is because experimental studies usually get the result (a picture of the sample fractured near the end or the value of the ultimate strength), corresponding to the process of fracture that has already finished when the end is crumpled.

To investigate or describe the phenomenon in question, it is necessary to have information on the initial stage of fracture (to identify the causes and mechanism of the corresponding phenomenon). In the complex picture of the sample fractured near the end (corresponding to the final stage of fracture), it is difficult to identify the processes at the initial stage of fracture.

A similar situation arises in an analysis of the results of experimental studies concerning **Problem 1** (*Fracture in composite materials under compression along the reinforcing elements*).

Basing on the above information and considerations, it seems appropriate and relevant to develop a theory that would describe the fracture of the ends in the continuum approximation or within the model of a piecewise homogeneous medium.

5.3 Theoretical Researches

As applied to Problem 3, the results of theoretical studies are presented in the publications listed in the Introduction (Sect. 5.1) in this chapter.

All of these results of theoretical studies are obtained *exclusively* **in the second direction (strict sequential approaches based on TLTDBS)**, the brief characteristics of which are given in Sect. 3.1.3. Below, in this paragraph, theoretical results are presented in a very brief form separately:

for the formulation of the General Concept;

for the model of the piecewise homogeneous medium;

for the continuum approximation (for the continuum model).

5.3.1 General Concept

In general, the composites of different structures are considered, which are modeled by *the orthotropic materials* in a continuum approximation. It is accepted at that: *these materials or samples from such materials have an end coinciding with one of the planes of orthotropy, and a normal compression load is applied to the end.* The above situation can be considered as a design scheme, which includes a sample (Fig. 5.1, after destruction) made of the unidirectional fibrous composite when the cross section is loaded with an external normal compression load.

General Concept. *In the above situation* (the internal structure of the composite allows to model in a continuum approximation this composite by an orthotropic material. The end is located in one of the planes of the material orthotropy. In the end, a normal compression load is applied – the compression is realized along one of the axes of symmetry of the material properties and the geometric shape of the structural element. *The initial stage (start) of fracture under the crumpling of the ends is a near-the-surface loss of stability near the loaded end.*

In analyzing the further development of the fracture mechanism under consideration, it is necessary to take into account its possible interaction with other fracture mechanisms.

The theoretical ultimate strength and the theoretical value of the ultimate shortening in near-the-surface fracture near the loaded by the normal compression end (fracture under the crumpling of ends) are the critical value of the load and the value of critical shortening calculated within the applied version of TLTDBS.

The formulated before **General Concept** makes it possible to develop the fracture mechanics when the ends are crumpled under compression, exploring the near-the-surface instability near the loaded end, both within the model of the piecewise homogeneous medium, and the continuum theory.

The studies under discussion accept that loading of the end is carried out by the "dead" loads, and therefore, the sufficient conditions of applicability of the static method of the research of the corresponding problems of TLTDBS are performed for the brittle and plastic fracture, according to Sect. 2.4.2. Thus, the studies of crumpling of the ends correspond fully to the generally accepted and rigorous method of research into the phenomenon of loss of stability—the analysis of the behavior of small perturbations within the linearized three-dimensional dynamic problems.

Note 5.1 It seems that *the determining factor providing* **the fracture in the form of crumpling ends** *is the implementation of* **certain boundary conditions at the ends**. The fact is that compression on the ends is realized:

in the case of samples—through the rather rigid bearing plate;
in the case of structural elements—through a fairly rigid mechanical interface.

In these two cases of the compression loading on the ends, *the same shortening at the end of reinforcing elements* (e.g., fibers or layers) and matrix (binding) is ensured due to the sufficient rigidity of the bearing plates and mechanical interfaces.

This situation corresponds to the case when the same normal displacements in the form of boundary conditions are given the entire surface of the end of the composite. If due to structural or technological features (e.g., polishing the surface of the bearing plate), very little friction or its complete absence at the end of the compression immersion is provided, then some parts of the composite at the end can slide freely enough along the plane of the bearing plate. In this situation, seemingly, the fracture *in the form of crumpling ends* arises.

Thus, for the occurrence of the phenomenon of fracture in the form of crumpling ends:

The "most favorable" are the boundary conditions on the surface of the end,

when the constant normal displacements and *zero tangent stresses* on the surface of the end are given; the "least favorable" conditions are the boundary conditions on the surface;

of the end, when the constant normal displacements and *zero tangent displacements* are given on the surface of the end.

It should be noted that the "least favorable" boundary conditions correspond to the fastening of the ends in the rigid holder (such as the results of the experiment presented in Figs. 3.9 and 3.10) when the fracture in the form of crumpling ends does not occur.

Note 5.2 In the study of the problems of fracture mechanics in the crumpling of the ends within the model of a piecewise homogeneous medium, the *two-dimensional* (plane) and *three-dimensional* (spatial) problems can be considered.

The plane problems (for the plane deformation) with the strict and consistent consideration and interpretation refer to the composites of the *tape* structure, in which the reinforcing elements are the parallel infinite tapes of constant cross section, directed perpendicular to the plane of the picture (as in 4.2 for Fig. 4.3a), and the *layered* composites. In an approximate interpretation, the results for plane problems (for the plane deformation) can be used in the analysis of phenomena that occur in unidirectional fibrous composites, similar to the approximate approach, which is described in a brief form in the final part of Sect. 4.3.1.

The spatial problems, in the sequential examination, refer to *the unidirectional fibrous* composites (the represented in Fig. 3.1 type).

The above information will be limited to the formulation and discussion of the **General Concept**, as well as when considering some moments related to the statement and analysis of the fracture mechanics when the ends of composite materials are crumpled in the case of compression.

5.3.2   On Studies Within the Model of Piecewise Homogeneous Medium

As an example of the study of the fracture of the ends, performed within the framework of the model of the piecewise homogeneous medium with the involvement of the TLTDBS's apparatus, the results of the article [7] considered in a brief form. There, the results are obtained for the simplest model of the fracture mechanics under the crumpling of the ends. Here, the designations of this article are changed to the designations adopted in this monograph.

The research is carried out following the **General Concept** set out in the previous subsection, for the brittle fracture under the end crumpling with the involvement of **theory 3** (*the second variant of the theory of small precritical deformations*) following the terminology of Sect. 2.2 for *the layered composites* (Fig. 3.2, the model of "infinitely long" layers) in the case of *a small concentration* of reinforcing elements (layers of filler), when the interconnection of two neighboring reinforcing elements (layers) *can be ignored* with loss of stability. At that, the analysis is done *only for one end* (un Fig. 5.2—for the top end).

The simplest design scheme, corresponding to the above statement of the problem, is presented in Fig. 5.2 concerning a plane problem (plane deformation) in the plane

Fig. 5.2

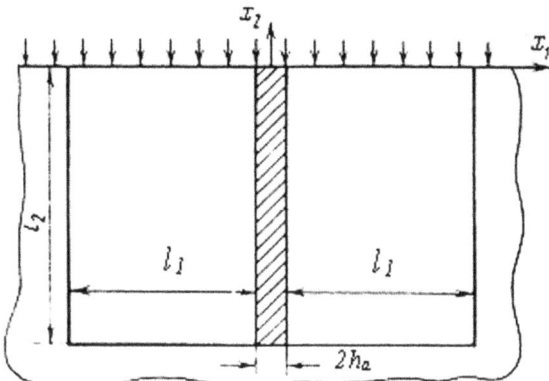

$x_1 0 x_2$ where the axis $0x_3$ is directed perpendicular to the plane of Fig. 5.2. Since the fact of crumpling the end is observed only near the end, attenuating at the distance from the end, and in this case, the situation is analyzed when the neighboring reinforcing elements do not interact with each other (due to their low concentration), the simplest design scheme from Fig. 5.2 is reduced to a lower semi-plane $x_2 \leq 0$ with one reinforcing element in the form of a semi-strip.

Thus, the study needs to be done for the part of the lower semi-plane containing the reinforcing element, setting the boundary conditions at $x_2 = 0$ and conditions of attenuation at and $x_1 \to \pm\infty$. The discussed part of the semi-plane is conditionally highlighted by the.

"wavy line" and $x_2 \to -\infty$ the boundary at $x_2 = 0$ where $2h_a$, as in Fig. 3.2, indicated the thickness of reinforcing element (layer).

In the analysis of *the brittle fracture under the crumpling of ends*, the properties of the materials of the filler and binder are modeled by a linearly elastic isotropic body. Therefore, the elasticity relations are used in the traditional form, taking into account the use of **theory 3** (*the second variant of the theory of small precritical deformations*).

$$\sigma_{ij}^a = \delta_{ij}\lambda_a\varepsilon_{nn}^a + 2\mu_a\varepsilon_{ij}^a; \quad \sigma_{ij}^m = \delta_{ij}\lambda_m\varepsilon_{nn}^m + 2\mu_m\varepsilon_{ij}^m;$$

$$2\varepsilon_{ij}^a = u_{i,j}^a + u_{j,i}^a; \quad 2\varepsilon_{ij}^m = u_{i,j}^m + u_{j,i}^m. \tag{5.1}$$

In (5.1), as in all this monograph, the designations are used for elastic constants in the form: λ_a, μ_a, E_a and ν_a or reinforcing elements (for filler); λ_m, μ_m, E_m and ν_m for the matrix (binder).

Consider in a brief form, the definition of the precritical state according to the above-mentioned statement. Let us limit ourselves to analyzing the compression of the end through the rigid supporting plate, which is conventionally shown in Fig. 5.2 by the vertical arrow system. In this case, according to Note 5.1, the surface of the end (in Fig. 5.2 for $x_2 = 0$) occurs as the filler and binder are shortened equally

$(\varepsilon_{22}^{a0} = \varepsilon_{22}^{m0})$. All the values of the precritical state, as in the whole present monograph, are noted additionally by the index "zero."

Accept also that at the end (at $x_2 = 0$ in Fig. 5.2) there is no friction between the sample and the support plate. According to Note 5.1, this condition corresponds to the "most favorable" boundary conditions for the occurrence of the phenomenon of fracture in the form of crumpling ends. Thus, to determine the precritical condition we get at the end (at $x_2 = 0$) the following boundary conditions

$$\varepsilon_{22}^{a0} = c, \sigma_{12}^{a0} = 0 \text{ for} |x_1| \le h_a \text{and} x_2 = 0;$$

$$\varepsilon_{22}^{m0} = c, \sigma_{12}^{m0} = 0 \text{ for} |x_1| \ge h_a \text{and} x_2 = 0; c = \text{const.} \tag{5.2}$$

Under boundary conditions in the form (5.2) at the end (at $x_2 = 0$ in Fig. 5.2) with taking into account the relations (5.1) for the plane deformation in the plane $x_3 = 0$ (Fig. 5.2, $\varepsilon_{33} \equiv 0$), the precritical state is determined in the analytical form in the following form

$$\sigma_{11}^{a0} = 0, \sigma_{12}^{a0} = 0, \sigma_{22}^{a0} = 4\mu_a(\lambda_a + \mu_a)(\lambda_a + 2\mu_a)^{-1}\varepsilon_{22}^{a0};$$

$$\sigma_{11}^{m0} = 0, \sigma_{12}^{m0} = 0, \sigma_{22}^{m0} = 4\mu_m(\lambda_m + \mu_m)(\lambda_m + 2\mu_m)^{-1}\varepsilon_{22}^{m0};$$

$$\varepsilon_{22}^{a0} = \varepsilon_{22}^{m0} = c = \text{const.} \tag{5.3}$$

Consider in a brief form the results studying the problem of stability following the above-mentioned statement of the problem concerning the precritical state in the form (5.3), following the article [7]. This problem has a local character near the end at $x_2 = 0$ in Fig. 5.2. The boundary conditions on the end surface for the problem of stability, matched with the conditions (5.2) for the precritical state, have the following form

$$\varepsilon_{22}^{a} = 0, \sigma_{12}^{a} = 0 \text{at} |x_1| \le h_a \text{and} x_2 = 0;$$

$$\varepsilon_{22}^{m} = 0, \sigma_{12}^{m} = 0 \text{at} |x_1| \ge h_a \text{and} x_2 = 0. \tag{5.4}$$

Due to the local nature of this problem of stability, the conditions of attenuation are also set when away from the end (under $x_2 \to -\infty$) and the conditions of attenuation at $x_1 \to \pm\infty$ in the case of the layered composites with a very low concentration of filler (the case when two adjacent layers of filler do not interact with each other at a loss of stability). These conditions are presented in the following form

$$u^a \to 0 \text{when} x_2 \to -\infty; u^m \to 0 \text{when} x_2 \to -\infty \text{or} x_1 \to \pm\infty. \tag{5.5}$$

For the problem of stability within relations (2.21)–(2.27) and (2.10)–(2.13) of the applied **Theory 3** (the second version of the theory *of* small precritical deformations) the conditions of continuity of stress and displacement vectors on $x_1 = \pm h_a$ with the expressions (2.10) and (2.23) are also formulated.

The problem of stability for the lower semi-plane $x_2 \leq 0$ (Fig. 5.2) within the above-mentioned theory with the constitutive equations in the form of (5.1), boundary conditions in the form (5.4) on the boundary of semi-plane at $x_2 = 0$ the conditions of attenuation in the form (5.5), and the above conditions of continuity of stress and displacement vectors when $x_1 = \pm h_a$ was studied by the numerical method, the brief information about which is presented in Sect. 4.4.2.

In the numerical study of this problem, the semi-infinite area (the lower half-plane) was replaced by the finite area in the form of a rectangle with sides $2(l_1 + h_a) \times l_2$, which is shown in Fig. 5.2. The values of quantities $l_1 \cdot h_a^{-1}$ and $l_2 \cdot h_a^{-1}$ were chosen from the condition of independence of the final result (the critical values of shortenings or stresses) on the values of these values. It was found that, with $l_1 \cdot h_a^{-1} \geq 10$ and $l_2 \cdot h_a^{-1} \geq 20$, the further increases in these values did not alter the final results. Therefore, all of the following results were obtained at the following values $l_1 \cdot h_a^{-1} = 10$ and $l_2 \cdot h_a^{-1} = 20$.

The results were obtained as concerning the determination of the modes of loss of stability for the problem, the design scheme for which is presented in Fig. 5.2, as in relation to the determination of the critical value of compression stresses σ_{22}^{a0} (in the filler) or σ_{22}^{m0} (in the matrix) through the critical value of shortening $\varepsilon_{22}^{a0} = \varepsilon_{22}^{m0}$ at the end. These results of determining the modes of loss of stability and critical values of loading parameters will be considered separately.

Consider first the results of determining *the modes of* loss *of* stability in the problem under discussion, the design scheme for which is presented in Fig. 5.2. It is convenient to characterize in this problem the mode of loss of stability by the horizontal displacements (displacements along the axis $0x_1$ in Fig. 5.2). In this case, the horizontal displacement of the points of the middle line of the reinforcing element, i.e., displacement $u_1^a(0, x_2)$, is particularly characteristic. Figure 5.3 shows a change in the immeasurable displacement $u_1^* = u_1^a(0, x_2) \cdot h_a^{-1}$ along the axis $0x_2^*$, where $x_{2*} = -x_2 \cdot h_a^{-1}$. Thus, Fig. 5.3 shows a change in the immeasurable horizontal displacement u_1^* of the points of the middle line of a reinforcing element when away from the end. The results are represented in Fig. 5.3 for the composites with $v_a = v_m = 0, 3$ for the following cases:

$E_a \cdot E_m^{-1} = 25$ as a solid line;

$E_a \cdot E_m^{-1} = 100$ in the form of a dotted line;

$E_a \cdot E_m^{-1} = 1600$ in the form of a bar-dotted line.

It follows from Fig. 5.3 that the attenuation of horizontal displacement u_1^* when away from the end has the form of a "rapidly attenuated wave" and almost complete attenuation is reached at $x_2^* \approx 15$, which corresponds to 7.5 thickness of filler. The maximum max u_1^* is achieved at $x_2^* \approx 5$ (2.5 thickness of filler) and is 5 thicknesses of filler.

It should be noted that the above numerical characteristics of the phenomenon of attenuation of horizontal displacements in the form of loss of stability relate only

Fig. 5.3

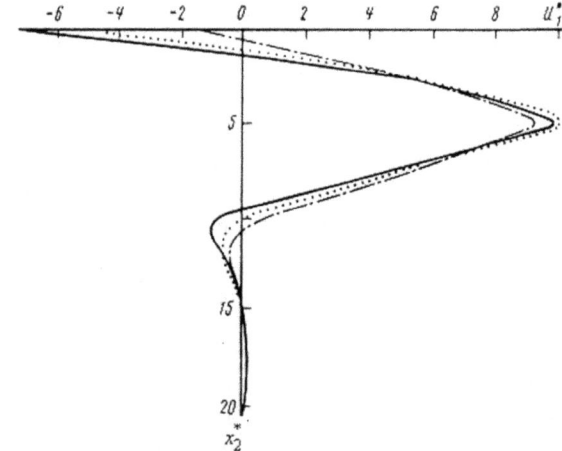

to the considered examples of composites, taking into account the statement of the problem discussed in the present subsection.

Consider below *the results on the determination of the critical values of the loading parameters* that correspond to the loss of stability. From the expressions (5.3) for the critical values, the relations follow

$$\left(\sigma_{22}^{a0}\right)^{\text{cri}} = E_a\left(1 - v_a^2\right)^{-1}\left(\varepsilon_{22}^{a0}\right)^{\text{cri}}; \ \left(\sigma_{22}^{m0}\right)^{\text{cri}}$$
$$= E_m\left(1 - v_m^2\right)^{-1}\left(\varepsilon_{22}^{m0}\right)^{\text{cri}}; \ \left(\varepsilon_{22}^{a0}\right)^{\text{cri}} = \left(\varepsilon_{22}^{m0}\right)^{\text{cri}}. \tag{5.6}$$

In (5.6), $\left(\varepsilon_{22}^{a0}\right)^{\text{cri}}$ and $\left(\varepsilon_{22}^{m0}\right)^{\text{cri}}$ are the critical shortenings for the reinforcing element and.

matrix at the end (Fig. 5.2), which are equal to each other because of the way of loading the end through the support plate.

Fig. 5.4 shows the dependence of the quantity

Fig. 5.4

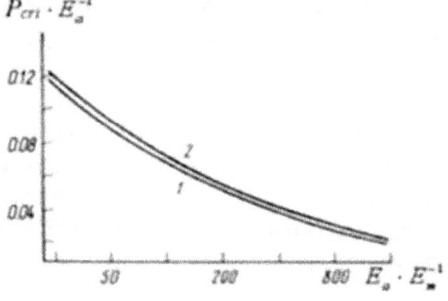

$$P_{kp} \cdot E_a^{-1} = \left| \left(\sigma_{22}^{a0} \right)^{\text{cri}} \right| \cdot E_a^{-1} \qquad (5.7)$$

on the parameter $E_a \cdot E_m^{-1}$ (curve 1 for $v_a = v_m = 0.3$; curve 2 for $v_a = 0.3$ and $v_m = 0.4$). A
logarithmic scale $\ln \left(E_a \cdot E_m^{-1} \right)$ is applied on the horizontal axis.

Note 5.3 Section 5.3.2 discusses in comparative detail the results outlined in the article [7] in relation to the following situation. The article [7], apparently for the first time, presents the results of the study of edge effects in the theory of **the stability of composites** within *the* model of piecewise homogeneous medium with the involvement of the TLTDBS apparatus.

In the case of **the problems of statics of composites**, the results of numerous studies of *the* edge effects in a three-dimensional statement involving a piecewise homogeneous medium model are outlined in a 12-volumes monograph [11, vol. 1, Section III, Chaps. 10, 11 and 12, pp. 313–445), which also provides a detailed setout of the developed numerical method of the study.

5.3.3 On Studies Within the Continuum Medium Model (Continuum Approximation)

In this subsection, in a fairly brief form, the results of the study of the phenomenon of fracture in the crumpling of the ends of samples or composite structural elements will be discussed. The composites are modeled by the homogeneous materials with average parameters, in relation to the *brittle* and plastic fracture.

Note that in Sect. 5.3.2, the corresponding results within the model of a piecewise homogeneous body are discussed only in relation to the brittle fracture.

These results are presented in a number of articles, which are listed in the Introduction (Sect. 5.1) in the present chapter. In addition, these results are fairly consistently set out in the monograph [4, vol. 2, Chap. 11, Sect. 5.4, pp. 529–551), which will be used in the present subsection with a short presentation of the corresponding results.

The basic provisions of the continuum theory of fracture in the form of crumpling ends under compression of the composites are reduced to the following provisions.

1. The setout in Sect. 5.3.1 **General Concept** is adopted, according to which the initial stage (start) of fracture in the crumpling of the ends is *a* near-the-surface loss *of* stability near the loaded end. The theoretical ultimate strength and *the* theoretical value of *the* limit shortening in near-the-surface fracture near the loaded by the normal compression end (fracture at *the* crumpling of the ends) is the critical value of the load and the value of critical shortening calculated within the framework of *the* applied version of TLTDBS.
2. The information and considerations are taken on the role of boundary conditions in the realization of fracture in the crumpled ends, which are outlined in Note 5.1.

Fig. 5.5

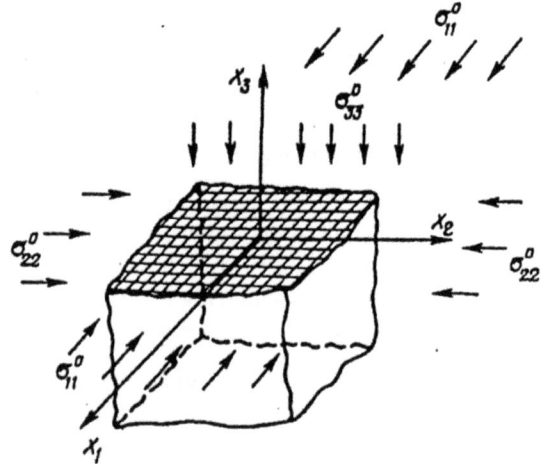

3. In the study of the phenomenon of crumpling ends, the influence of the lateral surfaces of the sample or structural elements, as well as the mutual interaction of two opposite ends (a long enough sample), will not be taken into account which makes it possible to carry out the research for the lower semi-space $x_3 \leq 0$ (Fig. 5.5), where the end is marked by the "grid."

4. In the study of surface instability near the loaded end, the precritical state will be considered homogeneous. If the precritical state is heterogeneous, then the local surface instability (as if highlighting the small neighborhood of this point) for each point of the end surface will be considered. Thus, we return anew to a homogeneous precritical state.

5. The external normal compressing load which is applied to the end at $x_3 = 0$ (Fig. 5.5) will be considered as the "dead" one. Thus, for the brittle and plastic fracture, the sufficient conditions of applicability of the statical method of research (Euler's method) are fulfilled according to Sect. 2.4.2.

6. The layered and fibrous composites will be considered. For the layered composites, the layers are assumed to be directed perpendicular to the surface of end $x_3 = 0$. As applied to the fibrous composites, the unidirectional or orthogonal-reinforced materials will be considered under condition that the direction of preferred reinforcing coincides with the axis $0x_3$ (Fig. 5.5); i.e., it is directed perpendicular to the end surface $x_3 = 0$.

7. In the continuum approximation, the above composites with the polymeric or metallic matrixes will be modeled with a homogeneous compressed orthotropic elastic or elasto-plastic material with the averaged constants, the axis of symmetry properties of which coincides with the axes of the chosen coordinate system (Fig. 5.5). In the case of transverse material, it will be assumed that the planes $x_3 = $ const are the planes of isotropy.

8. The only near-the-surface instability near the loaded end (in Fig. 5.5, the loaded end is marked with a "grid") will investigated when the stresses and displacements attenuate at $x_3 \to -\infty$.
9. The research will be carried out within the framework of one of the variants of TLTDBS—within the **theory 3** (the second variant *of* the theory of small precritical deformations) in accordance with the terminology of Sect. 2.2.

With the involvement of the main provisions 1–9.

the continuum fracture theory under the crumpling of ends for the brittle and plastic fracture is developed; the method of researching the plane and spatial problems is elaborated, based on the application of Fourier integrals; as well as the concrete results are obtained.

The above results are quite consistent and detailed, as already noted in the introductory part of this subsection, set out in the monograph [4, vol. 2, Chap. 11, Sect. 5.4, pp. 529–551). Below, in a very brief form, the final results only are given.

Consider in advance the following designations.

$\left(\Pi_3^-\right)_T^{cru}$ − : the theoretical ultimate strength in the uniaxial compression along the axis $0x_3$ (Fig. 5.5) under fracture in the form of a crumpled end;

$\left(\Pi_3^-\right)_T$ — The theoretical ultimate strength under uniaxial compression along the axis $0x_3$.

(without the appearance of fracture in the form of a crumpled end) under the internal fracture (away from the end). The same designation was used in the continuum fracture theory set out in Sect. 3.3.2.

As a result of rigorous analysis, the final result is obtained, which with the above designations can be presented in the following form

$$\left(\Pi_3^-\right)_T^{cru} < \left(\Pi_3^-\right)_T \tag{5.8}$$

The inequality (5.8) is consistent with the physical considerations and repeated the experimental studies testifying that the fracture of homogeneous materials tends to start, as a rule, on the surface of the material.

In this regard, it should be noted that although this chapter considers the composite materials, but in Sect. 5.3.3 the continuum approximation for the composites is considered and in the continuum approximation the composites are homogeneous materials.

Despite the inequality (5.8) is strongly proven, the presented in this inequality quantities differ slightly among themselves, which is also proven in the monograph [4, vol. 2, Chap. 11, Sect. 5.4, pp. 529–551) for the brittle and plastic fracture in the form of crumpling of the ends. To illustrate the above situation, let us look at an example for a case of the brittle fracture.

Example Consider a fibrous unidirectional composite in which the fibers are directed along the axis $0x_3$ (Fig. 5.5) and which in the cross-section plane (at $x_3 = $ const in Fig. 5.5) has a disordered structure. This composite material under brittle fracture is modeled by an elastic transversal-isotropic material in which the planes of isotropy are planes $x_3 = $ const (Fig. 5.5). In this modeling, the expression (11.276) is obtained in the monograph [4, vol. 2, Chap. 11, Sect. 5.4, pp. 542–543) that can be presented in the following form

$$\left(\Pi_3^-\right)_T^{cru} = \left(\Pi_3^-\right)_T \left[1 - \frac{G'2}{EE'}\left(1 - v^2\right)\left(1 - v'^2\frac{E}{E'}\right)\right]. \qquad (5.9)$$

In (5.9), the designations are additionally introduced:

E' is the Young modulus along the axis $0x_3$, E is the Young modulus in the plane of isotropy (when $x_3 = $ const); G' is the shear modulus under shear along the fibers $G' = G_{31} = G_{32}$.

For this fibrous unidirectional composite in the continuum approximation (as for a material with the reduced shear stiffness), the following conditions (5.10) are valid

$$G' << E'; G' \approx E. \qquad (5.10)$$

The relation

$$\left(\Pi_3^-\right)_T^{cru} \approx \left(\Pi_3^-\right)_T. \qquad (5.11)$$

follows from (5.9) and (5.10).

A similar relation occurs also in the plastic fracture, but the proof has a much more cumbersome form.

Corollary The relation (5.11) can be used in the approximate comparison of the value $\left(\Pi_3^-\right)_{exp}^{cru}$ (experimental ultimate strength under uniaxial compression along the axis $0x_3$ in the case of fracture in the form of crumpling ends) with the value of the quantity $\left(\Pi_3^-\right)_T^{cru}$, including two positions.

The first position is the visual confirmation the belonging the experimentally observed fracture to *the* fracture of the crumpled ends (such as the photograph in Fig. 5.1).

The second position is the comparison of the value of quantity $\left(\Pi_3^-\right)_{exp}^{cru}$ with the value of quantity $\left(\Pi_3^-\right)_{exp}$ according to the relation (5.11).

Consider now the proposed comparison procedure, which is presented in the foregoing **Conclusion**, in relation to the annealed and non-annealed boro-aluminum unidirectional composite VKA-1, which was discussed in **Example** in Sect. 3.3.4.1. In this case, the first position is implemented by analyzing the photo in Fig. 5.1; the second position is determined by an analysis of the results presented in the Table. 5.1.

Note that Table 5.1 corresponds to the monograph [4, vol. 2, p. 551, Table 11.8).

Note also that numerical values for $\left(\Pi_3^-\right)_{exp}^{cru}$ in Table 5.1 correspond to the averaged values for 32 samples in the case of annealed and for 14 samples in the case of non-annealed material.

Table 5.1

Material VKA – 1	Ultimate strength, MPa	
	$\left(\Pi_3^-\right)_{\exp}^{cru}$	$\left(\Pi_3^-\right)_T^{cru}$
Annealed	665	736
Non-annealed	1282	1467

More details of the above results are provided in the article [8].

The above information will be limited to a very brief analysis of the results on **Problem 3** (*End-Crush Fracture of Composite Materials Under Compression*), which are obtained in the Department of Dynamics and Stability of Continua of the S. P. Timoshenko Institute of Mechanics of the NASU.

References

1. Goldman, A.Y., Savelieva, N.F., Smirnov, V.I.: Issledovanie mekhanicheskikh svoistv tkanevykh stekloplastikov pri rastiazhenii i szhatii normalno k ploskosti armirovaniia (Research of mechanical properties of fiberglass fabrics under strain and compression normal to the plane of reinforcement). Mekhanika Polymerov **5**, 803–809 (1968)
2. Guz, A.N.: O kontinualnoi teorii razrusheniia kompozitnykh materialov pri smiatii tortsov (Continuum fracture theory of composite materials with crushed ends). Dokl. Akad. Nauk SSSR **298**(3), 565–570 (1988)
3. Guz, A.N.: Mekhanika razrusheniia kompozitnykh materialov pri szhatii (Fracture Mechanics of Composite Materials Under Compression). Naukova Dumka, Kyiv (1990)
4. Guz, A.N.: Osnovy mekhaniki razrusheniia kompozitov pri szhatii: V 2-kh tomakh (Fundamentals of the Fracture Mechanics of Composites Under Compression: In 2 volumes). Litera, Kyiv (2008)
5. Guz, A.N., Dekret, V.A.: Model korotkikh volokon v teorii ustoichivosti kompozitov (The Model of Short Fibers in the Theory of Stability of Composites). LAP LAMBERT Academic Publishing, Saarbrücken, Deutschland (2015)
6. Guz, A.N., Dekret, V.A.: Model volokon konechnykh razmerov v trekhmernoi teorii ustoichivosti kompozitnykh materialov (obzor) (Finite-fiber model in the three-dimensional theory of stability of composite materials (Review)). Prikladnaya Mekhanika **52**(1), 3–77 (2016)
7. Guz, A.N., Dekret, V.A.: Khrupkoe razrushenie kompozitnykh materialov pri smiatii tortsov (model kusochno-odnorodnoi sredy) (Brittle fracture of composite materials with crushed ends (model of piecewise homogeneous medium)). Dokl. Akad. Nauk SSSR **296**(4), 805–808 (1987)
8. Guz, A.N., Cherevko, M.A., Margolin, G.G., Romashko, I.M.: O razrushenii odnonapravlennykh boroaliuminievykh kompozitov pri szhatii (Fracture of unidirectional boron-aluminum composites under compression). Mekhanika kompozitnykh materialov **2**, 226–230 (1986)
9. Dekret, V.A.: Zastosuvannya metodu skinchennikh riznits do problemy pruzhnoï stiykosti (Finite differences method application to the problem of elastic stability). Dopovidi Akademii Nauk URSR, Ser. A. **7**, 537–539 (1973)
10. Dekret, V.A.: Ob odnom sposobe resheniya zadach trekhmernoy ustoychivosti kompozitnykh materialov lentochnoy struktury (On a method of solving problems of three-dimensional stability of a tape structure composite materials). Doklady Akademii nauk USSR, Ser.A. **2**, 31–33 (1989)
11. Guz, A.N. (ed.): Mekhanika kompozitov, v 12 tomah (Mechanics of Composites, in 12 volumes). (Vol. 1–4—Naukova Dumka, Vol. 5–12—A.S.K., Kyiv, 1993–2003), V.T. Golovach (ed.), T.1. Statika materialov. (Vol. 1. Statics of materials) (1993), N.A. Shulga (ed.),

T.2. Dinamika i ustoichivost materialov. (Vol.2. Dynamics and stability of materials) (1993), L.P. Khoroshun (ed.), T.3. Statisticheskaia mekhanika i effektivnye svoistva materialov (Vol.3. Statistical Mechanics and Effective Material Properties) (1993), A.N. Guz, S.D. Akbarov, (ed.), T.4. Mekhanika materialov s iskrivlennymi strukturami (Vol.4. Mechanics of materials with curved structures) (1995), A.A. Kaminsky (ed.), T.5. Mekhanika razrusheniia. (Vol.5. Fracture mechanics) (1996), N.A. Shulga, V.T. Tomashevsky (ed.), T.6. Tekhnologicheskie napriazheniia i deformatsii v materialakh. (Vol.6. Technological stresses and deformations in materials) (1997), A.N. Guz, A.S. Kosmodamiansky, V.P. Shevchenko (ed.), T.7. Kontsentratsiia napriazhenii. (Vol.7. Stress concentration) (1998), Ya.M. Grigorenko (ed.), T.8. Statika elementov konstruktsii. (Vol.8. Structural elements statics) (1999), V.D. Kubenko (ed.), T.9. Dinamika elementov konstruktsii. (Vol.9. Dynamics of structural elements) (1999), I.Yu. Babich (ed.), T.10. Ustoichivost elementov konstruktsii. (Vol.10. Stability of structural elements) (2001), Ya.M. Grigorenko, Yu.N. Shevchenko (ed.), T.11. Chislennye metody. (Vol.11. Numerical methods) (2002), A.N. Guz, L.P. Khoroshun (ed.), T.12. Prikladnye issledovaniia. (Vol.12. Applied research) (2003)

12. Guz, A.N.: Continuous theory of failure of composite materials with buckling at the ends (Brittle fracture). Sov. Appl. Mech. **23**(1), 52–60 (1987)

13. Guz, A.N.: Continuous theory of failure of composite materials with buckling at the ends (Plastic failure). Sov. Appl. Mech. **23**(5), 411–417 (1987)

14. Guz, A.N.: Continual theory of fracture of composite materials at bearing strain of end faces in compression. In: Proceedings of Conference Fracture of Engineering Materials and Structures, pp. 838–843. Elsevier, Singapore (1991)

15. Guz, A.N.: Fracture of fibrous composites at bearing strain in end faces in compression. In: Proceedings of 2nd International Symposium on Composite Materials and Structures, pp. 232–236. Beijing, China (1992)

16. Guz, A.N.: Fracture of fibrous composites at bearing strain in end faces in compression. In: Proceedings of ICCM/9 Composites and Applications, vol. VI, pp.613–618. Madrid, 12–16 July, (1993)

17. Guz, A.N.: Continual theory of fracture of composite materials at bearing strain in end faces in compression. In: Proceedings of ASM International Materials Congress, Material Park. Mechanisms and Mechanics of Composites Fracture, pp.201–207. Pittsburgh, Pennsylvania (1993)

18. Guz, A.N.: On failure propagation in composite materials in compression (Three-dimensional continual theory). In: Proceedings of ECF 11 Poitiers–Futuroscope, pp.1769–1774. Vol. III France, Sept. 3–6 (1996)

19. Guz, A.N.: Non-classical problems of composite failure. In: Proceedings of ICCST/1, pp.161–166. Durban, South. Africa, 18–20 June (1996)

20. Guz, A.N.: Non-classical problems of composite failure. In: Proceedings of ICF9 Advance in Fracture Research, Vol. 4, pp. 1911–1921. Sydney, Australia (1997)

21. Guz, A.N.: The fracture theory of composite at bearing strain in end faces. In: Report presented at the Conference Composite Construction and Innovation, pp.783–788. Innsbruck, Austria, Sept. 16–18 (1997)

22. Guz, A.N.: Some modern problems of physical mechanics of fracture. In: Cherepanov G.P. (ed.) FRACTURE. A Topical Encyclopedia of Current Knowledge, pp. 709–720. Krieger Publ. Company, Malabar, Florida (1998)

23. Guz, A.N.: Study and Analysis of Non-classical Problems of Fracture and Failure Mechanics and Corresponding Mechanisms. Lecture presented at Institute of Mechanics HANOI (1998)

24. Guz, A.N.: Description and study of some nonclassical problems of fracture mechanics and related mechanisms. Int. Appl. Mech. **36**(12), 1537–1564 (2000)

25. Guz, A.N.: In On study of nonclassical problems of fracture and failure mechanics and related mechanisms, pp. 35–68. ANNALS of the European Academy of Sciences Liège, Belgium (2006–2007)

26. Guz, A.N.: On study of nonclassical problems of fracture and failure mechanics and related mechanisms. Int. Appl. Mech. **45**(1), 1–31 (2009)

27. Kokhanenko, Yu.V.: Numerical solution of problems of the theory of elasticity and the three-dimensional stability of piecewise-homogeneous media. Sov. Appl. Mech. **22**(11), 1052–1058 (1986)
28. Kokhanenko, Yu.V.: Numerical study of the three-dimensional stability problems for laminated and ribbon-reinforced composites. Int. Appl. Mech. **37**(3), 317–340 (2001)

Part III
Other Non-Classical Problems of Fracture Mechanics

Chapter 6
Problem 4. Brittle Fracture of Materials with Cracks Taking into Account the Action of Initial (Residual) Stresses Along Cracks

This chapter, in a very brief form (in comparison with **Problem 1** and **Problem 2**, which are discussed in Part II of this monograph in Chaps. 3 and 4, respectively), presents the main results on the considered **Problem 4** which are obtained since 1980 in the Department of Dynamics and Stability of Continua of the S. P. Timoshenko Institute of Mechanics of the NASU. The presentation of these results is written in the style announced in the Introduction (Part I) to this monograph (without the excessively invoking aspects of a mathematical nature).

6.1 Introduction

The main results on this problem, obtained in the Department of Dynamics and Stability of Continua, are presented in the monographs [1, 2] (vol. 2), [3] (vol. 2), [4, 5] and in a number of other publications of a monographic nature, as well as in the review articles [6–10]. Initially, these results were published in the scientific articles, which are indicated in the bibliography for the above-mentioned monographs and review articles. From the list of publications given in this monograph, the articles [11–54] and materials on the reports at international conferences [55–59] are related to the scientific direction of the department.

It should be noted that the dissertation for the degree of Doctor of Physical and Mathematical Sciences (DSc) by V.L. Bogdanov also refers in part to the discussed scientific direction. On the whole, this dissertation is devoted to the combined study of **Problem 4** and **Problem 6**. The main results of this approach are presented in the monograph [4].

The discussed results in this chapter on the mechanics of brittle fracture of materials with cracks, taking into account the action of the initial (residual) stresses along the cracks, refer to the isotropic materials and orthotropic materials. At that, in the

A. N. Guz, *Eight Non-Classical Problems of Fracture Mechanics*,
Advanced Structured Materials 159,
https://doi.org/10.1007/978-3-030-77501-8_6

case of orthotropic materials, it is assumed that the plane cracks are located in the plane of symmetry of the material properties.

It is worth noting that the composite materials in the continuum approximation are also modeled by the orthotropic materials. Therefore, the results discussed in this chapter also apply in this sense to the composite materials.

Apparently, it should be emphasized that in the problem of constructing the mechanics of the interaction of initial (residual) stresses and stresses arising under the action of additional loads (these stresses can be conventionally called the additional stresses), three significantly different *situations or design schemes* are possible.

The first situation. *The initial (residual) stresses are significantly less than the additional stresses.* In this case, it is possible to ignore (discard) the initial (residual) stresses and carry out the research within the framework of the classical (linear or nonlinear) theory of elasticity relative to the problems of brittle fracture mechanics.

The second situation. *The initial (residual) stresses are of the same order as the additional stresses.* In this case, it is necessary to carry out the study of the general stress–strain state, without highlighting the initial (residual) and additional stresses. For the study under discussion, an exclusively *nonlinear* theory of elasticity of small or finite deformations should be used. Then the specific results concerning the mechanics of brittle fracture can be obtained only with the use of numerical methods and computational mechanics. With this approach, these results are apparently multiparametric, and their analysis, which is necessary for the mechanics of brittle fracture, is fraught with significant difficulties.

The third situation. *The initial (residual) stresses are significantly higher than the additional stresses.* The discussed situation is quite realistic, for example, for the composite materials. The fact is that the composite materials for specific structural elements are created in such a way that the reinforcing elements (fibers, etc.) are directed along the lines of force. Thus, the stresses along the reinforcing elements reach significantly higher values compared to the stresses in the perpendicular direction. In this case, the interfaces between the filler and the binder (matrix) are directed along the reinforcing elements (along the initial (residual) stresses). At that, the defects of various types are located in the interfaces, which lead to a significant decrease in adhesion strength (strength under the action of stresses in the direction perpendicular to the initial stresses, i.e., under the action of additional stresses). Thus, *in the third situation*, the additional stresses, being smaller than the initial (residual) stresses, can have a significant effect on the overall strength of the composite materials.

The above considerations indicate that *the third situation* is realistic and it is necessary to construct the mechanics of the interaction of the initial (residual) stresses and the additional stresses, despite the smallness of the second stresses in comparison with the first stresses when researching the mechanics of brittle fracture of materials with initial (residual) stresses. In this situation, the brittle fracture mechanics of such materials cannot be constructed using the basic relations and results of the classical linear theory of elasticity, since the singular part of the well-known Inglis–Muskhelishvili solution for the material with the crack does not include the initial (residual) stresses along the cracks.

In connection with the above considerations, for the construction of the mechanics of brittle fracture of materials with cracks, taking into account the action of the initial (residual) stresses along the cracks, to obtain results in a foreseeable form, the three-dimensional linearized theory of elasticity (TLTE) is used for the finite (large) and small initial deformations.

Following Note 2.2, *the three-dimensional linearized theory of elasticity (TLTE) for materials with initial (residual) stresses completely coincides with the three-dimensional linearized theory of elastic body stability (TLTEBS), if all stresses with the index "0" in the latter are considered known and corresponding the initial (residual) stresses.*

Note that Chap. 2 of this monograph sets out the main mathematical aspects of the three-dimensional linearized theory of deformable body stability. (TLTDBS), a special case of which is the three-dimensional linearized theory of elastic bodies stability (TLTEBS).

It should also be noted that the TLTEBS is presented in Chap. 2 of this monograph in a fairly general setting (*in the general form for the theories of large and small subcritical or initial deformations, for the compressible and incompressible hyperelastic bodies with an arbitrary structure of the elastic potential with the considered symmetry of properties, etc.*).

Following Note 2.2, the above form of generality also has the TLTE for the materials with initial (residual) stresses, with the involvement of which the studies were carried out on **Problem 4** in the Department of Dynamics and Stability of Continua of the S. P. Timoshenko Institute of Mechanics of the NASU.

Note 6.1 The results on the mechanics of brittle fracture of materials with cracks, taking into account the action of initial stresses along the cracks, which were obtained in the Department of Dynamics and Stability of Continua, were constructed *with the above General Statement.* This information also applies to the first publications of the department, which are the articles of the author of this monograph [15–17, 30–37] of 1980–1981. In other scientific centers around the world, mainly on the discussed **Problem 4**, the studies have been and are being carried out for the materials with elastic potentials of a particular specific structure. An example of this approach is the article [60] back of 1967, which was, apparently, the first publication in the world scientific literature on the mechanics of brittle fracture of materials with initial stresses along cracks and in which results were obtained for an incompressible material only with the elastic potential of the NeoHookean type.

6.2 Preliminary Discussion. Statement of Problems

A typical situation for the mechanics of brittle fracture of materials with initial (residual) stresses is shown in Fig. 6.1, where the designations are introduced:

$2a$—the crack width in the plane $y_3 = \text{const}$;

Fig. 6.1

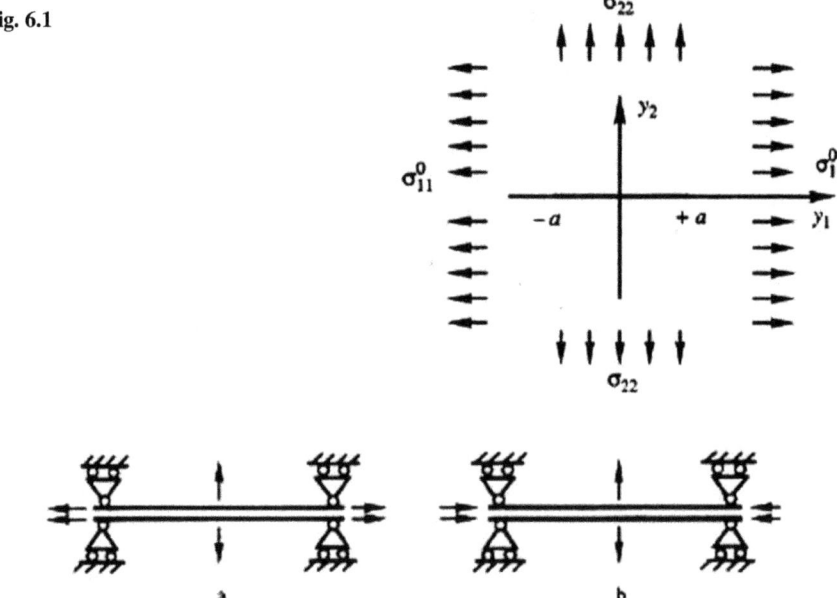

Fig. 6.2

σ_{11}^0—the initial (residual) stresses along the crack;

σ_{22}—the additional (acting or operational) stresses.

As applied to Fig. 6.1 within the framework of *the third situation*, discussed in the Introduction (Sect. 6.1) in this chapter, it is intuitively believed that the initial (residual) stresses σ_{11}^0 should have a certain effect on the brittle fracture process. To confirm this consideration, consider two "thought" experiments.

Two strings are shown in Fig. 6.2a that are placed parallel and tightly to one another, under the action of a tensile load along their axes.

To separate the strings, apply the forces in the middle of their length, directed perpendicular to the axes of the strings. Obviously, to separate the strings by a given deflection value, it is necessary to apply forces, the values of which should depend on the string tension, i.e., on the value of the tensile load.

Two rods are shown in Fig. 6.2b, which are located parallel and closely to one another, under the action of a compressive load along their axes. To separate the rods, in the middle of their length the forces are applied perpendicular to the rods' axes. Obviously, to separate the rods by a given deflection value, it is necessary to apply forces, the value of which should depend on the compressive load. When the compressive load reaches the Eulerian critical value, the system will be in a state of neutral equilibrium. In this state, it is enough to apply the negligible by value forces perpendicular to the axes of the rods to separate the rods.

Thus, upon preliminary consideration, the "thought" experiments clearly indicate that the initial (residual) stresses σ_{11}^0 (Fig. 6.1), corresponding to the horizontal loads in Fig. 6.2a, b, should influence the brittle fracture process. Apparently, the above mechanical effects obtained as a result of the "thought" experiments should also manifest themselves. Consequently, it can be assumed that concerning *the third situation* considered in the Introduction (Sect. 6.1) in this chapter, as a result of preliminary discussion (consideration), the expediency or necessity of developing the mechanics of brittle fracture of materials with initial (residual) stresses is proved to a certain extent.

Taking into account the above preliminary discussion, below **the rigorous statement** of the mechanics of brittle fracture of materials with cracks is considered, taking into account the action of the initial (residual) stresses along the cracks. At that, we will follow the corresponding approaches of the monographs [1, 2] (vol. 2), [3] (vol. 2), and [4], adhering to the presentation of the monograph [3] (vol. 2).

First of all, note that in the discussed mechanics of brittle fracture of materials with initial (residual) stresses along cracks, *the reference state corresponds to the second (unperturbed, initial) state in the terminology of* Chap. 2. In this regard, the Lagrangian coordinates ($m = 1, 2, 3$) are introduced in the reference (initial) state, which coincide with the Cartesian coordinates. All further presentation of the discussed mechanics is carried out in the above Lagrangian coordinates y_m. For example, all the necessary results for the plane problems are presented in Sects. 2.5.4 and 2.5.6 of Chap. 2 of this monograph (the general solutions, the representations of solutions of plane dynamic and static problems, etc.). The "prime" index in the various quantities indicates that these quantities refer to the results using the above Lagrangian coordinates.

Additional and more detailed information on the formation of the basic relations and results when using the above Lagrangian coordinates is given, for example, in the monograph [3] (vols. 1, 2).

The general statement of problems in the mechanics of brittle fracture of materials with initial (residual) stresses along cracks.

The isotropic and orthotropic hyperelastic materials with an arbitrary structure of the elastic potential are considered. In the case of orthotropic materials, it is assumed that the plane cracks are located in the planes of symmetry of material properties. These materials contain the plane cracks of various shapes, which are located in the parallel planes at $y_2 = $ const. One of the indicated planes (at $y_2 = 0$) is shown, for example, in Fig. 6.3.

The research is carried out in Lagrangian coordinates $y_m (m = 1, 2, 3)$, which are introduced in the second (initial, in the terminology of Chap. 2 of this monograph) state and coincide with Cartesian coordinates in this state. Thus, the initial (residual) state in this chapter is considered as the reference state. It is assumed that the initial (residual) stress–strain state is homogeneous, i.e., the expressions (2.78) hold. Additionally, it is assumed that the initial (residual) stresses act only along the planes in which the cracks are located. Thus, following Fig. 6.3, the initial (residual) stresses are taken

Fig. 6.3

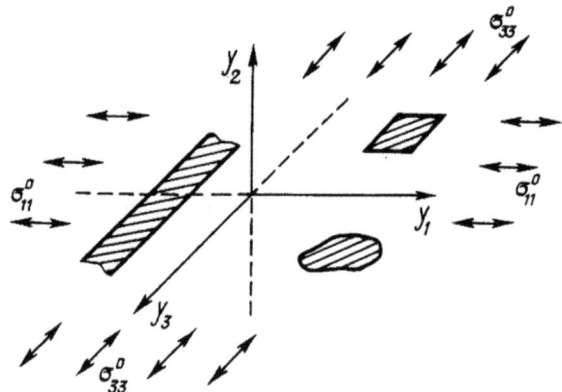

$$\sigma_{22}^0 = 0, \sigma_{11}^0 \neq 0, \sigma_{33}^0 \neq 0, \sigma_{ij}^0 = 0 \text{ for } i \neq j. \tag{6.1}$$

*The research is carried out in the unified general form for **theories 1, 2, and 3** in accordance with the terminology of* Chap. 2 *of this monograph with the involvement of the Lagrangian coordinates introduced in the initial (residual) state. For the two-dimensional problems, the mathematical apparatus described in* Sects. 2.6.4, 2.6.5, *and* 2.6.6 *of* Chap. 2 *of this monograph, and, for the three-dimensional problems, the general solutions set forth, for example, in the monograph* [3] (vol. 1, Chap. 1, Sect. 4) *are applied.*

Thus, it is assumed in this chapter that the following **assumptions 1–4** are always fulfilled, which are the main ones in the outlined mechanics of brittle fracture of materials with initial (residual) stresses along cracks and, naturally, determining the area of its application.

Assumption 1 In the initial (residual) stress–strain state, the material with cracks is loaded in such a way that *no initial (residual) stresses arise* on the planes in which the cracks are located.

Assumption 2 When the additional arbitrary loads (the additional to the initial (residual) stress–strain state) are applied to the material, the arising disturbances of the stress–strain state are *much less* than the corresponding values for the initial (residual) stress–strain state.

Assumption 3 The initial (residual) stress–strain state has such a structure that in the neighborhood of the crack, it can (with a sufficient degree of accuracy) be considered *locally homogeneous*.

Assumption 4 The solution to the linearized theory of elasticity, which arises in the framework of the considered variant of the mechanics of brittle fracture of materials with initial (residual) stresses along the cracks, *is unique*. This means that the conditions of the type (2.55) of Chap. 2 of this monograph are satisfied.

It should be noted that the first two assumptions are basic in the theory under consideration, the third assumption is auxiliary in the essence of the problem under consideration and is quite important in solving problems arising from this approach.

The third assumption is fulfilled when the first assumption is fulfilled if the initial (residual) stress–strain state does not change significantly at distances of the same order of value with the dimensions of the cracks.

Thus, the first and third assumptions ensure the existence of *a homogeneous* initial (residual) stress–strain state, for which the general solutions are constructed [3] (vol. 1, Chap. 1, Sect. 4).

The second assumption makes it possible to apply the linearized theory of elasticity to determine the additional (concerning the initial or residual state) stress–strain state. This assumption is only a consequence of the fact of considering *the third situation*, which is noted in the Introduction (Sect. 6.1) in this chapter, concerning the construction of the mechanics of interaction of initial (residual) stresses and stresses arising under the action of additional loads.

The fourth assumption ensures the uniqueness of the solution of the discussed linearized problem corresponding to the discussed mechanics of brittle fracture of materials with initial (residual) stresses along the cracks. Without fulfilling the above condition, it seems that the study of the problems under consideration is completely meaningless.

It should be noted that the second assumption in the discussed approach is violated near the crack tip since at the crack tip, the stress perturbations (stresses of the linearized theory) asymptotically tend to "infinity." A similar situation takes place in the classical linear mechanics of brittle fracture, which was discussed many times in the well-known publications, including the well-known monographs. Of course, at the crack tip, the "endless" stresses have no physical meaning. If to relate the behavior of stresses near the crack tip (amplitude values under the singularity) with the integral characteristics related to fracture criteria, then the useful, in some cases, necessary information can be obtained.

In the above sense, the mechanics of brittle fracture of materials with initial (residual) stresses along cracks, which is based on solutions of the linearized theory of elasticity for bodies with cracks, is also understood.

Note 6.2 Taking into account the General Statement of problems set out in this paragraph with the involvement of Lagrangian coordinates y_m ($m = 1, 2, 3$), which are introduced in the initial (residual, second, unperturbed) state, *the statement of specific problems* (for bodies of a specific shape with the cracks of a specific shape) of *the mechanics of brittle fracture of materials with initial (residual) stresses, which act along the cracks, completely coincides with the statement of the corresponding problems of the classical linear mechanics of brittle fracture.* With this approach, only the systems of basic differential equations and expressions for determining the t stress tensors components through the displacements, which are given in the monograph [3] (vol. 1, Chap. 1, Sect. 4), will differ. In the case of plane static and dynamic problems, the corresponding mathematical apparatus (using the complex potentials) is presented in Sects. 2.6.4, 2.6.5, and 2.6.6 of Chap. 2 of this monograph.

6.3 Plane and Anti-plane Statical Problems. Criteria of Fracture

In Sect. 6.3, the exact solutions are analyzed in a very brief form for the basic statical plane and anti-plane problems of the mechanics of brittle fracture of materials with the initial (residual) stresses that act along cracks, as applied to an infinite material with one plane (infinite in the direction of the axis $0y_3$) crack of finite width $2a$ placed (Fig. 6.4) in the plane $y_1 0y_3$ ($|y_1| \leq a$, $y_2 = 0$, $-\infty < y_3 < +\infty$).

The results were obtained for the cracks Mode I, Mode II, and Mode III, as well as for the wedging crack for the General Statement, which is described in the previous Sect. 6.2 near Fig. 6.3 with the involvement of the complex potentials presented in Sect. 2.6.5 of this monograph. The methods of the Riemann–Hilbert problem [61, 62] were used for the study. At that, in most of the considered problems, it was possible to restrict oneself to using the Keldysh–Sedov formula [63].

Initially, the main results were published in the articles of the author of the present monograph [15–17, 30–37] of 1980–1982. These results are outlined in the monograph [1] of 1983, which contains a list of the author's main publications for the indicated period.

6.3.1 Order of Singularity

Unfortunately, in the period 1980–1983 in the above publications, the question of the order of the singularity at the crack tip in the problems of the discussed mechanics of brittle fracture of materials with initial (residual) stresses was not fully solved.

Only in 1986, starting with the publication [19], **it was rigorously proved that the order of the singularity at the crack tip, calculated within the framework**

Fig. 6.4

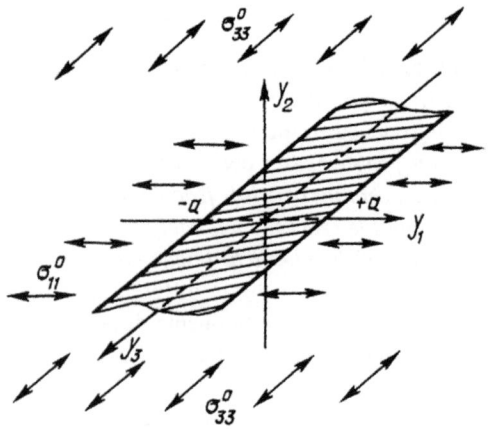

of the discussed variant of the mechanics of brittle fracture of materials with initial (residual) stresses along the cracks coincides with the order of the singularity within the framework of the classical linear brittle fracture mechanics. The above result is one of the main results for **Problem 4**, which is discussed in this chapter.

It should be noted that this result was obtained for **the General Statement**, which is set out in the previous Sect. 6.2 near Fig. 6.3, for the isotropic and orthotropic hyperelastic materials with an arbitrary structure of the elastic potential.

Taking into account the above results on the order of the singularity in the crack tip, all the exact solutions obtained are presented in the monograph [2] (vol. 2) for 1991. The most complete presentation of the discussed results obtained based on exact solutions is presented in the monographs [3] (vol. 2, Chap. 7, pp. 18–144) and [4] (Chap. 3, pp. 195–268).

6.3.2 Effects of Resonant Character

When loading the lower half-space $y_2 \leq 0$ with loads (6.1), which are considered in this chapter, a near-the-surface loss of stability can occur, which consists in the fact that the near-the-surface layers of the material of the lower half-space lose their stability by the modes, which decreases in value by moving from the boundary of the half-space (at $y_2 \to -\infty$). By using the mathematical apparatus of Sect. 2.6.5 of this monograph, the characteristic equations were presented in the [3] (vol. 2, Chap. 7) to determine the critical values of the loading parameters (6.1) corresponding to the occurrence of near-the-surface loss of stability, taking into account the notation of Sect. 2.5.5 for the case of unequal roots $\mu_1' \neq \mu_2'$ in the form

$$\mu_2' \gamma_{21}^{(2)} - \mu_1' \gamma_{21}^{(1)} = 0 \tag{6.2}$$

and for the case of equal roots $\mu_1' = \mu_2'$ in the following form

$$\gamma_{21}^{(2)} - \mu_1' \gamma_{21}^{(1)} \gamma_{22}^{(2)} = 0. \tag{6.3}$$

Equations (6.2) and (6.3) will be used below when analyzing the exact solutions for various cracks (Modes I, II, and III).

As examples, consider a number of the exact solutions for crack length $2a$ (Fig. 6.4).

So, for the Mode I crack, loaded with a load of intensity $g(y_1)$, for the case of unequal roots $(\mu_1' \neq \mu_2')$, the exact solution is given ([3] (vol. 2, pp. 75–79)) by the function

$$Z(z_j) = \frac{\mu_2' \gamma_{21}^{(2)}}{2\pi \left(\mu_2' \gamma_{21}^{(2)} - \mu_1' \gamma_{21}^{(1)}\right) \sqrt{z_j^2 - a^2}} \int_{-a}^{+a} \frac{g(t)\sqrt{a^2 - t^2}}{z_j - t} dt; \ j = 1, 2. \quad (6.4)$$

For the Mode II crack, loaded with an intensity load $h(y_1)$, for the case of equal roots $()\mu_1' = \mu_2'$, the exact solution is given ([3] (vol. 2, pp. 98–99)) by the formula

$$Z(z_1) = \frac{2i}{\pi \left(\gamma_{21}^{(2)} - \mu_1' \gamma_{21}^{(1)} \gamma_{22}^{(2)}\right) \sqrt{z_1^2 - a^2}} \int_{-a}^{+a} \frac{h(t)\sqrt{a^2 - t^2}}{z_1 - t} dt. \mu_1' = \mu_2'. \quad (6.5)$$

For the Mode III crack, loaded with a load of intensity $f(y_1)$, the exact solution is given ([3] (vol. 2, pp. 108–110)) by the formula

$$Z(z_3) = \frac{1}{\pi \sqrt{z_3^2 - a^2}} \int_{-a}^{+a} \frac{f(t)\sqrt{a^2 - t^2}}{z_3 - t} dt, \ z_3 = y_1 + \mu_3' y_2. \quad (6.6)$$

For the Mode I cracks, in the case of equal roots $\left(\mu_1' = \mu_2'\right)$ and Mode II cracks in the case of unequal roots $\left(\mu_1' \neq \mu_2'\right)$, the exact solutions are given by similar formulas (in structure) to formulas (6.4) and (6.5) [3] (vol. 2, Chap. 7).

From the analysis of the exact solutions in the form (6.4) and (6.5), as well as other exact solutions from [3] (vol. 2, Chap. 7) corresponding to *the Mode I and Mode II cracks*, it follows that the denominators of the discussed exact solutions *include the expressions corresponding to the left-hand sides of* Eqs. (6.2) and (6.3).

Thus, as the initial (residual) stresses tend to the values of the roots of Eqs. (6.2) and (6.3), *the exact solutions* of the type (6.4) and (6.5) corresponding to the Mode I and Mode II cracks *tend to "infinity."*

It follows from the analysis of the exact solution (6.6) corresponding to the Mode III cracks that for these cracks a phenomenon similar to the Mode I and Mode II cracks *is not observed.*

It should be noted that the asymptotic behavior of stresses and displacements near the crack tip was also determined based on the exact solutions of the type (6.4)–(6.6). It follows from the analysis of these results that for the individual stresses and displacements, the noted above for exact solutions of the type (6.4)–(6.6) regularities *remain.*

Thus, it can be assumed that, based on the results of the above analysis, the existence of the following regularity is proved.

*When the initial (residual) stresses tend to values corresponding to the near-the-surface instability of the half-space (to the roots of equations (6.2) and(6.3)), for the Mode I and Mode II cracks, **the phenomena of resonance character appear**, and for Mode III **such phenomena do not occur**.*

It is advisable to note that above (and, consequently, in the monographs [1] of 1983, [2] (vol. 2) of 1991, [3] (vol. 2) of 2008, and [4] of 2017, as well as earlier in the corresponding publications in periodicals), the above-formulated **regularity** was proved for *the hyperelastic isotropic and orthotropic materials with an arbitrary structure of the elastic potential,* based on the General Statement of problems, which is set out in Sect. 6.2 near Fig. 6.3.

The overwhelming majority of other authors consider the discussed regularity at present, proceeding from the results for the elastic potentials of a particular concrete structure.

6.3.3 Criteria of Fracture

The criteria for the brittle fracture of materials with initial (residual) stresses that act along cracks have been constructed starting with the article [17] of 1982, following the approach that was applied everywhere in the classical linear mechanics of brittle fracture of materials. In subsequent years, these criteria were discussed in several publications presented in the list of references to this monograph and were included in the monographs [1] of 1983, [2] (vol. 2) of 1991, [3] (vol. 2) of 2008, and [4] of 2017. Below, the criterion is given for the brittle fracture of materials with initial (residual) stresses along cracks as applied to the material with the Modes I, II, and III cracks, following the monograph [3] (vol. 2, p. 125–142) for the cases of unequal $\mu_1' \neq \mu_2'$ and equal $\mu_1' = \mu_2'$ roots.

So, in the case of unequal roots $\mu_1' \neq \mu_2'$, this criterion has the following form:

$$K_I^2 \left(\gamma_2^{(1)} \mu_2' \gamma_{21}^{(2)} - \gamma_2^{(2)} \mu_1' \gamma_{21}^{(1)} \right) + K_{II}^2 \left(\gamma_1^{(1)} - \gamma_1^{(2)} \right)$$
$$+ K_{III}^2 \left(\omega_{1331}' \omega_{2332}' \right)^{-\frac{1}{2}} i^{-1} \left(\mu_2' \gamma_{21}^{(2)} - \mu_1' \gamma_{21}^{(1)} \right) = 4\gamma i^{-1} \left(\mu_2' \gamma_{21}^{(2)} - \mu_1' \gamma_{21}^{(1)} \right).$$

(6.7)

In the case of equal roots $\mu_1' = \mu_2'$, this criterion has the following form:

$$K_I^2 \left[\left(\gamma_2^{(1)} + \gamma_2^{(2)} \right) \mu_1' \gamma_{21}^{(1)} - \gamma_2^{(1)} \gamma_{21}^{(2)} \right] + K_{II}^2 \left(\gamma_1^{(1)} \gamma_{22}^{(2)} - \gamma_1^{(2)} - \gamma_1^{(1)} \right)$$
$$+ K_{III}^2 \left(\omega_{1331}' \omega_{2332}' \right)^{-\frac{1}{2}} i^{-1} \left(\mu_1' \gamma_{21}^{(1)} \gamma_{22}^{(2)} - \gamma_{21}^{(2)} \right) = 4\gamma i^{-1} \left(\mu_1' \gamma_{21}^{(1)} \gamma_{22}^{(2)} - \gamma_{21}^{(2)} \right).$$

(6.8)

In expressions (6.7) and (6.8), the following notations are introduced:

K_I, K_{II} and K_{III}—stress intensity factors for the Mode I, Mode II, and Mode III cracks;

γ—the value of the specific surface energy per unit length along the axis $0y_3$. (Fig. 6.3);

All other notations in (6.7) and (6.8) correspond to the considered linearized theory and are given, for example, in monograph [3] (vol. 2).

It should be noted that the results (6.7) and (6.8) are given for a compressible body and due to the application of the General Statement of problems outlined in Sect. 6.2 near Fig. 6.3 and *refer to the isotropic and orthotropic materials with an arbitrary structure of the elastic potential.* Following Irwin, Orowan, and other authors, when calculating the energy loss for the fracture, it is possible to take into account in γ the specific irreversible energy associated with taking into account the plastic deformations in the local zones.

Consider the change in the fracture criteria in the form (6.7) and (6.8), when the values of the initial (residual) stresses in the form (6.1) verge toward the values corresponding to the occurrence of near-the-surface loss of stability of the half-space, i.e., to the values of the roots of Eqs. (6.2) and (6.3).

Recall that in this case, as noted in the previous section, the phenomena of *a resonant character* arise. Substituting (6.1) into (6.7) and (6.3) into (6.8), for the indicated case we obtain a fracture criterion for the unequal roots $\mu'_1 \neq \mu'_2$ in the following form:

$$K_I^2\left(\gamma_2^{(1)}\mu'_2\gamma_{21}^{(2)} - \gamma_2^{(2)}\mu'_1\gamma_{21}^{(1)}\right) + K_{II}^2\left(\gamma_1^{(1)} - \gamma_1^{(2)}\right) = 0 \qquad (6.9)$$

and the fracture criterion for the equal roots $\mu'_1 = \mu'_2$ in the following form:

$$K_I^2\left[\left(\gamma_2^{(1)} + \gamma_2^{(2)}\right)\mu'_1\gamma_{21}^{(1)} - \gamma_2^{(1)}\gamma_{21}^{(2)}\right] + K_{II}^2\left(\gamma_1^{(1)}\gamma_{22}^{(2)} - \gamma_1^{(2)} - \gamma_1^{(1)}\right) = 0. \quad (6.10)$$

From the fracture criteria in the forms (6.9) and (6.10) for one Mode I crack (in this case $K_{II} \equiv 0$) for the cases of unequal $\left(\mu'_1 \neq \mu'_2\right)$ and equal $\left(\mu'_1 = \mu'_2\right)$ roots, the fracture criterion is obtained in the following form:

$$K_I^2 = 0. \qquad (6.11)$$

Note that the fracture criterion in the form (6.11) refers to the additional loads in the direction perpendicular to the crack plane (in the direction of the axis $0y_2$ in Fig. 6.4).

These loads are linearly included in K_I.

Note also that criterion (6.11) refers to the Mode I crack when the initial (residual) stresses in the form (6.1) verge toward the values corresponding to *the resonant character phenomena* (to the roots of Eqs. (6.2) and (6.3)).

In this case, it is sufficient to apply the insignificant additional loads (at $K_I \to 0$) perpendicular to the crack, so that, following the criterion in the form (6.11), failure occurs. The corresponding effect was also discussed in a "thought experiment" concerning the compression of two rods (Fig. 6.2b)*when the compressive load along the rods axis verge towards the value of the Euler critical value.*

We restrict ourselves to the above information when considering in this section the static plane and anti-plane problems of the mechanics of brittle fracture of materials, taking into account the initial (residual) stresses that act along the cracks.

Note 6.3 This section contains a sufficient amount of information in an analytical (mathematical) form, which does not correspond to the style of writing this review, announced in the Introduction (part I of this monograph). The noted situation is because *all the results of* Sect. 6.3 *are based on the exact solutions obtained and on their corresponding analysis.*

6.4 Spatial Statical Problems

In this section, the information in an annotated form is considered about the results on the spatial static problems of the mechanics of brittle fracture of materials with initial (residual) stresses that act along the cracks. These results were obtained within the framework of the General Statement of problems, which is presented near Fig. 6.3, and refers to the situation shown in Fig. 6.3. It is appropriate to note that, apparently, the first publications in this direction with the discussed approach were the articles [36, 37]. In subsequent years, several articles were published in this direction, the main ones of which are presented in the list of references to this monograph.

The results are also, in part, presented in the review articles [6–9] devoted to the non-classical problems of fracture mechanics. The results obtained on the spatial statical problems of the discussed mechanics of brittle fracture of materials with initial (residual) stresses along cracks are included in the monographs [1] (Chap. 4) of 1983, [2] (vol. 2, Chaps. 4 and 5) of 1991 and [4] (Chaps. 4 and 5 and this chapter) of 2017. In the bibliography lists of the above-mentioned monographs, the main articles on the considered **Problem 4**, published before the publication of the corresponding monograph, are presented.

6.4.1 On Statement of Spatial Statical Problems in Mechanics of Brittle Fracture of Materials with Initial (Residual) Stresses Acting Along Cracks

The research is carried out in a General Statement, which is set out near Fig. 6.3. At that, **assumptions 1–4**, outlined in the final part of Sect. 6.2, and three additional assumptions are accepted.

Additional assumption 1 The initial (residual) stresses are considered (in contrast to Fig. 6.3) in the following form:

$$\sigma_{33}^0 = 0; \ \sigma_{11}^0 = \sigma_{22}^0 \neq 0; \ \sigma_{ij}^0 = 0 \text{ for } i \neq j. \tag{6.12}$$

Note that expressions (6.12) are a special case of expressions (6.1) and correspond to an axisymmetric initial (residual) stress–strain state, where in this case the axis $0y_3$ is the axis of symmetry.

Additional assumption 2 The hyperelastic isotropic and transversely isotropic materials with an arbitrary structure of the elastic potential are considered. At that, in the case of the transversely isotropic bodies, it is assumed that the planes $y_3 = $ const are the isotropic planes.

Additional assumption 3 The plane cracks are located in planes $y_3 = $ const, and thus the initial (residual) stresses act only along the planes in which the cracks are located.

It should be noted that in this section concerning the spatial problems, the location of the coordinate axes has been changed in comparison with Fig. 6.3, which thus only applies to the plane and anti-plane problems. This change corresponds to the traditionally accepted arrangement of the axes in the plane and anti-plane, as well as in the spatial problems of fracture mechanics.

6.4.2   To the Method of Research of Spatial Statical Problems

Considering Note 6.2, these problems are reduced to the mixed problems for the harmonic potential, as in the classical linear brittle fracture mechanics. Using the various integral transforms and existing methods, the various types of one-dimensional integral equations are obtained.

The above results were obtained for the hyperelastic isotropic and transversely isotropic materials with an arbitrary structure of the elastic potential. Subsequently, a numerical study of the obtained one-dimensional integral equations for the hyperelastic isotropic and transversely isotropic materials with a specific structure of the elastic potential and analysis of mechanical effects, which are characteristic of the mechanics of brittle fracture of materials with initial (residual) stresses along cracks, is carried out.

6.4.3   Concrete Results (Using Both Exact Solutions and Computer Methods) Obtained for the Following 11 Design Schemes

1.   The internal disk-shaped circular crack (Mode I crack, radial shear crack, torsion);
2.   The outer disk-shaped circular crack (Mode I crack, radial shear crack, torsion);
3.   The general problem for shear in the case of an internal disk-shaped circular crack (exact solutions, asymptotic representation near the crack tip);
4.   The general problem for shear in the case of an external disk-shaped circular crack (exact solutions, asymptotic representations near the tip of the crack);

5. The crack in the form of a plane elliptical disk (Mode I crack, shear crack, exact solutions, asymptotic representations near the crack tip);
6. The general non-axisymmetric problem for a crack in the form of a circular disk (Mode I crack, shear crack, exact solutions, asymptotic representations near the crack tip);
7. The general non-axisymmetric problem for a crack in the form of a circular disk near the half-space boundary (asymptotic stress distribution in the neighborhood of the crack edge, stress intensity factors);
8. The axisymmetric problems for a crack in the form of a circular disk near the boundary of a half-space (Mode I crack, shear crack, torsion, asymptotic distribution near the crack edge, stress intensity factors);
9. The non-axisymmetric and axisymmetric problems for the interacting parallel coaxial cracks in the form of circular disks *under normal loading on their edges* (two parallel coaxial cracks, a periodic system of parallel coaxial cracks, asymptotic representation near the crack edge, stress intensity factors);
10. The axisymmetric problems for interacting parallel coaxial cracks in the form of circular disks *under the action of radial shear* (two parallel coaxial cracks, periodic system of parallel coaxial cracks, asymptotic representation near the crack edge, stress intensity factors);
11. The problems for interacting parallel coaxial cracks in the form of circular disks under *the action of torsional forces* (two parallel coaxial cracks, a periodic system of parallel coaxial cracks, asymptotic representation near the crack edge, stress intensity factors).

Note 6.4 The specific results presented above in an annotated form for the spatial statical problems of the mechanics of brittle fracture of materials with initial (residual) stresses acting along cracks are apparently presented in the most complete form in the monograph [4] of 2017.

6.4.4 On Phenomena of Resonant Character for Spatial Statical Problems of Non-classical Problem 4 of Fracture Mechanics

As in the plane statical problems considered in Sect. 6.3 and the spatial problems, considered in an annotated form in Sect. 6.4, *there are resonant effects* when the initial (residual) stresses tend to the values corresponding to the near-the-surface instability of the half-space. In the case of the spatial statical problems, the near-the-surface instability should be considered concerning the initial (residual) stresses in the form (6.12). In this case, the effects of a resonant character occur for the opening mode cracks (Mode I) and other cracks. The more detailed information is presented in the monographs [1] of 1983, [2] (vol. 2) of 1991, and [4] of 2017.

6.5   On Dynamical Plane and Anti-plane Problems in Mechanics of Brittle Fracture of Materials with Initial (Residual) Stresses Along Cracks

In this section, the information on the plane and anti-plane dynamical problems of the brittle fracture mechanics of materials with initial (residual) stresses that act along cracks is considered in an annotated form. At that, *only* the results for a moving crack are considered. The research is carried out within the framework of the General Statement set forth after Fig. 6.3, and all assumptions and comments set out in the final part of Sect. 6.2 are applied. As applied to the plane and anti-plane problems, an infinite material with one plane (infinite in the direction of the axis $0y_3$, Fig. 6.4) crack of the finite width $2a$ is considered, which is located in the plane $y_1 0y_3$ (Fig. 6.4) ($|y_1| \leq a$; $y_2 = 0$; $-\infty < y_3 < +\infty$).

With the above brief statement, **an additional assumption** is made, which is as follows:

The above crack (Fig. 6.4) *moves uniformly and rectilinearly at a constant velocity* $v = $ const *along the axis* $0y_1$. The study is carried out *for a subsonic regime of motion, i.e. the conditions* (2.132) *(part I of this monograph) are satisfied, taking into account the notation given below of the expression* (2.132), *for the materials with initial (residual) stresses.*

The studies were carried out using *the complex potentials* for the plane and anti-plane dynamical problems of the brittle fracture mechanics of materials with initial (residual) stresses that act along cracks. These complex representations are presented in a very short form in Sect. 2.6.6 (Part I of this monograph).

On the mechanics of moving cracks in the materials with initial (residual) stresses along cracks in the historical aspect, apparently, the first publication in the world scientific literature was the article by the author of this monograph, which was published in the journal "Prikladnaya Mekhanika" (1982, 18, No. 2, pp. 60–67). In the English translation, the above article was published in the journal "Soviet Applied Mechanics "(1982, 18, No. 2, pp. 137–143). As the initial (residual) stresses *tend to zero*, the above results, which are the exact solution, turn into the exact solution [64], published in 1951. Moreover, the article [64] in the historical aspect was the first publication in the world scientific literature on the theory of moving cracks in the classical linear mechanics of brittle fracture.

The results on the mechanics of moving cracks in materials with initial (residual) stresses along cracks, published before 1983, are presented in the monograph [1], the results up to 1991—in the monograph [2] (vol. 2), and the results up to 2008—in the monograph [3] (v. 2).

Additionally, it should note the modern review [10] on the discussed scientific direction, which was published in 2010.

In the most General Statement, the main results on the mechanics of moving cracks in materials with initial (residual) stresses that act along cracks are presented in the articles [42–45]. As the initial (residual) stresses *tend to zero*, the results presented in articles [42–45] also include the results for the moving cracks of the classical linear mechanics of brittle fracture, which were first presented in the articles [64] of 1951 and [65] of 1960.

Above in this section, in a very brief form as applied to the mechanics of moving cracks in materials with initial (residual) stresses that act along the cracks, the following are considered:

The statement of problems with an indication of the general approach to their solution;

The aspects of a historical character related to the discussed scientific direction.

Below, in a very brief form are also considered:

The brief description of the solution method;

The specific results obtained;

The effect of a resonant character is found in the discussed results.

Method of solving:

Application of the complex representations outlined in Sect. 2.5.6 (Part I of this monograph);

The attraction of methods of reduction to the Riemann–Hilbert problem [61, 62];

Application of the Keldysh–Sedov formula [63], which made it possible to obtain;

In the main, the exact solutions of all the problems under consideration.

The concrete results obtained (using the exact solutions) for the following design schemes:

The moving Mode I crack for cases of the unequal and equal roots (exact solution, asymptotic distribution near the leading crack tip, analysis of the effect of initial (residual) stresses);

The moving Mode II crack for cases of the unequal and equal roots (exact solution, asymptotic distribution near the leading crack tip, analysis of the effect of initial (residual) stresses);

The moving Mode III crack (exact solution, asymptotic distribution near the leading crack tip, analysis of the influence of initial (residual) stresses).

In the most complete form, the above specific results are presented in the monograph [3] (vol. 2, Chap. 10, Sects. 1, 2, 3, and 4, pp. 318–392) and initially in the articles [42–45].

On the resonance effects for the plane and anti-plane dynamical problems of non-classical

Problem 4 *of the fracture mechanics is applied to moving cracks.*

For the Mode I and Mode II cracks, the resonant effects *occur* when the crack propagation velocity *tends to the Rayleigh wave velocity* in an elastic material *with the considered initial (residual) stresses.*

In the above case, for the Mode III cracks, the resonance effects *do not occur* (are absent).

We restrict ourselves to the information set out above in this chapter for a brief discussion of the results on **Problem 4** (*The brittle fracture of materials with cracks taking into account the action of the initial (residual) stresses along cracks*), which were obtained in the Department of Dynamics and Stability of Continua Media of the S. P. Timoshenko Institute of Mechanics of the NASU.

Below, in Sect. 6.6, we will additionally provide, as examples, some information on the repetition by other authors of the above results related to **Problem 4**, as well as to other **Problems** considered in this monograph.

6.6   Repetition of Results

The results for **Problem 4** (*The brittle fracture of materials with cracks taking into account the action of the initial (residual) stresses along the cracks*), presented in a brief form in this chapter, were obtained in a unified general form for the compressible and incompressible isotropic and orthotropic (for the spatial problems—transversely isotropic) materials **with an arbitrary structure of the elastic potential**.

As already noted in the final part of Sect. 6.1, in other scientific centers (especially by English-speaking researchers), the research on **Problem 4** has been and is being carried out for the materials **with elastic potentials of a particular special structure**.

In connection with the above, the following three positions should be noted.

The first position. The published results for materials *with elastic potentials of a specific structure* for the problems that are studied and indicated in this chapter concerning materials *with an arbitrary structure of elastic potential* (isotropic and orthotropic materials) **cannot be recognized as new scientific results**. In any case, **the references** to results in this chapter **are mandatory.**

The second position. The absence of the above **references** cannot be justified by the "inaccessibility of the Russian-language publications." The point is that the main results on **Problem 4** presented in this chapter were also published in the journal "Prikladnaya Mekhanika," which was translated into English and published as "Soviet Applied Mechanics" in 1966–1991, since 1992 was translated into English as "International Applied Mechanics" and is currently published by the Springer Group.

The third position. Before submitting an article to the journal for publication, it is advisable to carry out an appropriate information search in the discussed scientific direction. With a modern accessible international database system, this information search is quite easy to realize.

Adherence of the above three positions in the preparation of new publications on **Problem 4**, as well as on other non-classical problems of the fracture mechanics discussed in this monograph, makes it possible to form a new publication that meets the ethical standards generally accepted in the world scientific community. The non-conformity to comply with the above three positions in the preparation of new publications on the issues under discussion, apparently, not only takes such publications and corresponding actions beyond the ethical standards adopted in the world scientific community, but also transfers such actions to *the sphere of legal responsibility.*

The expediency of the above approach is confirmed by three examples with which the author of this monograph met and which are discussed below.

Example 1 In 1996, an article by E. Soos was published in the "International Journal of Engineering Science," which was included in the list of references to this monograph as an article [66]. The specified article of E. Soos *violates all possible ethical norms accepted in the world scientific community.* In this regard, the author of this monograph prepared a small article, which, along with the attached copy of the correspondence with E.Soos, was sent to the editorial office of "International Journal of Engineering Science," in which E. Soos was a member of the editorial board. The prepared article was not published in the "International Journal of Engineering Science." Therefore, this article was published in the journal "Prikladnaya Mekhanika" and was included in the list of references to this monograph as the article [40]. In subsequent years, the materials of the article by E. Soos [66], with some modification, were included in the monograph [67] of 2003. In order not to clutter up this book with repeated analysis of the article by E. Soos [66], we will give only brief information from the Introduction to the article [40].

"... The article [66] is characterized by:

(1) The comparatively detailed presentation of the mathematical apparatus I have outlined for the considered variant of the mechanics of brittle fracture of materials with initial stresses, sometimes without appropriate references;

(2) The harsh criticism, even using the term "paradox," of my results (as will be shown below, this criticism arose as a result of a complete lack of understanding by the author of the article [66] of the statements of the problems he is trying to deal with);

(3) The publication by the author of article [66] new and correct, in his opinion, results (as will be shown below, these results completely coincide with my results "ten years ago," if we replace some of the notation introduced in [66]). The whole complexity of the situation lies in the fact that the author of article [66] was familiar with my results "ten years ago" (this will be documented below). ...

Documents

1. From *E.Soos letter* dated 03/15/1995.

"Dear Professor Guz, 1995.03.15
Since 1992 I study intensively your fundamental books, published in the period 1971–1990, concerning general, three-dimensional theory of the stability of solid bodies ...".

2. In response to the letter from E.Soos dated 03/15/1995 by mail, my monograph

[3] (vol. 2) of 1991 was sent to him, in which my results "ten years ago" are presented in sufficient detail.

After the above, the article by E. Soos [66] was published.

Thus, from the information presented above, it follows that E. Soos and his article [66] *go far beyond the three above positions.*

The additional more detailed information on the discussed issue can be obtained from article [40].

Example 2 In 2002, an article by E. Radi, D. Bigoni, and D. Capuati was published in the "International Journal of Solids and Structures," which was included in the list of references to this monograph as an article [68].

In [68], the asymptotic stress distribution near the crack tip was obtained for Mode I and Mode II for the materials with initial stresses along the crack in the framework of an incompressible elastic body model with elastic relations of a particular form for a linearized theory of a simplified type.

The indicated results [68] are a special case of more general results, which are indicated in Sect. 6.3 and were published in 1980–1982. (i.e., 20 years before [68]). Thus, the results of [68] *are not new*, and at the same time, there are *no references* to the above results "twenty years ago" in comparison with [68] for 2002. Hence in [68], *the first position* is violated.

On page 3971 of article [68], in a footnote, the authors indicate that publications in Russian are not available. At the same time, the publications of "twenty years ago," noted in Sect. 6.3, *are also presented* in the journal "Soviet Applied Mechanics" *in English*. Hence, *the second position* also takes place in a negative sense.

It follows from the above that the corresponding information search *was not carried out*, otherwise the authors of [68] would have learned about the noted results "twenty years ago." Therefore, *the third position* also takes place in a negative sense.

Thus, it follows from the information above that the authors of the article [68] and their article [68] *go beyond the three above positions.* The additional information on this issue concerning the article [68] can be obtained from the article [48] of 2003, where the corresponding analysis was carried out. On this issue, a letter was sent to the "International Journal of Solids and Structures," which the editors of this journal published in 2003 in the form of a very abbreviated note [46].

Note 6.5 In Examples 1 and 2 of this, two cases of repetition of the results on **Problem 4** described in Sect. 6.3 and obtained earlier in the Department of Dynamics and Stability of Continua of the S. P. Timoshenko Institute of Mechanics of the NASU.

Moreover, the results in the form of repetition are a very special case of the results of Sect. 6.3. The repetition of the results briefly discussed in this monograph **Problems 1–8** also takes place concerning other **Problems**, not only concerning **Problem 4**.

Below, as an example of repetition of results on **Problems 1–8**, consider Example 3 related to repetition of results on **Problem 1** (*Fracture in composite materials under compression along reinforcing elements*), which are described in Part II of this monograph.

Example 3 In 2006, the *"International Journal of Solids and Structures"* published an article by Y. B. Fu and Y. T. Zhang (Int. J. Solids Struct., 2006, vol. 43, P. 3306–3323), which is not included in the listed literature for this monograph. In the discussed article, Fu and Zhang repeated (even in a less general form) the results obtained by the author of this monograph *more than 35 years ago* (in comparison with 2006).

In this regard, a letter was sent to the editorial office of the "International Journal of Engineering Science" *that was not published*. This letter was fully published in the article (Guz A.N. // Prikladnaya Mekhanika.—2006.—42, No. 11.—P. 3–29) on p. 23–24.

The main analysis of Fu and Zhang's article is presented in the first part of this letter. In this regard, below is shown only the first part of the letter, in which the references to publications are indicated following the list of references to this monograph (in the part III).

"... In the paper Fu and Zhang of 2006, a continuum theory of the fracture of composites under compression is presented, which is based on the criterion of the loss of elliptic properties by a system of differential equations of the three-dimensional linearized theory of stability of elastic bodies. This continuum theory allows one to determine the theoretical ultimate compressive strength of composites.

It should be noted that *more than 35 years ago*, this continuum theory of fracture of composites, also based on the criterion of the loss of elliptic properties, was proposed in the papers by Guz [69] of 1969 and [70] of 1971. Guz's approach (1969, 1971) along with the definition of theoretical ultimate compressive strength allows also us to determine the surfaces along which the fracture propagates, i.e. is more general than Fu and Zhang's (2006) results. In subsequent years, the numerous results obtained based on the Guz' approach (1969, 1971) were published in the periodicals. These results are summarized on pages 78–144 in Guz's monograph [71] of 1990, which is presented in the LIBRARY OF CONGRESS (USA) (Access: Jefferson or Adams Bldg General or Area Studies Rms). In a brief form, these results are also presented on pp. 107–146 of vol. 5 of the multivolume (in 12 volumes) monograph [72] of 1993–2003. This monograph is also presented in the LIBRARY OF CONGRESS (USA) (Access: Jefferson or Adams Bldg General or Area Studies Reading Rms, Call number: TA.418.9.C6M435 1993).

In English, the results obtained based on Guz's approach (1969, 1971) are presented in numerous publications in periodicals and reported at numerous scientific conferences. For example, let us point out the publications of Guz [73] of 1982, [56] of 1985, [74] of 1986, [75] of 1987, [76] of 1989, [77] of 1993, [78] of 1996, [58] of 1997, [79] of 1998, and [7] of 2000. These results in the form of an article by Guz [80] of 1998 were included in the **encyclopedia** (FRACTURE. A Topical Encyclopedia of Current Knowledge. Edited by Genady P. Cherepanov.—Krieger Publ. Company, Florida, USA.—1998.—870 p.). The results of the analysis of the phenomenon of loss of ellipticity properties are also presented on pages 413–430 in the monograph Guz [81] of 1999 ...".

It follows from the information presented above in a copy of a part of the letter from the author of this monograph to the editorial office of the journal "International Journal of Engineering Science," that the authors Fu and Zhang and their 2006 paper *go beyond the three positions* indicated at the beginning of Sect. 6.6.

We restrict ourselves to the above information in **Examples 1–3** when analyzing the repetition by other authors of the results presented in this monograph and obtained on **Problems 1–8** of the non-classical fracture mechanics in the Department of Dynamics and Stability of Continua of the S. P. Timoshenko Institute of Mechanics of the NASU.

6.7   On Increasing the Objectivity of Citation

In the previous section, we analyzed three examples of considering more specific results (in comparison with the results of other authors that were published *earlier*) on the non-classical problems of fracture mechanics *without corresponding citation of previously obtained more general results of other authors*. Thus, all three discussed examples are characterized *by a lack of objective citation*.

Unfortunately, a similar situation occurs relatively often in other scientific fields. It is advisable to note that, apparently, there would be no need to consider *the three positions* formulated at the beginning of Sect. 6.6, if a *sufficiently high level of citation objectivity* had been maintained. In this case, the authors of the articles analyzed in Sect. 6.6 obviously could not publish their articles, since it would follow from the literature review or the Introduction to these articles that their articles proposed for publication contain results that are a special case the previously published results.

The need to increase the objectivity of citation, or rather the level of objectivity of citation, also follows from the situation that has developed recently in the process of evaluating the results of scientific work. Currently, the bibliometric methods based on *the analysis of citation* (articles, journals, monographs, individual scientists, etc.) are mainly used to evaluate the individual articles, individual journals, individual monographs, and the work of individual scientists. In this regard, it follows that *the citation should have an objective character*.

The analysis of the objectivity of citation was carried out in a short form in the article (Guz A.N. // Prikladnaya Mekhanika.—2006.—42, No. 11.—P. 3–29) on pages 22–29, where the **Examples 2 and 3** are also considered in more detail, brief information about which is presented in Sect. 6.6 of this monograph; as a result of this analysis, the conclusions were formulated that are presented below.

Conclusions

1.   The problem of *ensuring the objectivity of citation* in publications in periodicals (journals) exists and is actual.
2.   In the process of formation of the information scientific space, the problem of ensuring the objectivity of citation *is the weakest link*.

3.   The approaches to ensuring the objectivity of citation in periodicals (and other) publications *have not yet been developed*.

In the above article, a Proposal of the general character was formulated, referring to the entire system of publications of scientific research results in periodicals and contributing to increasing the objectivity of citation. This proposal is written again below.

"**Proposal.** In the opinion of the author of this monograph, the most active way can be to contribute to ensuring the objectivity of citation in publications in scientific and scientific and technical journals, if to provide in the structure of published articles the necessary and mandatory points. So, for example, at present in the overwhelming majority of scientific and scientific-technical journals in all published articles the following items are presented in English: Abstract or Summary; Key words.

As a possible action of a general nature, ensuring the objectivity of citation and the novelty of the published results, we can propose to introduce the following paragraph into the structure of articles:

The novelty of the results and objective citation in this article is provided by the database ().

The authors of each published article in parentheses above place the name of the international database within which the authors carried out an information search. In this case, *the authors of the published article are naturally responsible for the results of the information search.* Of course, it is preferable to introduce the above-formulated paragraph in English. It is obvious that the adoption of the above-formulated Proposal will to a certain extent complicate the work of the authors of published articles when preparing an article for submission to the journal.

Nevertheless, the adoption of the above-formulated Proposal will significantly increase the responsibility of the authors of publications *for the novelty of the results and objective citation*, which is extremely important for the world scientific community in the formation of the information scientific space."

The above Proposal (Guz A.N. // Prikladnaya Mekhanika.—2006.—42, No. 11.—P. 3–29) was discussed in July 2007 in the international Internet journal.

iMechanica, web of mechanics and mechanicians

http://www.imechanica.org

It is entitled "Objective citation—a proposal from the Timoshenko Institute." The discussion was organized by Prof. Andrew Norris (Mechanical & Aerospace Engineering, Rutgers University, NJ, USA). It should be noted that the journal (6.14) is electronic only and uses the server of the Harvard School of Engineering and Applied Sciences. Information about the discussion and its analysis are presented in the article (Guz A.N., Ruschitsky J.J. // Prikladnaya Mekhanika.—2009.—45, No. 3.—P. 3–22). The participants in the above discussion in the journal (6.14) expressed various considerations, but from substantive considerations, the following general conclusion can be drawn—*the discussed problem takes place, but its solution is not simple.*

The author of **Proposal (6.13)** is most impressed by the considerations of Prof. E. Norris on this issue, which are presented in English at the beginning of page 11 of the above article (Guz A.N., Ruschitsky J.J.). Below on the same page 11, a translation into Russian of the discussed considerations of Prof. E. Norris is shown. For the information of readers, it is necessary to note that in the Russian translation, the term "statement" corresponds to the Proposal (6.13).

Prof. E. Norris stated:

> What can you expect from adding another hurdle? But I think this kind of statement would be worth including. It would make the author stop and think and give the reader confidence in the article. It also reduces the burden of the reviewer - I must say - reduces the reviewer's responsibility to be still a literary detective. And that could improve the problem of attracting competent reviewers.

The above information, inspiring optimism and hope, will be limited when discussing the problem of increasing the level of *ensuring the objectivity of citation* in scientific publications.

Yet the main conclusion is that in the issue under discussion, the consistent actions and expected results *are yet to come*. Although the author of this monograph is convinced that **Proposal (6.13)** will make an appropriate contribution to the solution of the problem under discussion.

Note 6.6 The material presented in this section relates not only to **Problems 1–8** considered in this monograph but also *to all publications of scientific results.* The expediency of including the material of Sect. 6.7 in this chapter, devoted to the analysis of the results for **Problem 4** (*Brittle fracture of materials with cracks, taking into account the action of the initial (residual) stresses along the cracks*), is justified by the fact that Sect. 6.6 is also included in this chapter, in which the analysis of **Examples 1–3** with a clear violation of the objectivity of citation is performed.

References

1. Guz, A.N.: Mekhanika khrupkogo razrusheniia materialov s nachalnymi napriazheniiami (Mechanics of Brittle Fracture of Materials with Initial Stresses). Naukova Dumka, Kyiv (1983)
2. Guz, A.N. (ed.): Neklassicheskie problemy mekhaniki razrusheniya, v 4 tomah, 5 knigah (Non-classical problems of fracture mechanics, in 4 volumes, 5 books) (Naukova Dumka, Kyiv, 1990–1993) A.A. Kaminsky (ed.), T.1. Razrushenie viazkouprugikh tel s treshchinami (Vol. 1. Fracture of viscoelastic bodies with cracks) (1990) A.N. Guz (ed.), T.2. Khrupkoe razrushenie materialov s nachalnymi napriazheniiami, (T.2. Brittle fracture of materials with initial stresses) (1991) A.A. Kaminsky, D.N. Gavrilov (ed.), T.3. Dlitelnoe razrushenie polimernykh i kompozitnykh materialov s treshchinami (T.3. Long-term fracture of polymer and composite materials with cracks) (1992) A.N. Guz, M.Sh. Dyshel, V.M. Nazarenko (ed.), T.4, kniga 1. Razrushenie i ustoichivost materialov s treshchinami (V.4, book 1. Fracture and stability of materials with cracks) (1992) A.N. Guz, V.V. Zozulya (ed.), T.4, kniga 2. Khrupkoe razrushenie materialov

pri dinamicheskikh nagruzkakh (V.4, book 2. Brittle fracture of materials under dynamic loads) (1993)

3. Guz, A.N.: Osnovy mekhaniki razrusheniia kompozitov pri szhatii: V 2-kh tomakh (Fundamentals of the fracture mechanics of composites under compression: In 2 volumes). (Litera, Kyiv, 2008) T. 1. Razrushenie v strukture materiala. (Fracture in structure of materials) T. 2. Rodstvennye mekhanizmy razrusheniia. (Related mechanisms of fracture)

4. Bogdanov, V.L., Guz, A.N., Nazarenko, V.M.: Obieedinennyi podkhod v neklassicheskikh problemakh mekhaniki razrusheniia (A Unified Approach in Non-classical Problems of Fracture Mechanics). LAP LAMBERT Academic Publishing, Saarbrücken, Deutschland (2017)

5. Guz, A.N., Bogdanov, V.L., Nazarenko, V.M.: Fracture of Materials Under Compression Along Cracks. Springer Nature Switzerland AG (2020)

6. Guz, A.N.: Study and Analysis of Non-classical Problems of Fracture and Failure Mechanics and Corresponding Mechanisms. Lecture presented at Institute of Mechanics. HANOI (1998)

7. Guz, A.N.: Description and study of some nonclassical problems of fracture mechanics and related mechanisms. Int. Appl. Mech. 36(12), 1537–1564 (2000)

8. Guz, A.N.: In On study of nonclassical problems of fracture and failure mechanics and related mechanisms. ANNALS of the European Academy of Sciences, pp. 35–68. Liège, Belgium (2006–2007)

9. Guz, A.N.: On study of nonclassical problems of fracture and failure mechanics and related mechanisms. Int. Appl. Mech. 45(1), 1–31 (2009)

10. Guz, A.N.: Mekhanika dvizhushchikhsia treshchin v materialakh s nachalnymi (ostatochnymi) napriazheniiami (obzor) (Mechanics of moving cracks in materials with initial (residual) stresses (review)). Prikladnaya Mekhanika 47(2), 3–75 (2011)

11. Bogdanov, V.L.: Osesimetrichna zadacha pro pripoverkhnevu trishchinu normalnogo vidrivu v kompozitnomu materiali z zalishkovimi napruzhenniami (Axisymmetric problem on the near-surface crack of normal rupture in a composite material with residual stresses). Matematychni Metody ta Fizyko-Mekhanichni Polya 50(2), 45–54 (2007)

12. Bogdanov, V.L.: Neosesimmetrichnaia zadacha o periodicheskoi sisteme diskoobraznykh treshchin normalnogo otryva v tele s nachalnymi napriazheniiami (Nonaxisymmetric problem of a periodic system of penny-shaped cracks of normal rupture in a body with initial stresses). Matematychni Metody ta Fizyko-Mekhanichni Polya 50(4), 149–151 (2007)

13. Bogdanov, V.L.: O kruchenii predvaritelno napriazhennogo materiala s dvumia parallelnymi soosnymi treshchinami (On torsion of a prestressed material with two parallel coaxial cracks). Dopovidi NAN Ukrainy 11, 59–66 (2008)

14. Bogdanov, V.L.: Neosesimetrichna zadacha pro dvi paralelni spivvisni trishchini normalnogo vidrivu v materiali z pochatkovimi napruzhenniami (Non-axisymmetric problem of two parallel coaxial cracks of normal ruption in a material with initial stresses). Dopovidi NAN Ukrainy 8, 49–59 (2010)

15. Guz, A.N.: K linearizirovannoi teorii razrusheniia khrupkikh tel s nachalnymi napriazheniiami (On the linearized fracture theory of brittle bodies with initial stresses). Dokl. Akad. Nauk SSSR 252(5), 1085–1088 (1980)

16. Guz, A.N.: Treshchiny otryva v uprugikh telakh s nachalnymi napriazheniiami (Rupture cracks in elastic bodies with initial stresses). Dokl. Akad. Nauk SSSR 254(3), 571–574 (1980)

17. Guz, A.N.: O kriterii khrupkogo razrusheniia materialov s nachalnymi napriazheniiami (Criterion for brittle fracture of materials with initial stresses). Dokl. Akad. Nauk SSSR 262(2), 285–288 (1982)

18. Guz, A.N.: O kriterii khrupkogo razrusheniia pri szhatii materialov s defektami (Criterion of brittle fracture upon compression of materials with defects). Dokl. Akad. Nauk SSSR 285(4), 828–831 (1985)

19. Guz, A.N.: O poriadke osobennosti v konchike treshchiny v materialakh s nachalnymi napriazheniiami (Singularity order at the crack tip in materials with initial stresses). Dokl. Akad. Nauk SSSR 289(2), 310–313 (1986)

20. Guz, A.N.: Dvizhushchiesia treshchiny v kompozitnykh materialakh s nachalnymi napriazheni-iami (Moving cracks in composite materials with initial stresses). Mekhanika kompozitnykh materialov **37**(44352), 695–708 (2001)
21. Babich, V.M., Guz, A.N., Nazarenko, V.M.: Disk-shaped normal-rupture crack near the surface of a semiinfinite body with initial stresses. Sov. Appl. Mech. **27**(7), 637–643 (1991)
22. Bogdanov, V.L.: On a circular shear crack in a semi-infinite composite with initial stresses. Mater. Sci. **43**(2), 321–330 (2007)
23. Bogdanov, V.L.: Effect of residual stresses on fracture of semi-infinite composite with cracks. J. Mech. Adv. Mater. Struct. **15**(6), 453–460 (2008)
24. Bogdanov, V.L.: Influence of initial stresses on fracture of composite materials containing interacting cracks. J. Math. Sci. **165**(3), 371–384 (2010)
25. Bogdanov, V.L.: Nonaxisymmetric problem of the stress-strain state of an elastic half-space with a near-surface circular crack under action of loads along it. J. Math. Sci. **174**(3), 341–366 (2011)
26. Bogdanov, V.L.: Influence of initial stresses on the stressed state of a composite with a periodic system of parallel coaxial normal tensile cracks. J. Math. Sci. **186**(1), 1–13 (2012)
27. Bogdanov, V.L.: On the interaction of a periodic system of parallel coaxial radial-shear cracks in a prestressed composite. J. Math. Sci. **187**(5), 606–618 (2012)
28. Bohdanov, V.L.: Mutual influence of two parallel coaxial cracks in a composite material with initial stresses. Mater. Sci. **44**(4), 530–540 (2008)
29. Bohdanov, V.L.: Influence of initial stresses on the fracture of a composite material weakened by a subsurface mode III crack. J. Math. Sci. **205**(5), 621–634 (2015)
30. Guz, A.N.: Theory of cracks in elastic bodies with initial stresses. Formulation of problems, tear cracks. Sov. Appl. Mech. **16**(12), 1015–1023 (1980)
31. Guz, A.N.: Theory of cracks in prestressed elastic bodies. Shear cracks and limiting cases. Sov. Appl. Mech. **17**(1), 1–8 (1981)
32. Guz, A.N.: Theory of cracks in prestressed highly elastic materials. Sov. Appl. Mech. **17**(2), 11–21 (1981)
33. Guz, A.N.: Theory of cracks in elastic bodies with initial stresses (stiff materials). Sov. Appl. Mech. **17**(4), 311–315 (1981)
34. Guz, A.N.: Theory of cracks in elastic bodies with initial stresses (cleavage problem). Sov. Appl. Mech. **17**(5), 405–411 (1981)
35. Guz, A.N.: Theory of cracks in elastic bodies with initial stresses (three-dimensional static problems). Sov. Appl. Mech. **17**(6), 499–513 (1981)
36. Guz, A.N.: Three-dimensional problem for a disk-shaped crack in an elastic body with initial stress. Sov. Appl. Mech. **17**(11), 963–970 (1981)
37. Guz, A.N.: General three-dimensional static problem for cracks in an elastic body with initial stress. Sov. Appl. Mech. **17**(12), 1043–1050 (1981)
38. Guz, A.N.: Mechanics of the brittle failure of materials with initial stress. Sov. Appl. Mech. **19**(4), 293–307 (1983)
39. Guz, A.N.: Criterion of brittle fracture near stress raisers in composites in compression. Sov. Appl. Mech. **22**(12), 1148–1154 (1986)
40. Guz, A.N.: On the development of brittle-fracture mechanics of materials with initial stresses. Int. Appl. Mech. **32**(4), 316–323 (1996)
41. Guz, A.N.: Order of singularity in problems of the mechanics of brittle fracture of materials with initial stresses. Int. Appl. Mech. **34**(2), 103–107 (1998)
42. Guz, A.N.: Dynamic problems of the mechanics of the brittle fracture of materials with initial stresses for moving cracks. 1. Problem statement and general relationships. Int. Appl. Mech. **34**(12), 1175–1186 (1998)
43. Guz, A.N.: Dynamic problems of the mechanics of the brittle fracture of materials with initial stresses for moving cracks. 2. Cracks of normal separation (Mode I). Int. Appl. Mech. **35**(1), 1–12 (1999)
44. Guz, A.N.: Dynamic problems of the mechanics of the brittle fracture of materials with initial stresses for moving cracks. 3. Transverse-shear (Mode II) and longitudinal-shear (Mode III) cracks. Int. Appl. Mech. **35**(2), 109–119 (1999)

45. Guz, A.N.: Dynamic problems of the mechanics of the brittle fracture of materials with initial stresses for moving cracks. 4. Wedge problems. Int. Appl. Mech. **35**(3), 225–232 (1999)
46. Guz, A.N.: Comments on «Effects of prestress on crack-tip fields in elastic incompressible solids». Int. J. Solids Struct. **40**(5), 1333–1334 (2003)
47. Guz, A.N.: On some nonclassical problems of fracture mechanics taking into account the stresses along cracks. Int. Appl. Mech. **40**(8), 937–942 (2004)
48. Guz, A.N., Guz, I.A.: On publications on the brittle fracture mechanics of prestressed materials. Int. Appl. Mech. **39**(7), 797–801 (2003)
49. Guz, A.N., Klyuchnikov, Yu.V.: Three-dimensional static problem for an elliptical crack in an elastic body with initial stress. Sov. Appl. Mech. **20**(10), 898–907 (1984)
50. Guz, A.N., Nazarenko, V.M., Bogdanov, V.L.: Fracture under initial stresses acting along cracks: approach, concept and results. Theor. Appl. Fract. Mech. **48**, 285–303 (2007)
51. Guz, A.N., Nazarenko, V.M., Bogdanov, V.L.: Combined analysis of fracture under stress acting along cracks. Arch. Appl. Mech. **83**(9), 1273–1293 (2013)
52. Guz, A.N., Nazarenko, V.M., Nikonov, V.A.: Torsion of a pre-stressed halfspace with a disc-shaped crack at the surface. Sov. Appl. Mech. **27**(10), 948–954 (1991)
53. Kluchnikov, Yu.V.: Three-dimensional static problem for an external disk-shaped crack in an elastic body with initial stresses. Sov. Appl. Mech. **20**(2), 118–122 (1984)
54. Nazarenko, V.M., Bogdanov, V.L., Altenbach, H.: Influence of initial stress on fracture of a halfspace containing a penny-shaped crack under radial shear. Int. J. Fract. **104**, 275–289 (2000)
55. Guz, A.N.: Foundations of mechanics of brittle fracture of materials with initial stresses. In Proceedings of 6th ICF6 (India, 1984), pp. 1223–1230
56. Guz, A.N.: Three-dimensional stability theory of deformed bodies. Internal instability. Sov. Appl. Mech. **21**(11), 1023–1034 (1985)
57. Guz, A.N.: The study and analysis of non-classical problems of fracture and failure mechanics. In: Abstracts of IUTAM Symposium of Nonlinear Analysis of Fracture, p. 19. Cambridge, 3–7 Sept, 1995
58. Guz, A.N.: Non-classical problems of composite failure. In: Proceedings of ICF9 Advance in Fracture Research, vol. 4, pp.1911–1921. Sydney, Australia (1997)
59. Guz, A.N.: On the singularities in problems of brittle fracture mechanics in case of initial (residual) stresses along the cracks. In: Proceedings of the 3rd International Conference on Nonlinear Mechanics, pp. 219–223. Shanghai, China (1998)
60. Kurashide, M.: Circular crack problem for initially stressed neo-Hookean solid. ZAMM. **49**(2), 671–678 (1969)
61. Galin, L.A.: Kontaktnye zadachi teorii uprugosti (Contact Problems of Elasticity Theory). Fizmatgiz, Moscow (1953)
62. Muskhelishvili, N.I.: Nekotorye osnovnye zadachi matematicheskoy teorii uprugosti (Some Basic Problems of the Mathematical Theory of Elasticity). Nauka, Moscow (1966)
63. Keldysh, M.V., Sedov, L.I.: Effektivnoe reshenie nekotorykh kraevykh zadach dlya garmonich-eskikh funkciy (Efficient solution of some boundary problems for harmonic functions). Dokl. Akad. Nauk SSSR **16**(1), 7–10 (1937)
64. Yoffe, E.: The moving Griffith crack. Phil. Mag. **4**(330), 739–750 (1951)
65. Craggs, I.W.: On the propagation of a crack in an elastic-brittle materials. J. Mech. Phys. Solids. **8**(1), 66–75 (1960)
66. Soos, E.: Resonance and stress concentration in a prestressed elastic solid containing a crack. An apparent paradox. Int. J. Engng. Sci. **34**(3), 363–374 (1996)
67. Cristescu, N.D., Craciun, E.M., Soos, E.: Mechanics of Elastic Composites. CRC Press (2003)
68. Radi, E., Bigoni, D., Capuani, D.: Effect of pre-stress on crack tip fields in elastic incompressible solids. Int. J. Solids and Struct. **39**, 3971–3996 (2002)
69. Guz, O.M.: Pro vyznachennia teoretychnoi hranytsi mitsnosti na stysk armovanykh materialiv (Determining the theoretical compressive strength of reinforced materials). Dopovidi Akademii Nauk URSR, Ser. A. **3**, 236–238 (1969)
70. Guz, A.N.: O postroenii teorii prochnosti odnonapravlennykh armirovannykh materialov na szhatie (On construction of the strength theory of unidirectional reinforced materials on compression). Probl. Prochn. **3**(3), 37–40 (1971)

71. Guz, A.N.: Mekhanika razrusheniia kompozitnykh materialov pri szhatii (Fracture Mechanics of Composite Materials Under Compression). Naukova Dumka, Kyiv (1990)
72. Guz, A.N. (ed.): Mekhanika kompozitov, v 12 tomah (Mechanics of Composites, in 12 volumes). (Vol. 1–4 – Naukova Dumka, Vol. 5–12 – A.S.K., Kyiv, 1993–2003) V.T. Golovach (ed.), T.1. Statika materialov. (Vol. 1. Statics of materials) (1993) N.A. Shulga (ed.), T.2. Dinamika i ustoichivost materialov. (Vol.2. Dynamics and stability of materials) (1993) L.P. Khoroshun (ed.), T.3. Statisticheskaia mekhanika i effektivnye svoistva materialov (Vol.3. Statistical Mechanics and Effective Material Properties) (1993) A.N. Guz, S.D. Akbarov, (ed.), T.4. Mekhanika materialov s iskrivlennymi strukturami (Vol.4. Mechanics of materials with curved structures) (1995) A.A. Kaminsky (ed.), T.5. Mekhanika razrusheniia. (Vol.5. Fracture mechanics) (1996) N.A. Shulga, V.T. Tomashevsky (ed.), T.6. Tekhnologicheskie napriazheniia i deformatsii v materialakh. (Vol.6. Technological stresses and deformations in materials) (1997) A.N.Guz, A.S.Kosmodamiansky, V.P.Shevchenko (ed.), T.7. Kontsentratsiia napriazhenii. (Vol.7. Stress concentration) (1998) Ya.M.Grigorenko (ed.), T.8. Statika elementov konstruktsii. (Vol.8. Structural elements statics) (1999) V.D. Kubenko (ed.), T.9. Dinamika elementov konstruktsii. (Vol.9. Dynamics of structural elements) (1999) I.Yu. Babich (ed.), T.10. Ustoichivost elementov konstruktsii. (Vol.10. Stability of structural elements) (2001) Ya.M. Grigorenko, Yu.N.Shevchenko (ed.), T.11. Chislennye metody. (Vol.11. Numerical methods) (2002) A.N. Guz, L.P. Khoroshun (ed.), T.12. Prikladnye issledovaniia. (Vol.12. Applied research) (2003)
73. Guz, A.N.: Continuum theory of fracture in the compression of composite materials with metallic matrix. Sov. Appl. Mech. **18**(12), 1045–1052 (1982)
74. Guz, A.N.: Continuous theory of failure of composite materials under compression in the case of a complex stresses state. Sov. Appl. Mech. **22**(4), 301–315 (1986)
75. Guz, A.N.: Continuous theory of failure of composite materials with buckling at the ends (Plastic failure). Sov. Appl. Mech. **23**(5), 411–417 (1987)
76. Guz, A.N.: Construction of a theory of failure of composites in triaxial and biaxial compression. Sov. Appl. Mech. **25**(1), 29–33 (1989)
77. Guz, A.N.: Continual theory of fracture of composite materials at bearing strain in end faces in compression. In: Proceedings of ASM International Materials Congress, Material Park. Mechanisms and Mechanics of Composites Fracture, pp.201–207. Pittsburgh, Pennsylvania (1993)
78. Guz, A.N.: On failure propagation in composite materials in compression (Three-dimensional continual theory). In: Proceedings of ECF 11 Poitiers–Futuroscope, vol. III, pp. 1769–1774. France, Sept. 3–6, 1996
79. Guz, A.N.: Conditions of hyperbolicity and mechanics of failure of composites in compression. ZAMM **78**(1), 427–428 (1998)
80. Guz, A.N.: Some modern problems of physical mechanics of fracture. In: Cherepanov, G.P. (ed.). FRACTURE. A Topical Encyclopedia of Current Knowledge, pp. 709–720. Krieger Publ. Company, Malabar, Florida (1998)
81. Guz, A.N.: Fundamentals of the Three-Dimensional Theory of Stability of Deformable Bodies. Springer, Berlin Hiedelberg New York (1999)

Chapter 7
Problem 5. Brittle Fracture in the Form of Separation into the Slender Parts of Composite Materials Under Tension or Compression Along Reinforcing Elements

In this chapter, in a very brief form (even in comparison with **Problem 4**, which was discussed in the previous chapter), the main results on the **Problem 5** are presented, obtained on the themes of the Department of Dynamics and Stability of Continua of the S. P. Timoshenko Institute of Mechanics of the NASU since 1983. The presentation of these results is written in the style announced in the introduction (Part I) to this monograph (without excessively invoking aspects of a mathematical nature).

7.1 Introduction

The main results on this problem were obtained in the Department of Dynamics and Stability of Continua of the S. P. Timoshenko Institute of Mechanics of the National Academy of Sciences of Ukraine and subsequently developed in Azerbaijan and Turkey.

The first results are presented in publications of 1983–1986 [4–6, 10, 11, 20–22, 32–35, 51, 52]. The first publications were, apparently, the articles [10, 11] by the author of this monograph. The main results are presented in the monographs [12] (Chap. 7) of 1990, [19] (vol. 4) of 1995, [38] of 2000, [13] (vol. 2, Chap. 11) of 2008, and in other monographic publications. Moreover, the monograph [38] in English was published in two versions—in hardcover (ISBN 0-7923-6477-5) and in softcover (ISBN 1-4020-0383-8).

The results discussed here were included in the review papers [37, 39, 40], which are devoted exclusively to the presentation and discussion of **Problem 5**, and also in the review papers [53–57], which are devoted to the presentation and discussion of all questions pertaining to all non-classical problems of fracture mechanics. The list of references to this monograph includes the articles [1–7, 10, 11, 20–36, 41–52], which present the main results related to **Problem 5**. Additional information

A. N. Guz, *Eight Non-Classical Problems of Fracture Mechanics*,
Advanced Structured Materials 159,
https://doi.org/10.1007/978-3-030-77501-8_7

about publications on this issue can be obtained from the lists of references to the monographs [12, 19] (vol. 4), [38], and [13] (vol. 2).

In this scientific direction, a dissertation for the degree of Doctor of Physical and Mathematical Sciences (DSc) was prepared and defended by S. D. Akbarov.

The fracture mechanism considered in this chapter, as already noted in Sect. 1.3 of (Part I of this monograph), is called "separation into the slender parts." This fracture mechanism is not observed for the homogeneous materials (which conditionally include metals and alloys when they are studied at the level of continuum representations) and is characteristic of the composite materials with a clearly marked direction of preferential reinforcement.

This mechanism consists of dividing the material into separate parts along the direction of action of the compressive load with condition that *the compressive load* is directed along the direction of reinforcement in the unidirectional composites or one of the directions of reinforcement in the composites with a clearly marked direction of preferential reinforcement.

It should be noted that some authors associate the appearance of the phenomenon of "separation into the slender parts" *under compression* with the fact that in this case (within the framework of the continuum concepts) *the transverse elongation* (due to Poisson's ratio) reaches its limiting values.

In this regard, note also that the phenomenon of "separation into the slender parts" is also observed *under tension* along the direction of preferential reinforcement. Such an example is shown in Fig. 0.18 of Introduction (p. 64) of the book [13] (vol. 1), which is borrowed from the monograph [18] (p. 46) according to the list of references to the monograph [13]. The above example of fracture by "separation into the slender parts" *under tension* cannot be explained based on the noted approach within the framework of the continuum representation (*the transverse elongation or shortening reaches its limiting values*), since *under tension*, the transverse compression occurs, and the composite fractures along the direction of the tensile load (perpendicular to the transverse shortening).

Thus, we conclude that in the discussed fracture mechanism by "separation into the slender parts", apparently, the more complex fracture mechanism exists in the microstructure of the composite.

It is also appropriate to note that even in the case of a seemingly successful application of the concept of ultimate transverse elongation in the case of *compression* of unidirectional composites, *the fracture mechanism in the microstructure still remains unclear* since the concept of ultimate transverse elongation operates with the integral characteristics (from the point of view of microstructure mechanics) and the force does not work in the transverse direction.

Taking into account the information presented in the introduction (Sect. 7.1), below in this chapter the information is briefly considered on the following issues concerning **Problem 5**:

Consideration of experimental studies;
Explanation of the fracture mechanism by "separation into the slender parts".
Based on model concepts;

Information on analytical and numerical studies concerning the discussed.

Explanation of the fracture mechanism by "separation into the slender parts."

In presenting the above results, we will follow the review articles [55] of 2000, [56] of 2006, and [57] of 2009.

7.2 Experimental Researches

The main publications on the results of experimental studies refer to the case of fracture by "separation into the slender parts" *under compression.* So, the nature of the fracture of a unidirectional boron-aluminum composite *under uniaxial compression* along the direction of reinforcement (along the boron fibers) is shown in Fig. 7.1.

Note that Fig. 7.1 corresponds to Fig. 11 from the review articles [55, 57]. A description of the considered in Fig. 7.1 boron–aluminum composite is presented in the **example** in the final part of Sect. 3.3.4.1 (part II of this monograph).

It should be also noted that for the realization of fracture in the form of "separation into the slender parts," which is shown in Fig. 7.1, the additional techniques were taken to exclude the fracture in the form of crushing of the ends, which is shown in Fig. 5.1 and which is the subject of research in Chap. 5 (part II of this monograph).

The nature of the fracture of unidirectional fiberglass under uniaxial *compression* along the direction of reinforcement (along the glass fibers) is shown in Fig. 7.2.

Note that Fig. 7.2 corresponds to Fig. 12 of the review articles [55, 57].

Figure 7.2 is also presented on p. 110 of the monograph [17] from the bibliography to the two-volume monograph [13]. The monograph [17] also contains a description of the corresponding experimental studies.

Typical for the fracture shown in Figs. 7.1 and 7.2 is the complete fracture along planes and surfaces that are located along the reinforcement elements (along the fibers) for the unidirectional fiber composites. Consequently, it is quite logical to expect that the discussed type of fracture occurs as a result of the action of forces directed perpendicular to the reinforcing elements.

Fig. 7.1

Fig. 7.2

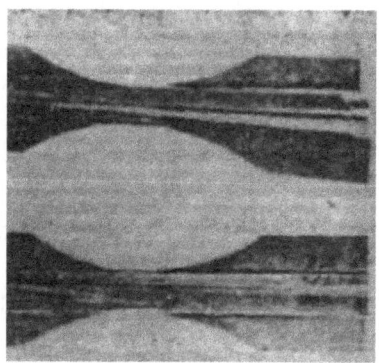

The nature of the fracture of fiberglass LTW (longitudinal-transverse winding) under uniaxial *compression* along the direction of reinforcement (along the glass fibers in one of the characteristic layers) is shown in Fig. 7.3.

Note that Fig. 7.3 corresponds to Fig. 13 of the review articles [55, 57]. Also, note that Fig. 7.3 is presented on page 81 of the monograph [8] from the bibliography to the two-volume monograph [13].

In the specimens, the fractured view of which is represented in Fig. 7.3, the longitudinal-transverse winding was carried out along the horizontal and vertical axes on Fig. 7.3. Thus, in Fig. 7.3, the fracture occurred also on the planes and surfaces that are arranged along the reinforcement elements in one of the characteristic layers.

The examples of fracture by "separation into the slender parts," which are presented in Figs. 7.1, 7.2 and 7.3, refer to the different composites and are obtained concerning the case of uniaxial *compression*.

In the introduction (Sect. 7.1) to this chapter, an example of "separation into the slender parts" failure is indicated, which refers to the case of uniaxial *tension*. This example is considered below in more detail.

The corresponding fracture is shown in Fig. 7.4, which is borrowed from page 46 of the monograph [18] according to the bibliography for the two-volume monograph [13]. Following the description on page 46 of the monograph [18] from the list [13], the character of the *tensile fracture* of the composite along the direction of

Fig. 7.3

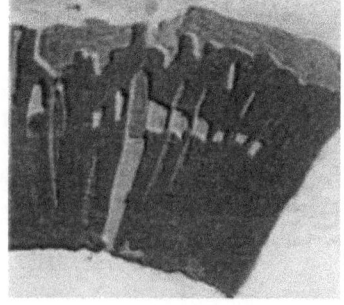

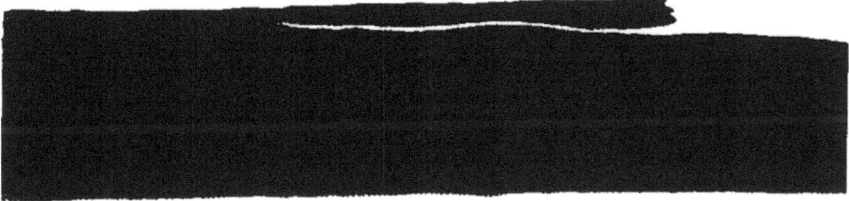

Fig. 7.4

reinforcement is shown in Fig. 7.4. At that, the fracture (separation into separate longitudinal parts) began in the presence of a local single curvature in the composite structure.

It follows from the analysis of the results of experimental studies presented in Figs. 7.1, 7.2, 7.3 and 7.4 that the following can be considered the main specific features of fracture by "separation into the slender parts:"

1. This type of fracture occurs in compression and some cases in tension of the unidirectional fibrous and layered composites under the action of a load along the fibers and layers.
2. This type of fracture manifests itself or consists in the general (not local) fracture along planes and surfaces that are placed parallel to the reinforcing elements.
3. This type of fracture is realized if the additional techniques are adopted that exclude the implementation of fracture of another type.

The above results of the analysis of experimental studies on fracture by "separation into the slender parts" (in the form of three specific features) are used in the next section when explaining the mechanism of the discussed type of fracture.

7.3 Explanation of Mechanism of Fracture in the Form of "Separation into the Slender Parts"

The explanation discussed in this section is presented in a rather short form, corresponding to the review articles [55] of 2000 and [57] of 2009. Here, the results of the author of this monograph and his pupils are used, which are presented in the articles indicated in the introduction of Sect. 7.1).

It should be noted that the considered fracture mechanism by "separation into the slender parts" occurs under the action of a load along the reinforcing elements, and the fracture is realized along the planes and surfaces that are located along the reinforcing elements.

The occurrence of this fracture mechanism both under compression along the reinforcing elements and in some cases also under tension along the reinforcing elements makes it possible to exclude from the considered fracture mechanism the various *phenomena associated with loss of stability*. Thus, under the uniaxial tension–compression along the reinforcing elements, the following fracture mechanism is most probable.

> ***The considered fracture by "separation into the slender parts" can occur only due to the internal forces (stresses) that act perpendicular to the direction of the external load and arise as a result of the influence of the microstructure of the composite, or due to the corresponding shear stresses.***

Thus, it is necessary to determine the mechanism of the *occurrence of these stresses in the microstructure of the composite and to find out the possibility of these stresses in the microstructure reaching the limiting values corresponding to the indicated type of fracture, provided that the stresses along the direction of reinforcement (along the direction of the applied external loads) do not reach their limit values.*

It is well known from numerous studies that there are various types of curvature in the microstructure of composite materials. Thus, almost always the possibility exists of the appearance of different stresses in the microstructure of the composite.

Taking into account the above information and considerations, it was for the first time proposed in the works of the author and his pupils to explain the fracture mechanism by "separation into the slender parts" *due to internal stresses that arise as a result of curvatures in the microstructure and act on the areas, the normal to which coincides with the normal to the curved interface.*

Let us consider the proposed approach concerning the micro-element in Fig. 7.5, including a curved interface between the reinforcing element and the matrix. All values related to the reinforcing element are marked with the index "a," and all values related to the matrix are marked with the index "m."

In Fig. 7.5, the following designations are introduced:

H—the rise;

Λ—the half-wave length of curvatures (e.g., periodic curvature is considered);

n and τ—the unit vectors of the normal and tangent to the curved interface;

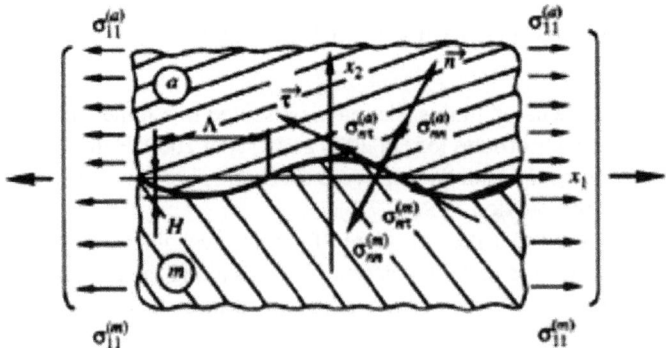

Fig. 7.5

$\sigma_{11}^{(a)}$ and $\sigma_{11}^{(m)}$—the stresses along the direction of the external load, which are balanced with the external load;

$\sigma_{nn}^{(a)}$, $\sigma_{n\tau}^{(a)}$, $\sigma_{nn}^{(m)}$, and $\sigma_{n\tau}^{(m)}$—the stresses in the reinforcing element and matrix, which are applied in the curved boundary and are self-balanced within each period of curvature.

It should be noted that the last stresses change sign after a half-period of curvature along the axis $0x_1$ (Fig. 7.5). Consequently, under a tensile or compressive external load, *the tensile stresses always arise in individual sections of the curved boundary,* which can lead to the exhaustion of the adhesive strength.

Similar considerations apply to the shear stresses.

Consider now some estimates of a quantitative nature. So, for the not very large curvatures, the conditions are fulfilled

$$H \ll \Lambda. \tag{7.1}$$

Obviously, for the composites reinforced in the direction of the axis $0x_1$ (Fig. 7.5), under the considered loading, the inequality (7.2) is hold

$$\left| \sigma_{11}^{(m)} \right| \ll \left| \sigma_{11}^{(a)} \right| \tag{7.2}$$

(due to a natural inequality of the following type:

$$E_m \ll E_a, \tag{7.3}$$

where E_a and E_m are Young's moduli of reinforcing elements and matrix) with the same elongation of the reinforcing element and matrix in the direction of the axis $0x_1$ (Fig. 7.5).

In connection with the considered type of loading for the elastic models (brittle fracture), the following relations are also valid:

$$\sigma_{11}^{(a)} \approx E_a \cdot k_a; \, \sigma_{11}^{(m)} \approx E_m \cdot k_m, \tag{7.4}$$

where k_a and k_m are the coefficients.

By virtue of the condition of continuity of the stress vector on the curved boundary, the following relations also take place:

$$\sigma_{nn}^{(a)} = \sigma_{nn}^{(m)}; \, \sigma_{n\tau}^{(a)} = \sigma_{n\tau}^{(m)}. \tag{7.5}$$

By virtue of conditions (7.5) at the interface for the stresses $\sigma_{nn}^{(m)}$ and $\sigma_{n\tau}^{(m)}$, it is no longer possible to obtain estimates of the type of the second expression (7.4). Therefore, for these stresses at the interface, despite the existence of inequality (7.3), it is no longer possible to obtain an estimate of the type (7.2).

Thus, at the interface, the stresses $\sigma_{nn}^{(m)} \sigma_{n\tau}^{(m)}$ and $\sigma_{n\tau}^{(m)}$ can be much larger in value than can be obtained based on an estimate of the type of the second expression (7.4).

Taking into account the above estimates and considerations, it is necessary to clarify the following situation. Assume that the stresses $\sigma_{11}^{(a)}$ and $\sigma_{11}^{(m)}$ (Fig. 7.5) are less than the corresponding ultimate strengths separately for the reinforcing elements and the matrix

$$\left|\sigma_{11}^{(a)}\right| < \Pi_1^{\pm(a)}; \ \left|\sigma_{11}^{(m)}\right| < \Pi_1^{\pm(m)}. \tag{7.6}$$

The conditions (7.6) ensure that no fracture of the composite occurs due to the rupture of the reinforcing elements and the matrix under the action of stresses directed along the direction of the external load.

Note that conditions (7.6) can be fulfilled due to restrictions on the value of the external load. Thus, *it is necessary to prove that there are reasonable limits of variation of the quantities H and Λ (Fig. 7.5), satisfying condition (7.1), under which conditions (7.6) are also satisfied, and for stresses $\sigma_{nn}^{(m)}$ and $\sigma_{n\tau}^{(m)}$ at the interface the following conditions are satisfied*:

$$\left|\sigma_{nn}^{(m)}\right| = A_p^+ \text{or} \left|\sigma_{n\tau}^{(m)}\right| = A_c^+, \tag{7.7}$$

where A_p^+ and A_c^+ are limits of adhesion tensile and shear strength as applied to the interface between the reinforcing elements and the matrix.

It should also be noted that for most engineering composites, due to the presence of various defects at the interface between the reinforcing elements and the matrix, the following conditions are usually met:

$$A_p^+ < \Pi_1^{+(m)}. \tag{7.8}$$

The necessary conditions. *Thus, for the proposed fracture mechanism by "separation into the slender parts," it is necessary to prove that there are reasonable limits of variation of the parameters H and Λ (Fig. 7.5), at which, taking into account the fulfillment of conditions (7.1) and (7.6), the stresses $\sigma_{nn}^{(m)}$ and $\sigma_{n\tau}^{(m)}$ can occur (at the interface) no less than the values of the stresses $\sigma_{11}^{(m)}$ (Fig. 7.5).*

Note that when solving this problem *in the case of applying the continuum theory, it is necessary to construct such a theory that would make it possible to determine the self-balanced (within each curvature) stresses. Usually, the continuum theories allow one to determine the stresses on the areas, the dimensions of which significantly exceed the dimensions of the curvatures.*

Note 7.1 It is advisable to note that the above-formulated **necessary conditions** for explaining the fracture mechanism by "separation into the slender parts" were obtained within the framework of *the model concepts* for any considered pair of materials (matrix, filler). Thus, the fulfillment of the **necessary conditions** for a specific pair of materials (matrix and filler, taking into account their mechanical and strength properties) indicates that the fracture mechanism by "separation into the slender parts" for the pair of materials under discussion can arise if, in a certain volume of the composite created from this pair of materials, **the periodic distortions exist** in the structure of the composite, characterized by the values of the parameter $H \Lambda^{-1}$ (Fig. 7.5) in a certain interval.

Thus, to clarify the possibility of satisfying *the sufficient conditions* for explaining the fracture mechanism by "separation into the slender parts," it is necessary, first of all, to show that certain parts of the material *with periodic curvatures in the structure* in the *real* composites exist.

Apparently, two cases of the existence of the periodic curvatures in the structure of the composite can be distinguished.

The first case is when the presence of periodic curvatures in the structure of the composite is an indispensable property of these materials, which is taken into account in the technology of their creation.

The second case is when the presence of periodic curvature in the structure of the composite is an undesirable consequence of technological processes associated either with their imperfection or caused by the influence of the initial (residual) stresses, which are also called the technological stresses.

Consider an example for *the first* above case, shown in Fig. 7.6, which is placed on page 21 of monograph [18] in the bibliography of the two-volume monograph [13]. This figure shows the structure of the composite (spatially cross-linked fiberglass), where the sinusoidal curvatures of the reinforcing elements are clearly visible, which are not associated with the technology errors or with the influence of technological stresses, and are an indispensable property of this composite.

Consider an example for *the second* above case, shown in Fig. 7.7, which corresponds to Fig. 4.12a on p. 103 of the monograph [14] in the bibliography of the two-volume monograph [13]. This figure shows the section of the cut of a thick-walled cylindrical shell made of epoxyphenolic fiberglass, obtained by winding on a mandrel. This figure clearly shows the local periodic curvatures of the reinforcing elements, obtained as a result of the loss of stability during cooling from a temperature of $160°$ C under the action of technological stresses.

Fig. 7.6

Fig. 7.7

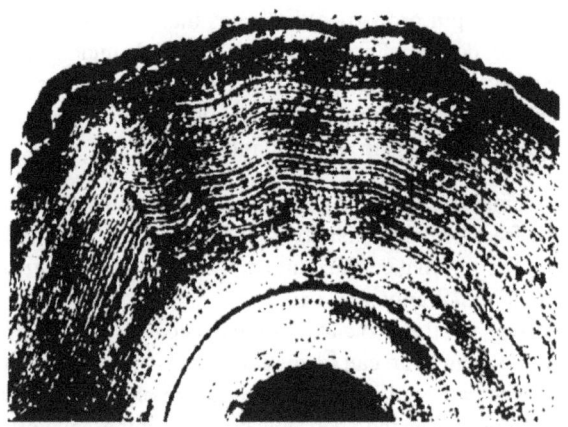

Thus, as it follows from the examples given in Figs. 7.6 and 7.7 that in *the real* (not in model) composites (Fig. 7.6) and structural elements (Fig. 7.7), at least the parts of the material with periodic curvatures in the structure exist. Consequently, taking into account the information presented in this section, the conclusion can be formulated that the explanation of the fracture mechanism in the form of "separation into the slender parts" proposed and described in Sect. 7.3 which seems to be quite realistic.

We will restrict ourselves to the information outlined in this section when discussing the proposed explanation of the fracture mechanism in the form of "separation into the slender parts." Additional information can be obtained from the monograph [38] (e.g., pp. 367–371) and the monograph [13] (vol. 2, for example, pp. 479–482).

7.4   On Development of Fundamentals of Mechanics of Composites with Curved Structures

It follows from the previous section that to explain the fracture mechanism in the form of "separation into the slender parts" and to determine the parameters of the material structure and the value of the load corresponding to the occurrence of the discussed type of fracture, it is necessary to have sufficiently accurate information about the stress–strain state of composites with curved structures.

7.4.1 Introduction

Therefore, in this section, the basic information on the development of the foundations of the mechanics of composites with curved structures is given in a very brief form, which made it possible to carry out analytical and numerical studies of the stress–strain state of such composites.

Additional information can be obtained in the monographs [12] (Chap. 7) of 1990, [19] (vol. 4) of 1995, and [13] (vol. 2, Chap. 11) of 2008, and also in the monograph [38] in English, which is fully devoted to the issue under discussion.

Since this chapter deals with **Problem 5** (*Brittle fracture in the form of "separation into the slender parts" under stretching or compressing the composite materials along reinforcing elements*), the *model of the linear elastic body* is used to study the brittle fracture.

It is worth noting that in the mechanics of composites with curved structures, classification of curvatures is carried out, highlighting the large-scale and small-scale curvatures.

The large-scale curvatures are usually understood to mean the curvatures, the geometric parameters (boom, wavelength) of which are comparable to the minimum dimensions of a structural element or specimen.

The statement of problems for composites and structural elements made of them in the case of the large-scale curvatures of the filler, methods for solving problems, and the results of studying problems are presented, for example, in Russian in the article [9] of 1982, from the list of references to the monograph [15] or in English in the article [16] from the list of references to the monograph [58].

The small-scale curvatures are usually understood to mean curvatures, the geometric parameters of which are significantly less than the minimum dimensions of the considered structural element or sample. In this section, all the discussed approaches and the results obtained on their basis concerning the analysis of the fracture mechanism in the form of "separation into the slender parts" refer to the mechanics of composites with the small-scale curvatures of reinforcing elements, which obviously follows from Figs. 7.5, 7.6 and 7.7.

The results on the construction of the mechanics of composites with curved structures, which are discussed in this section, were obtained, as in other parts of the mechanics of composites, using the generally accepted *two approaches*:

the second approach is based on the continuum approaches;
the first approach is based on a piece-wise homogeneous body model.

It is well known and generally accepted that *the first approach* is the most rigorous and accurate.

Within the framework of the continuum approaches (as applied to the studies of Sect. 7.4 on the mechanics of composites with curvatures in the structure), *the theories were developed that make it possible to determine the stresses on the areas, the sizes of which are commensurate with or slightly less than the half-periods of curvatures in the composite structure.*

It should be noted that the continuum theories previously developed by other authors make it possible to determine the stresses on areas *that are much larger than the half-periods of curvature* in the composite structure, which *makes it impossible to apply them* to the study of the fracture mechanism in the form of "separation into the slender parts."

Taking into account the above information of an introductory nature, below in this section, the information in a very brief form is considered for the composites with curvatures separately for the continuum theories and results based on them, as well as for a piece-wise homogeneous body model and results based on it.

7.4.2   Continuum Theories and Results Based on Them

It should be noted that in the scientific direction discussed in this section, the first results were published in the articles [10, 11, 51, 52] of 1983 by the author of this monograph. At that, in [10], apparently, a continuum theory for the composites with curvatures was proposed, which made it possible to determine the stresses on areas whose dimensions are of the same order of value or somewhat less than half-periods of curvatures.

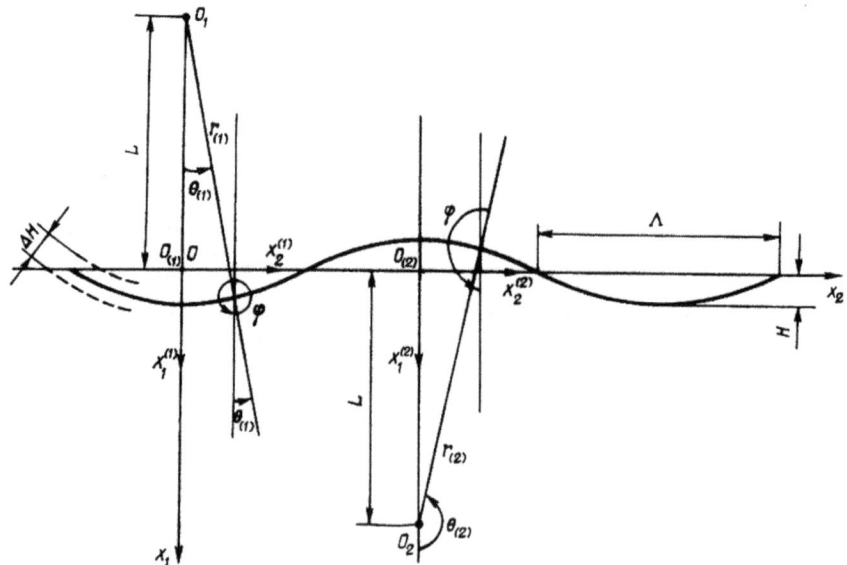

Fig. 7.8

Figure 7.8 presents, in part, the design scheme of the article [10] to take into account the effect of the curvature of reinforcing elements. The construction corresponds to the continuum theory presented in sufficient detail in the monograph [12] of 1990 on pages 494–510 and is also given in the monographs [19] (v.4) of 1995 on pages 25–40, [38] of 2000 on the pp. 13–28, and [13] (vol. 2) of 2008 on pp. 440–464.

Following the above sources, the basic relations of the continuum theory [10] for composites with periodic curvatures (Fig. 7.8) are presented in the Cartesian coordinates $x_j(j = 1, 2, 3)$ concerning the plane problem in the plane (Fig. 7.8), where the axis $0x_3$ is directed perpendicular to the plane $(x_1 0 x_2)$ from Fig. 7.8. In this case, we will use the designations for the quantities L, H, and Λ, which are also indicated in Fig. 7.8, where the axis $0x_2$ is directed along the reinforcing elements.

The main relations are considered for geometrically linear problems. In this case, the elasticity relations have the following form

$$\sigma_{ij} = \mu_{ij\alpha\beta}\frac{\partial u_\alpha}{\partial x_\beta}; i, j, \alpha, \beta = 1, 2, 3, \tag{7.9}$$

where for the components of the elastic coefficient tensor the relation holds (7.10)

$$\mu_{ij\alpha\beta} = \mu_{ij\alpha\beta}^{(0)} + \varepsilon\left(\sin \pi \Lambda^{-1} x_2\right)\mu_{ij\alpha\beta}^{(1)} + \varepsilon^2 \frac{1}{2}\left(1 - \cos 2\pi \Lambda^{-1} x_2\right)\mu_{ij\alpha\beta}^{(2)}. \tag{7.10}$$

In (7.10) and below, $\mu^{(0)}$ denotes the tensor of elastic coefficients of a linearly elastic orthotropic body with *rectilinear* orthotropy, the components of which are determined from the well-known expression (7.11)

$$\mu_{ij\alpha\beta}^{(0)} = \delta_i^j \delta_\alpha^\beta A_{i\beta} + \left(1 - \delta_i^j\right)\left(\delta_i^\alpha \delta_j^\beta + \delta_i^\beta \delta_j^\alpha\right)G_{ij}. \tag{7.11}$$

The components of tensors $\mu^{(1)}$ and $\mu^{(2)}$ are determined through the elasticity coefficients $A_{i\beta}$ and G_{ij} by expressions, which are given, for example, in the monograph [13] (vol. 2) on pages 454–456.

It should be noted *that the continuum theory [10]for the composites with small-scale curvatures is constructed with an accuracy of ε^2*, where this parameter is determined by the expression

$$\varepsilon = \frac{\Lambda}{\pi L} \approx \frac{8}{\pi}\frac{H}{L} \ll 1, \tag{7.12}$$

and the coefficients of elasticity $A_{i\beta}$ and G_{ij} in all relations of this theory are the averaged coefficients of elasticity for the considered composite **without curvatures** (within the framework of the model of an orthotropic body with rectilinear orthotropy).

The equations in the displacements of the discussed continuum theory are for the material occupying the region D of the form (7.13)

$$L_{ij}u_\alpha + \varepsilon \frac{\partial}{\partial x_i}\left[(\sin \pi \Lambda^{-1}x_2)\mu_{ij\alpha\beta}^{(1)}\frac{\partial}{\partial x_\beta}\right]u_\alpha + \qquad (7.13)$$

$$+\varepsilon^2 \frac{\partial}{\partial x_i}\left[\frac{1}{2}(1 - \cos 2\pi \Lambda^{-1}x_2)\mu_{ij\alpha\beta}^{(2)}\frac{\partial}{\partial x_\beta}\right]u_\alpha = 0, \; x_k \in D,$$

where the following notation is introduced for the differential operators $L_{j\alpha}$ (differential operators of the Lamé equations for an orthotropic body with the rectilinear orthotropy)

$$L_{j\alpha} = \left[A_{j\alpha} + (1 - \delta_j^\alpha)G_{j\alpha}\right]\frac{\partial^2}{\partial x_j \partial x_\alpha} + \delta_j^\alpha(1 - \delta_i^j)G_{ij}\frac{\partial^2}{\partial x_i^2} - \rho\delta_j^\alpha\frac{\partial^2}{\partial \tau^2}. \quad (7.14)$$

The boundary conditions in stresses on a part S_1 of the surface have the usual form

$$N_i\sigma_{ij} = P_j, \; x_k \in S_1 \qquad (7.15)$$

taking into account expressions (7.9) and (7.10).

It follows from expressions (7.13) that the basic equations of the discussed theory [10] are equations with periodic coefficients, which should be expected, taking into account the fragment from the design scheme, which is shown in Fig. 7.8.

Results based on continuum theory [10].

Let us dwell on a short list of some interesting, in the author's opinion, results obtained based on the continuum theory [10].

1.   When constructing, discussing, and analyzing this theory, the following statement was proved

     Statement. For insignificant curvatures in the structure of the composite (for small values ε from (7.12)) when constructing the theory in the first approximation (up to ε), one should take into account *not the change in the values of the elastic constants (within the framework of the model of an orthotropic body with rectilinear orthotropy), but a change in the type of symmetry of an anisotropic body.*

     Proof of the Statement, in essence, is represented by expressions (11.34) and (11.35) on page 453 of the monograph [13] (vol. 2); the actuality of the statement under discussion is commented on p. 454 of the above monograph.

2.   The *unique exact solution* was obtained (It remains the only one to this day)— the problem on the pure shear in the plane x_20x_3 (Fig. 7.8). This *exact solution* is presented in detail for the first time on pp. 13–14 of the article [51] of 1983.

3.   The analysis of vibrations of the composites with curved structures is carried out in the article [11] using the variational methods.

4.   A general method for solving the problems of the theory [10] was developed in the article [51] based on the representation of the basic relations in the form of series in a small parameter (7.12).

5.   Using the above method, the solution [52] of the problems of quasi-uniform states of the tension–compression and shear were obtained.
6.   Based on the above-mentioned results and following the approach described in relative detail in Sect. 7.3, the conditions [52] (p. 14) were obtained for the geometric parameters of curvatures, when the fracture may occur in the form of "separation into the slender parts." For the fiberglass, these conditions are given below

$$H > 0.16\Lambda. \tag{7.16}$$

The above results are presented in the articles [10, 11, 51, 52] of 1983. They are obtained within the framework of the continuum theory [10] of the composites with curved structures, in subsequent years were included in the corresponding sections of the monographs [12] of 1990, [19] (vol.4) of 1995, [38] of 2000, and [13] (vol. 2) of 2008.

The further development of the continuum theories for the composites with curved structures and the study of the corresponding specific problems are presented in the articles of S.D. Akbarov and his pupils, which are partly presented in the introduction to this chapter (Sect. 7.1) and are also partly included in the bibliography to this monograph. It should be noted that these studies were carried out mainly in Azerbaijan and Turkey, and the corresponding results are included in the monographs [19] (vol.4) of 1995 and [38] of 2000.

7.4.3   Model of a Piece-Wise Homogeneous Medium and Results Based on It

In this chapter, **Problem 5** (*Brittle fracture in the form of "separation into the slender parts" under tension or compression of composite materials along reinforcing elements*) as in **Problem 1** (*Fracture in composite materials under compression along reinforcing elements*) and in **Problem 3** (*Fracture in the form of end-crush fracture under compression of composite materials*), considered, respectively, in Chaps. 3 and 5 (part II of this monograph), the studies were carried out using two well-known and generally accepted approaches in the mechanics of composites.

In the first approach, the model of a piece-wise homogeneous medium is used, and in the second approach, the composite is modeled by the homogeneous anisotropic material with the averaged characteristics (continuum theories).

A short and consistent description of these approaches is presented in the final part of Sect. 1.4.1 (part I of this monograph).

As applied to **Problem 5**, discussed in this chapter, a summary of the results obtained using the second approach (continuum theory) is presented in the previous Sect. 7.4.2. The present Sect. 7.4.3 provides very brief information on the results for **Problem 5** on the layered and unidirectional fiber composites obtained using the first approach (piece-wise homogeneous medium model).

First of all, let us dwell **on the statement of problems and the method of their study**.

Within the framework of the model of linear elastic isotropic, in some cases orthotropic, bodies, the static problems are considered (for the layered composites—the plane and spatial, and the unidirectional fibrous composites—the spatial) problems, concerning the composites with curved structures.

At that, the conditions of continuity of the stress vectors and displacements are set at the boundaries—the interface between the filler and the binder.

Figure 7.9 shows a fragment of a layered composite, where all the quantities related to the matrix are marked with the index (1), and all the quantities related to the filler are marked with the index (2). With the middle surface of the filler, which in Fig. 7.9 is indicated by the number 1, the Cartesian coordinate system (x_1, x_2, x_3) is associated. It is assumed that the middle surface of the filler layer is given by the equation

$$x_2 = F(x_1, x_3) = \varepsilon f(x_1, x_3), \tag{7.17}$$

where ε is a small parameter satisfying the condition

Fig. 7.9

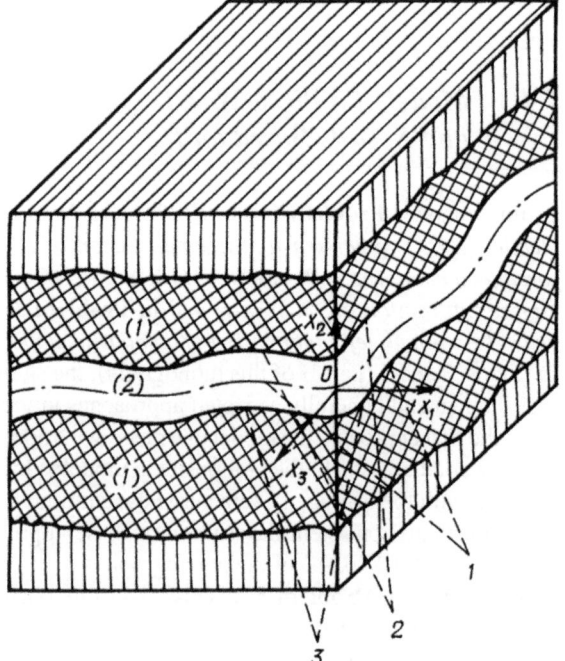

Fig. 7.10

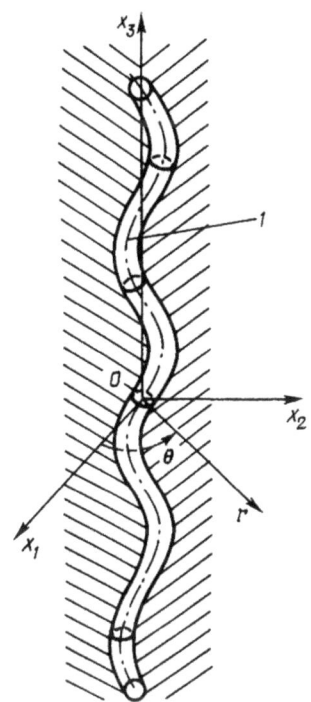

$$\varepsilon < 1. \tag{7.18}$$

The conditions of continuity of the stress vectors and displacements are formulated on the lower (in Fig. 7.9 indicated by the number 3) and upper (in Fig. 7.9 indicated by the number 2) interface surfaces, *into which, due to dependencies (7.17), a small parameter ε will enter.*

Figure 7.10 shows the fragment of a unidirectional fiber composite in which, for example, one fiber is highlighted. Further, also, for example, the case is considered when the middle line of the fiber or the line of centers of transverse normal (concerning the curved line) sections (in Fig. 7.10 this line is indicated by the number 1) is a plane curve and is located in a plane $x_2 0 x_3$. In this case, this centerline can be represented by the equation

$$x_2 = F(x_3) = \varepsilon \delta(x_3), \ x_1 = 0, \tag{7.19}$$

where ε is a small parameter satisfying the condition (7.18).

The conditions of continuity for the stress vectors and displacements are formulated on the curved surface of a circular cylinder, which is the interface between the filler and the matrix. The section of this interface by the plane $x_1 = 0$ in Fig. 7.10 is shown by two dark lines, between which is the middle line of the fiber, shown in Fig. 7.10 in the dash-dotted line. By virtue of expressions (7.19), the conditions of continuity formulated in the above way include a small parameter ε satisfying condition (7.18).

The method for solving the problems of determining the stress–strain state of the composites with curved structures (the plane and spatial problems for layered composites, the spatial problems for unidirectional fiber composites) **within the framework of a piece-wise homogeneous medium model** *consists in representing all the sought quantities in the form of series in a small parameter* ε, *which satisfies condition* (7.18). As a result, we arrive at each of the approximations to *the corresponding problem for the composite with non-curved structures*. The results from the previous approximations are included in the right-hand sides of the corresponding conditions of continuity of the stress and displacement vectors for the considered approximation.

Note 7.2 In the above solution method, the formulation of continuity conditions for stress and displacement vectors is carried out on curved interfaces. It is taken into account that the layers have a constant thickness and the fibers have a constant circular cross section, which is counted in normal cross sections concerning the curved midlines and mid-surfaces. To implement this procedure, a parametric representation of lines and surfaces is used, which complicates the construction of a solution procedure in an analytical form. It should be noted that the provision on the constancy of the thickness of the layers and the shape of the cross section of the fibers as applied to the filler (reinforcing elements) corresponds to a certain extent to structural composites. The discussed research method for composites with curved structures was proposed by S.D. Akbarov and the author of this monograph concerning layered composites in articles [4, 32] for 1984, [5, 34] for 1985 and in several others, and as applied to unidirectional fiber composites in articles [192] for 1984, [5, 35] for 1985, [6] for 1986 and in several others. In the above articles [4–6, 32–35] for 1984–1986, the results obtained for composites with curved structures within the framework of the model of a piece-wise homogeneous medium were included in the following years in the corresponding sections of monographs [12] for 1990, [19] (vol. 4) for 1995, [38] for 2000, and in [13] (v. 2) for 2008. Further development of the mechanics of composites with curved structures within the framework of the model of a piece-wise homogeneous medium and the study of the corresponding specific problems is presented in the articles of S.D. Akbarov and his students, which are partially indicated in the Introduction to this chapter (Sect. 7.1) and also, partially, included in the list of references to this monograph (part III); It should be noted that these studies were carried out mainly in Azerbaijan and Turkey, and the corresponding results are included in a comparatively detailed form in monographs [19] (vol. 4) for 1995 and [38] for 2000.

It should be noted that in the above monographs [12, 19 vol. 4, 38, 13 vol. 2] and in the articles indicated in the Introduction to Sect. 7.1), the numerous specific scientific results obtained within the framework of the piece-wise homogeneous model of the medium using **the method for solving problems**, the brief information about which is presented above in Sect. 7.4.3.

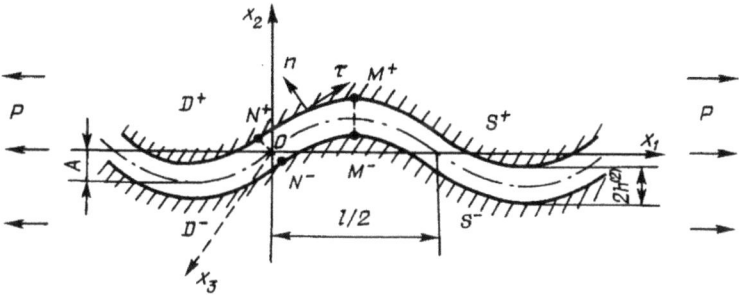

Fig. 7.11

Without being able to present even brief information about the main results, below, as examples, the brief information about two interesting results is given.

Examples of results within the framework of the piece-wise homogeneous medium model.

Let us consider two examples for the composites with curved structures *in the case of low filler concentrations.* In this case, the problems are reduced to an infinite elastic space (matrix, all values for which are marked with the subscript "*m*") with one infinite in one direction *curved reinforcing element* (filler, all values for which are marked with the subscript "*a*") under tension "at infinity" along the reinforcing element.

When studying the phenomena inside the composite following **Note 3.8** (Part II of this monograph) "at infinity" (along the reinforcing elements), *the same* tensile or compressive deformations are set *for the matrix and the reinforcing element.*

Example 7.1 *The layered composite with curved layers. A plane problem.*

All additional designations are shown in Fig. 7.11. The plane deformation in the plane $x_1 0 x_2$ is considered. Hence, it is assumed that the curvature of the layers does not depend on the coordinate x_3.

Note that Fig. 7.11 corresponds to the cross section of the material at $x_3 = 0$ and is a special case of the situation shown in Fig. 7.9.

For the study, **the method of solving the problems** is used, brief information about which is presented above in this subsection.

Restrict ourselves to considering the case when the equation of the curved middle surface of the reinforcing element can be represented in the following form

$$x_2 = A \sin 2\pi l^{-1} x_1 = \varepsilon l \sin 2\pi l^{-1} x_1; \varepsilon = A \cdot l^{-1}. \qquad (7.20)$$

According to Fig. 7.11, in (7.20), introduced the following designations:
A–rise;
l–the wavelength of the curvature shape.
According to (7.17), the case (7.20) corresponds to the notation

$$F(x_1, x_2) = A \sin 2\pi l^{-1} x_1; \ f(x_1, x_2) = l \sin 2\pi l^{-1} x_1; \ \varepsilon = A \cdot l^{-1}. \qquad (7.21)$$

The studies were carried out taking into account three approximations by ε. In this case, the normal $\sigma_{nn}^{\pm}$ and shear $\sigma_{n\tau}^{\pm}$ stresses were determined on the interface lines (lines $S^{\pm}$ in Fig. 7.11) of the material properties of the binder and filler (interface) at points $N^{\pm}$ and $M^{\pm}$ in Fig. 7.11.

It should be noted that the indicated stresses $\sigma_{nn}^{\pm}$ and $\sigma_{n\tau}^{\pm}$ the interfaces are self-balanced within each period of curvature (Fig. 7.11).

The calculations were carried out for the following parameter values

$$v_a = v_m = 0.3; \ E_a \cdot E_m^{-1} = 20; \ 50; \ 100; \ 150. \qquad (7.22)$$

From the analysis of the results for the above cases (7.22), the following conclusion was made.

At $E_a \cdot E_m^{-1} \geq 50$, even at relatively small values ε(7.20) $\varepsilon \equiv Al^{-1} \geq 0.025$ **the values of the normal self-balanced stresses applied to the interface can exceed the normal stresses in the matrix** $\sigma_{11}^{(m)}$ **(in the absence of curvature in the structure)**, which are balanced with external forces p (Fig. 7.11).

The above results can be used to substantiate the proposed explanation of the fracture mechanism in the form of "separation into the slender parts." The previously discussed results should be supplemented with an analysis involving the numerical values of the ultimate strength of the considered binder (matrix) and filler (reinforcing elements).

These results are presented in the article [4] of 1984. This article also outlines other interesting findings that follow from the analysis. The considered results are fully included in the corresponding sections of the monographs [12, 19 vol. 4, 38, 13 vol. 2], which also presents the numerous results for the composites with different curvatures in the structure, taking into account *the interaction of these curvatures*.

Example 7.2 *Fibrous unidirectional composite with curved fibers. Spatial problem.*

All additional designations are shown in Fig. 7.12. The studies are carried out in the Cartesian coordinates (x_1, x_2, x_3) and in coordinates of a circular cylindrical coordinate (r, θ, x_3) system, which are also indicated in Fig. 7.12.

Note that Fig. 7.12 corresponds to Fig. 7.10. For the study, the method of solving problems is used, brief information about which is presented above in this subsection.

As in the case in Fig. 7.10, it is assumed in Example 7.2, that the line of centers of the fiber cross sections is a plane curve, which is located in the plane $x_1 = 0$ in Fig. 7.12.

Restrict ourselves to considering the case when the equation of the line of centers of cross sections of a curved fiber can be represented in the following form

$$x_2 = A \sin 2\pi l^{-1} x_3 = \varepsilon l \sin 2\pi l^{-1} x_3, \ x_1 = 0; \ \varepsilon = A \cdot l^{-1}. \qquad (7.23)$$

Fig. 7.12

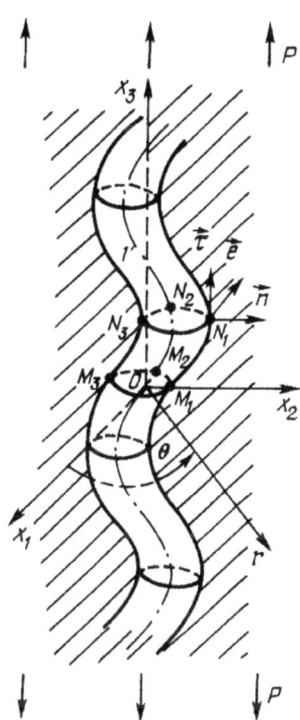

According to Fig. 7.12 in (7.23), the following designations are introduced:

A—the rise;

l—the wavelength of the curvature form concerning the line of centers of the cross sections of the curved fiber.

Following the notation (7.19), the notation holds

$$F(x_3) = A \sin 2\pi l^{-1} x_3, \; \delta(x_3) = l \sin 2\pi l^{-1} x_3, \; x_1 = 0; \; \varepsilon = A \cdot l^{-1}. \qquad (7.24)$$

The studies were carried out taking into account three approximations by ε. The normal σ_{nn} and tangential $\sigma_{n\tau}$ and $\sigma_{n\theta}$ stresses at the interface between the properties of the binder and filler, which is a curved cylindrical surface, were determined. The numerical values of these stresses were determined for points M_1, M_2, and M_3, as well as N_1, N_2, and $N_{3,}$, which are marked in Fig. 7.12, with the uniaxial loading "at infinity" by the forces of intensity, which are directed along the axis $0x_3$ (Fig. 7.12).

It should be noted that the stresses σ_{nn}, $\sigma_{n\tau}$, and $\sigma_{n\theta}$ at the interface are self-balanced within each bending period (Fig. 7.12).

The calculations were carried out for the following parameter values

$$v_a = v_m = 0.3; \; E_a \cdot E_m^{-1} = 20; 50; 100; 150; 200. \qquad (7.25)$$

From the analysis of the results for the above cases (7.25), the following conclusion was made.

For $E_a \cdot E_m^{-1} \geq 100$, even with relatively small values ε (7.23) $\varepsilon = A \cdot l^{-1} \geq 0,020$, the values of self-balanced normal and tangential stresses applied to the interface can exceed normal stresses in the matrix $\sigma_{33}^{(m)}$ (in the absence of curvatures in the structure), which are balanced with external forces p (Fig. 7.12).

The above results can be used to substantiate the proposed explanation of the fracture mechanism in the form of "separation into the slender parts." The previously discussed results should be supplemented with an analysis involving the numerical values of the ultimate strength of the considered bind (matrix) and filler (fibers).

These results are presented in the article [35] of 1985. This article also presents other interesting conclusions that follow from the above analysis. The considered results are included in full in the corresponding sections of the monographs [12, 19 vol. 4, 38, 13 vol. 2], where the numerous results are also given for the composites with different curvatures in the structure, including taking into account *the interaction of these curvatures.*

Note 7.3 When carrying out a comparative analysis of the results that are discussed in the framework of the continuum theory in Sect. 7.4.2, and the results that are discussed in the framework of the model of a piece-wise homogeneous medium in Sect. 7.4.3, it is necessary to take into account the following situation.

Within the framework of the continuum theory in Sect. 7.4.2, the small parameter ε is introduced by expression (7.12), which, due to the notation in Fig. 7.8 *differs* from expressions (7.20) and (7.21), by which a small parameter ε. is introduced in Sect. 7.4.3 within the framework of the model of a piece-wise homogeneous medium. Hence, *for the same curvature, the numerical values of the small parameter for the two discussed approaches differ from each other.*

Note 7.4 This chapter discusses the various issues related to brittle fracture of type of "separation into the slender parts." In this regard, in these studies, *the model of a linearly elastic body* is exclusively used, including for the constructed foundations of the mechanics of composites with curved structures.

Taking into account the above results, in several articles, beginning with [20] of 1985, as well as [1, 22] of 1986, the fundamentals of the mechanics of composites with curved structures were built, when the linear viscous model was used for the matrix (binder). These studies were continued in Azerbaijan and Turkey, where several articles were published. These studies (when using the linear viscoelasticity model for the matrix), apparently, have independent significance and are not intended to explain the mechanism of brittle fracture in the form of "separation into the slender parts."

We restrict ourselves to the above information in this chapter with a short outlining of the results on **Problem 5** (*Brittle fracture in the form of "separation into the slender parts" under tension or compression of composite materials along reinforcing elements*).

References

1. Akbarov, S.D.: Vliianie reologicheskikh parametrov materiala matritsy na raspredelenie samouravnoveshennykh napriazhenii v mnogosloinom kompozite s iskrivlennymi strukturami (Effect of rheological parameters of material of matrix on distribution of self-equilibrated stresses in multilayered composite with curved structures). Mekhanika Kompozitnykh Materialov. **61**(4), 617 (1986)
2. Akbarov, S.D.: K mekhanike kompozitnykh materialov s lokalnymi iskrivleniiami v strukture (Towards mechanics of composite materials with local structure curvatures). Prikladnaya Mekhanika. **23**(1), 119–122 (1987)
3. Akbarov, S.D.: O raspredelenii samouravnoveshennykh napriazhenii v mnogosloinom kompozitnom materiale s iskrivlennymi strukturami (On distribution of self-balanced stresses in a multilayer composite material with curved structures). Mat. metody i fiz.-mekh. polia. **26**, 83–89 (1987)
4. Akbarov, S.D., Guz, A.N.: O napriazhennom sostoianii v kompozitnom materiale s iskrivlennymi sloiami s maloi kontsentratsiei napolnitelia (On stress state in a composite material with curved layers with a low filler concentration). Mekhanika kompozitnykh materialov **6**, 990–996 (1984)
5. Akbarov, S.D., Guz, A.N.: K mekhanike kompozitnykh materialov s iskrivlennymi strukturami (Towards mechanics of composite materials with curved structures). Dokl. Akad. Nauk SSSR **281**(1), 37–41 (1985)
6. Akbarov, S.D., Guz, A.N.: Ob odnom effektie v mekhanike razrusheniia kompozitnykh materialov (On one effect in fracture mechanics of composite materials). Dokl. Akad. Nauk SSSR **290**(1), 23–26 (1986)
7. Akbarov, S.D., Guz, A.N.: Raspredelenie napriazhenii v mnogosloinom kompozitnom materiale s iskrivlennymi strukturami (model kusochno-odnorodnogo tela) (Stress distribution in a multilayer composite material with curved structures (piecewise uniform body model)). Mekhanika kompozitnykh materialov **4**, 592–599 (1987)
8. Babich, IYu., Guz, A.N.: O primenimosti podkhoda Eilera k issledovaniiu ustoichivosti deformirovaniia anizotropnykh nelineino-uprugikh tel pri konechnykh dokriticheskikh deformatsiiakh (On applicability of Euler's approach to the study of anisotropic nonlinear elastic bodies deformation stability under finite subcritical deformations). Dokl. Akad. Nauk SSSR **202**(4), 795–796 (1972)
9. Bogdanov, V.L., Guz, A.N., Nazarenko, V.M.: Obieedinennyi podkhod v neklassicheskikh problemakh mekhaniki razrusheniia (A Unified Approach in Non-classical Problems of Fracture Mechanics). LAP LAMBERT Academic Publishing, Saarbrücken, Deutschland (2017)
10. Guz, A.N.: O kontinualnoi teorii kompozitnykh materialov s melkomasshtabnymi iskrivleniiami v strukture (The continuous theory of composite materials with small-scale curvatures in the structure). Dokl. Akad. Nauk SSSR **268**(2), 307–313 (1983)
11. Guz, A.N.: O teorii kolebanii kompozitnykh materialov s melkomasshtabnymi iskrivleniiami v strukture (The vibration theory of composite materials with small-scale curvatures in the structure). Dokl. Akad. Nauk SSSR **270**(4), 824–827 (1983)
12. Guz, A.N.: Mekhanika razrusheniia kompozitnykh materialov pri szhatii (Fracture Mechanics of Composite Materials Under Compression). Naukova Dumka, Kyiv (1990)
13. Guz, A.N.: Osnovy mekhaniki razrusheniia kompozitov pri szhatii: V 2-kh tomakh (Fundamentals of the Fracture Mechanics of Composites Under Compression: In 2 volumes). Litera, Kyiv, (2008). T. 1. Razrushenie v strukture materiala. (Fracture in Structure of Materials)
14. Guz, A.N., Kuliev, G.G.: K postanovke zadach ustoichivosti deformirovaniia tonkikh tel s treshchinami (On Stating Problems of Deformation Stability for Thin Bodies with Cracks). Doklady Akademii nauk USSR, Ser.A. **12**, 1085–1088 (1976)
15. Guz, A.N., Rushchitskyi, Ya.Ya., Guz, I.A.: Vvedenie v mekhaniku nanokompozitov (Introduction to Mechanics of Nanocomposites). S. P. Timoshenko Institute of Mechanics, Kyiv (2010)

16. Guz, A.N., Cherevko, M.A.: K mekhanike razrusheniia voloknistogo kompozitnogo materiala pri szhatii (On fracture mechanics of a fibrous composite material under compression). Dokl. Akad. Nauk SSSR **268**(4), 806–808 (1981)

17. Dovzhyk, M.V.: Razrushenie materiala s dvumya diskoobraznymi treschinami pri szhatii vdol treschin dlya malykh rasstoyaniy mezhdu treschinami (Fracture of a material compressed along two closely spaced penny-shaped cracks). Prikladnaya Mekhanika **49**(1), 100–108 (2013) T. 2. Rodstvennye mekhanizmy razrusheniia. (Related mechanisms of fracture)

18. Menshykov, A.V., Guz, I.A.: Zavisimost kofficientov intensivnosti napryazheniy sdviga ot sily treniya pri garmonicheskom nagruzhenii krugovoy treschiny (The dependence of the shear stress intensity factors on the friction force under the harmonic loading of a circular crack). Problemy Mashinostroeniya **9**(3), 65–71 (2006)

19. Guz, A.N. (ed.): Mekhanika kompozitov, v 12 tomah (Mechanics of Composites, in 12 volumes). (Vol. 1–4—Naukova Dumka, Vol. 5–12—A.S.K., Kyiv, 1993–2003), V.T. Golo-vach (ed.), T.1. Statika materialov. (Vol. 1. Statics of materials) (1993), N.A. Shulga (ed.), T.2. Dinamika i ustoichivost materialov. (Vol.2. Dynamics and stability of materials) (1993), L.P. Khoroshun (ed.), T.3. Statisticheskaia mekhanika i effektivnye svoistva materialov (Vol.3. Statistical Mechanics and Effective Material Properties) (1993), A.N. Guz, S.D. Akbarov, (ed.), T.4. Mekhanika materialov s iskrivlennymi strukturami (Vol.4. Mechanics of mate-rials with curved structures) (1995), A.A. Kaminsky (ed.), T.5. Mekhanika razrusheniia. (Vol.5. Fracture mechanics) (1996), N.A. Shulga, V.T. Tomashevsky (ed.), T.6. Tekhnologich-eskie napriazheniia i deformatsii v materialakh. (Vol.6. Technological stresses and deforma-tions in materials) (1997), A.N. Guz, A.S. Kosmodamiansky, V.P. Shevchenko (ed.), T.7. Kontsentratsiia napriazhenii. (Vol.7. Stress concentration) (1998), Ya.M.Grigorenko (ed.), T.8. Statika elementov konstruktsii. (Vol.8. Structural elements statics) (1999), V.D. Kubenko (ed.), T.9. Dinamika elementov konstruktsii. (Vol.9. Dynamics of structural elements) (1999), I.Yu. Babich (ed.), T.10. Ustoichivost elementov konstruktsii. (Vol.10. Stability of struc-tural elements) (2001), Ya.M. Grigorenko, Yu.N.Shevchenko (ed.), T.11. Chislennye metody. (Vol.11. Numerical methods) (2002), A.N. Guz, L.P. Khoroshun (ed.), T.12. Prikladnye issledovaniia. (Vol.12. Applied research) (2003)

20. Akbarov, S.D.: A method of solving problems in the mechanics of composite materials with curved viscoelastic layers. Sov. Appl. Mech. **21**(3), 221–225 (1985)

21. Akbarov, S.D.: Normal stresses in a fiber composite with curved structures having a low concentration of filler. Sov. Appl. Mech. **21**(11), 1065–1069 (1985)

22. Akbarov, S.D.: Stress state in a viscoelastic fibrous composite with curved structures and low fiber concentration. Sov. Appl. Mech. **22**(6), 506–513 (1986)

23. Akbarov, S.D.: Stress distribution in multi-layered composite with small-scale antiphase curvatures in structure. Sov. Appl. Mech. **23**(2), 107–111 (1987)

24. Akbarov, S.D.: Stress state in a laminar composite material with local warps in the structure. Sov. Appl. Mech. **24**(5), 445–452 (1988)

25. Akbarov, S.D.: Distribution of self-balanced stresses in a laminated composite material with antiphase locally distorted structures. Sov. Appl. Mech. **24**(6), 560–566 (1988)

26. Akbarov, S.D.: Solution of problems of the stress-strain state of composite materials with curvilinearly anisotropic layers. Sov. Appl. Mech. **25**(1), 12–20 (1989)

27. Akbarov, S.D.: The distribution of self-equilibrated stresses in fibrous composite materials with twisted fibers. Mech. Comp. Materials. **3**, 803–812 (1990)

28. Akbarov, S.D.: On the crack problems in composite materials with locally curved layers. Mech. Comp. Materials. **6**, 750–759 (1994)

29. Akbarov, S.D.: On the determination of normalized non-linear mechanical properties of composite materials with periodically curved layers. Int. J. Solid Struct. **32**(21), 3229–3243 (1995)

30. Akbarov, S.D., Aliev, S.A.: Stress state in laminar composite material with partial distortion in structure. Sov. Appl. Mech. **26**(12), 1127–1132 (1990)

31. Akbarov, S.D., Djamalov, Z.R.: Influence of geometric non-linearly calculation of stress disturbation in laminar composites with curved structures. Mech. Comp. Mater. **6**, 799–812 (1992)

32. Akbarov, S.D., Guz, A.N.: Method of solving problems in mechanics of composite materials with bent layers. Sov. Appl. Mech. **20**(4), 299–304 (1984)
33. Akbarov, S.D., Guz, A.N.: Method of solving problems in mechanics of fiber composites with curved structures. Sov. Appl. Mech. **20**(9), 777–784 (1984)
34. Akbarov, S.D., Guz, A.N.: Model of a piecewise-homogeneous body in the mechanics of laminar composites with fine-scale curvatures. Sov. Appl. Mech. **21**(4), 313–318 (1985)
35. Akbarov, S.D., Guz, A.N.: Stress state of a fiber composite with curved structures with a low fiber concentration. Sov. Appl. Mech. **21**(6), 560–565 (1985)
36. Akbarov, S.D., Guz, A.N.: Continuum theory in the mechanics of composite materials with small-scale structural distorsion. Sov. Appl. Mech. **27**(1), 107–117 (1991)
37. Akbarov, S.D., Guz, A.N.: Mechanics of composite materials with curved structures (survey). Composite laminates. Sov. Appl. Mech. **27**(6), 535–550 (1991)
38. Akbarov, S.D., Guz, A.N.: Mechanics of Curved Composites. Kluwer Academic Publisher, Dordrecht Boston London (2000)
39. Akbarov, S.D., Guz, A.N.: Mechanics of curved composites (piecewise-homogeneous body model). Int. Appl. Mech. **38**(12), 1415–1439 (2002)
40. Akbarov, S.D., Guz, A.N.: Mechanics of curved composites and some related problems for structural members. Mech. Advan. Mater. Struc. **11 Pt.II**(6), 445–515 (2004)
41. Akbarov, S.D., Guz, A.N., Djamalov, Z.R., Movsumov, E.A.: Solution of problems involving the stress state of composite materials with curved layers in the geometrically nonlinear statement. Int. Appl. Mech. **28**(6), 343–346 (1992)
42. Akbarov, S.D., Guz, A.N., Mustafaev, S.M.: Mechanics of composite materials with anisotropic distorted layers. Sov. Appl. Mech. **23**(6), 528–533 (1987)
43. Akbarov, S.D., Guz, A.N., Yahnioglu, N.: Mechanics of composite materials with curved structures and elements of constructions (review). Int. Appl. Mech. **34**(11), 1067–1078 (1998)
44. Akbarov, S.D., Guz, A.N., Zamanov, A.D.: Natural vibrations of composite materials having structures with small-scale curvatures. Int. Appl. Mech. **28**(12), 794–800 (1992)
45. Akbarov, S.D., Verdiev, M.D., Guz, A.N.: Stress and deformation in a layered composite material with distorted layers. Sov. Appl. Mech. **24**(12), 1146–1153 (1988)
46. Akbarov, S.D., Kosker, R., Ucan, Y.: Stress distribution in a composite material with a row of antiphase periodically curved fibers. Int. Appl. Mech. **42**(4), 486–488 (2006)
47. Akbarov, S.D., Maksudov, F.G., Panakhov, P.G., Seyfullayev, A.I.: On the crack problems in composite materials with curved layers. Int. J. Eng. Sci. **32**(6), 1003–1016 (1994)
48. Akbarov, S.D., Mustafaev, S.M.: Distribution of self-balanced stresses in composite materials with curved curvilinearly anisotropic layers. Sov. Appl. Mech. **27**(12), 1225–1227 (1991)
49. Akbarov, S.D., Sisman, T., Yahnioglu, N.: On the fracture of the unidirectional composites in compression. Int. J. Eng. Sci. **35**(12/13), 1115–1136 (1997)
50. Akbarov, S.D., Yahnioglu, N.: Stress distribution in a strip fabricated from a composite material with small-scale curved structure. Int. Appl. Mech. **32**(9), 684–690 (1996)
51. Guz, A.N.: Mechanics of composite materials with a small-scale structural flexure. Sov. Appl. Mech. **19**(5), 383–392 (1983)
52. Guz, A.N.: Quasiuniform states in composites with small-scale curvatures in the structure. Sov. Appl. Mech. **19**(6), 479–489 (1983)
53. Guz, A.N.: Some modern problems of physical mechanics of fracture. In: Cherepanov, G.P. (ed.) FRACTURE. A Topical Encyclopedia of Current Knowledge, pp. 709–720. Krieger Publ. Company, Malabar, Florida (1998)
54. Guz, A.N.: Study and Analysis of Non-classical Problems of Fracture and Failure Mechanics and Corresponding Mechanisms. Lecture presented at Institute of Mechanics HANOI (1998)
55. Guz, A.N.: Description and study of some nonclassical problems of fracture mechanics and related mechanisms. Int. Appl. Mech. **36**(12), 1537–1564 (2000)
56. Guz, A.N.: In On study of nonclassical problems of fracture and failure mechanics and related mechanisms. ANNALS of the European Academy of Sciences, pp. 35–68. Liège, Belgium (2006–2007)

57. Guz, A.N.: On study of nonclassical problems of fracture and failure mechanics and related mechanisms. Int. Appl. Mech. **45**(1), 1–31 (2009)
58. Guz, A.N., Rushchitskii, J.J.: Short Introduction to Mechanics of Nanocomposites. Scientific & Academic Publishing Co., LTD, USA (2013)

Chapter 8
Problem 6. Fracture Under Compression Along Parallel Cracks

In this chapter, in a very brief form, the main results on **Problem 6** are obtained on the themes of the Department of Dynamics and Stability of Continua of the S. P. Timoshenko Institute of Mechanics of the NASU since 1981. The presentation of these results is written in the style announced in the Introduction (Part I) to this monograph (without excessively invoking aspects of a mathematical nature).

8.1 Introduction

In 2014, a review article [1] was published, prepared by the author of this monograph. The article [1] presents an analysis of the results on **Problem 6** and related problems that are obtained by *scientists around the world*, including the results obtained on **Problem 6** in the Department of Dynamics and Stability of Continua. In presenting the specific results in the article [1], considerable attention is paid to the aspects of a historical nature related to the formation and development of the discussed scientific direction. In this regard, a more detailed understanding of the considered scientific direction can be obtained from the article [1], the list of references to which includes 214 publications.

Taking into account the above information, below in this chapter, we *will not pay* the proper attention to information of a historical nature and the analysis of publications of *scientists around the world* on the problems under consideration, which, in the opinion of the author of this monograph, was already implemented in [1].

Thus, in this chapter, the statement of problems and the obtained *only* in the Department of Dynamics and Stability of Continua results on **Problem 6** are considered. According to the style of the article [1], *the scientists are indicated* who obtained the main results for each section of the considered scientific direction. Following the

© The Author(s), under exclusive license to Springer Nature Switzerland AG 2022   289
A. N. Guz, *Eight Non-Classical Problems of Fracture Mechanics*,
Advanced Structured Materials 159,
https://doi.org/10.1007/978-3-030-77501-8_8

above approach, the list of references to this monograph includes the following articles [1–88], in which the main results on **Problem 6** were initially obtained on the
themes of the Department of Dynamics and Stability of Continua. The first publications, apparently, were the articles [5, 35–37, 89] of 1981–1982 of the author of this
monograph. Besides, these results were included in the materials of reports [90–117]
at the international scientific conferences and review articles [1, 33, 55, 56], which
are devoted exclusively to the presentation and discussion of **Problem 6** as well as to
the review articles [118–122], which are devoted to the presentation and discussion
of all questions related to all non-classical problems of fracture mechanics.

The main results, originally published in the above articles and materials of the
international conferences, were presented in subsequent years in the monographs
[123] (pp. 240–279) of 1983, [124] (pp. 361–490) of 1990., [125] (vol. 4, pp. 10–
250) of 1992, [126] (vol. 5, pp. 174–233) of 1996, [127] (vol. 2, p. 145–317) of 2008,
[128] of 2017, and [129] of 2020. Moreover, the largest volume of information is
apparently presented in the monographs [128] and [129].

On the considered scientific direction of the department, three dissertations for the
degree of Doctor of Physical and Mathematical Sciences (DSc) were prepared and
defended: V. M. Nazarenko, I. A. Guz, and V. L. Bogdanov. It should be noted that I.
A. Guz's dissertation also included the results on **Problem 1** (*Fracture in composite
materials under compression along reinforcing elements*). V. L. Bogdanov's dissertation also included the results on **Problem 4** (*Brittle fracture of materials with cracks
taking into account the action of the initial (residual) stresses along the cracks*),
since he developed a unified approach that allows research on **Problems 4 and 6**.

The fracture mechanics of materials under compression along parallel planes, in
which the plane cracks of arbitrary shape are located, is a special section of the fracture
mechanics, in which the description of fracture mechanisms cannot be carried out
within the framework of classical fracture mechanics. The above conclusion follows
from the subsequent discussion in the final part of Sect. 8.1.

The object of research is the isotropic and orthotropic materials in which the plane
cracks of arbitrary shape are located in the parallel planes. It is assumed in the case
of the orthotropic materials that the parallel planes in which the cracks are located
coincide with the planes of symmetry of the material properties.

It is worth noting that composite materials in the continuum approximation are
also modeled by the orthotropic materials. Therefore, the results discussed in this
chapter also apply in this sense to the composite materials. When compressed by
a uniformly distributed load along the planes in which the cracks are located, the
homogeneous stress–strain state appears in the above materials with cracks (there is
no singular part in the corresponding exact solution). In this regard, in this situation,
the following result for the stress intensity factors is obtained

$$K_I = 0; \; K_{II} = 0; \; K_{III} = 0. \tag{8.1}$$

It should be noted that *result* (8.1) *takes place (for the considered loading and
location of cracks)* **in the case of the arbitrary nonlinear models of deformable
bodies taking into account elastic, plastic, and viscous deformations.** Thus, *in the*

above general case, the Griffith–Irwin criterion and the critical crack opening (CCO) criterion do not work, like all other criteria of the classical fracture mechanics.

In connection with the above, conclude that, as applied to the fracture mechanics in the case of compression along the parallel planes in which the cracks are located, **it is necessary** to involve the approaches that are different from the approaches of the classical fracture mechanics. Hence, **Problem 6** is the typically non-classical problem in mechanics.

8.2 General Statement of Problems: General Concept and Basic Approaches

In this section, the brief information is presented which is related to the general issues of the statement of the discussed problems, the formation of the basic concept (development of fracture criteria), and the general characteristics of the developed basic approaches to the study of these problems.

8.2.1 General Statement of Problems

As already noted in the final part of Sect. 8.1, the compression of the material along the parallel planes, in which the plane cracks of arbitrary shape are located, is considered. The research is carried out for the isotropic and orthotropic materials. It is assumed in the case of orthotropic materials that the parallel planes in which the cracks are located coincide with the planes of symmetry of the material properties.

A similar situation occurs when the laminated composite is compressed along the layers for the case of layers *of the constant thickness* when the plane cracks of various shapes are located at the interface between the binder (matrix) and filler (reinforcing layers).

As already noted in the final part of Sect. 8.1 for the homogeneous isotropic and orthotropic materials, and, obviously, for the above situation in the case of layered composites, the expressions (8.1) hold and, therefore, in these situations, *the classical fracture mechanics does not work.* Thus, the research should be carried out within the framework of the non-classical problem of fracture mechanics. Below, separately for the two cases described above, the design schemes are presented for studying the main classes of problems.

In connection with the above, come to the conclusion that, as applied to the fracture mechanics in the case of compression along the parallel planes in which the cracks are located, **it is necessary** to involve approaches that are different from the approaches of the classical fracture mechanics. Hence, **Problem 6** is the typically non-classical problem in mechanics.

Fig. 8.1

The design schemes for isotropic and orthotropic materials: The main design schemes are shown in Fig. 8.1a–d concerning the plane problems in the plane $x_1 0 x_2$, where the axis $0x_3$ is directed perpendicular to the plane of the figure.

In the case of spatial problems, Fig. 8.1a–e can be considered as a section. In Fig. 8.1, the cracks are depicted by "bold" lines, which are infinite and of constant width in the direction of the axis $0x_3$ in the case of plane deformation. Figure 8.1a shows a general view of the location of the cracks. Figure 8.1b shows the simplest case of one crack when cracks in adjacent planes are located at such a distance that their mutual influence can be ignored. Figure 8.1c presents the simplest case of interaction of two identical cracks located in two neighboring parallel planes. Figure 8.1d shows the case of a larger number of the identical cracks located in the parallel planes. This situation is modeled by a periodic row (along the vertical axis) of the identical cracks located one above the other, when it is necessary to take into account the mutual influence of two neighboring cracks in the periodic row (and, therefore, all cracks).

It should be noted that Fig. 8.1a–d shows the design schemes for the study of material with cracks, which are located "inside" the material, as evidenced by the designation of the volume selected inside the material by the wavy lines. In this regard, the cracks shown in Fig. 8.1a–d can be called the "internal" cracks. Along with the "internal" cracks in the material, there are the cracks near the material surface, *which interact with the material surface when fractured*. Therefore, these cracks can be called "near-the-surface" cracks.

Figure 8.1e shows the simplest case of one near-the-surface crack located in a plane parallel to the surface of the material. The boundary of the material or the boundary surface (in Fig. 8.1e, the material occupies the lower half-plane) is indicated by a "bold" straight line.

It is advisable to note that the marked in Fig. 8.1c–e by the "dense" hatching parts of the material are discussed below in 8.2.3.

The design schemes for laminated composite materials: The layered composites are considered that are formed by the layers of filler and binder of *the constant thickness*, in the interface of which the plane cracks of arbitrary shape are located. The filler and binder layers are considered the isotropic or orthotropic materials. In the case of the orthotropic materials, it is assumed that one of the planes of symmetry of the material properties coincides with the plane of separation of the properties of the filler and the binder (interface).

Taking into account the above conditions, the main design schemes for the layered composites are shown in Fig. 8.2a–e concerning the plane problems in a plane $x_1 0 x_2$, where the axis $0 x_3$ is directed perpendicular to the plane of the figure. In the case of the spatial problems, Fig. 8.2a–e should be considered as a section.

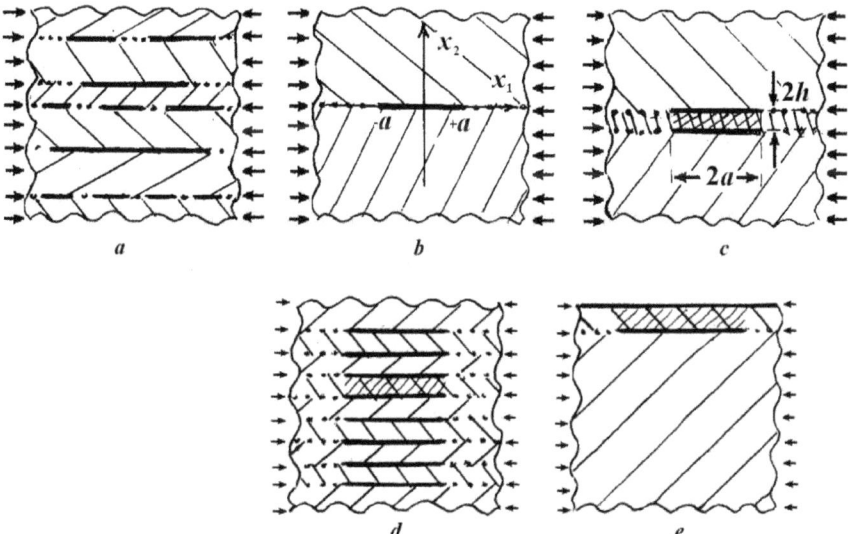

Fig. 8.2

It is advisable to note that all the information, designations, and considerations given near Fig. 8.1 concerning the homogeneous materials apply equally to the design schemes in Fig. 8.2 concerning the layered composites or the specified data should be somewhat specialized.

So, in Fig. 8.2, the layers of filler and binder are indicated by hatching at different angles; the "bold" lines indicate the cracks at the interface; the dots denote the parts of interfaces without cracks.

Additionally, for the layered composites, the main design schemes for which are shown in Fig. 8.2 (additionally in comparison with homogeneous materials, the main design schemes for which are presented in Fig. 8.1), *the following condition is adopted.*

When studying the phenomena inside the composite, including the phenomena in the near-the-surface layers of the composite, following Note 8.3 (Part II of this monograph) "at infinity" (along the reinforcing elements), the same tensile or compressive deformations are given *for the matrix and the reinforcing elements.*

It is advisable to note that concerning the layered composites with the "dense" shading, the marked parts of the material in Fig. 8.2a–e are discussed below in Sect. 8.2.3.

We restrict ourselves to the above information when discussing the design schemes for the homogeneous (in the continuum approximation) materials (Fig. 8.1a–e) and layered composite materials (Fig. 8.2). Additional information can be obtained from the review article [1] and monographs [123, 124, 125 vol. 4, book 1, 126 v. 5, 127 vol. 2, 128].

8.2.2   General Concept

It follows from the information and considerations given in the concluding part of Sect. 8.1 that *under compression along a system of the plane cracks located in parallel planes*, **in the homogeneous** isotropic and orthotropic materials and *under compression along layers*, at the interfaces of which the plane cracks are located, **in the laminated composites with layers of constant thickness,** the classical Griffith-Irwin fracture mechanics **does not work**.

It is advisable to note that it is assumed in the above two cases for the orthotropic homogeneous materials and orthotropic layers concerning the layered composites that the planes in which the cracks are located are parallel to the plane of symmetry of the material properties. Thus, for the cases shown in Figs. 8.1 and 8.2, *the compression is carried out along the axes and planes of symmetry of the material properties, taking into account the presence of cracks.*

To carry out the research on fracture mechanics concerning the above non-classical problem of fracture mechanics (classified as **Problem 6** in this monograph), the General Concept was formed by analogy with the corresponding situation in the mechanics of thin-walled structural elements (rods, beams, plates, and shells).

So, in the mechanics of thin-walled structural *elements under compression along the axes of symmetry* (material properties and geometric shape), *the phenomenon of loss of stability of the equilibrium state arises,* which in most cases subsequently leads to the *exhaustion of the bearing capacity.*

It should be emphasized, as noted above, that *in the discussed cases* (Figs. 8.1 and 8.2) *of fracture mechanics of materials under compression*, the compression along the axes of symmetry is also analyzed.

The above information and considerations justify the formed General Concept, which can be formulated as follows.

General Concept
In the situations shown in Figs. 8.1a–e and 8.2a–e, the beginning (start) of the material fracture process is determined by the local loss of stability of the equilibrium state of the material that surrounds the cracks. In this case, the theoretical ultimate strength corresponds to the value of the critical load at loss of stability, which is calculated within the framework of the accepted concrete theory of stability of deformable bodies.

It is advisable to note that the above-stated **General Concept** refers *to the arbitrary models* (elastic, plastic, viscoelastic, etc., media) of deformable bodies concerning the isotropic and orthotropic (the cracks are located in the plane of symmetry of material properties) materials. It is only necessary in the case of a specific model to apply the stability criteria generally accepted in the related classes of problems.

Note 8.1   In the design schemes of Fig. 8.1g and Fig. 8.2g (a large number of equally spaced identical cracks, which is modeled by an infinite periodic row of cracks), *the local* fracture near the cracks can turn into *the general* fracture, since in this case (due to the periodic location of the cracks) a kind of "hinge" appears *throughout the total thickness* of the material.

The above **General Concept** seems to be such a natural generalization that at present, apparently (at least in the opinion of the author of this monograph), it is not possible to strictly establish in which publication it was first consistently presented.

It should be noted that the above-stated **General Concept** determines only the general direction of research of the analyzed phenomena. Within the framework of the discussed concept, various approaches and methods are being developed, the priority publications for the development of which are established, which will be done below in this chapter.

We will restrict ourselves to the above information when discussing the **General Concept**. At that, its implementation will actually be considered when presenting the specific results below in the subsequent sections of this chapter.

8.2.3   General Approaches

For **Problem 6**, the various approaches have been proposed, and, within the framework of these approaches, numerous methods for studying the specific problems are developed. A certain understanding of the discussed approaches and research methods can be obtained from the results presented in the review article [1] of 2014. Nevertheless, at present, two fundamentally different general approaches exist, a very brief discussion of which will be carried out in this section.

It should be noted that the numbering of approaches (first and second) is determined not by the generality or severity of the approaches, but by the sequence of their formation in the historical aspect.

8.2.3.1   The First General Approach: Beam Approximation or Beam Approach

The beam approach consists of the selection of a part of the material located between two neighboring cracks in the parallel planes or between the crack and the boundary surface of the material, and in the application of the applied stability theories for the selected part of the material within the framework of the mechanics of thin-walled systems using the hypotheses of plane sections, Kirchhoff–Love, Timoshenko type, and other hypotheses. In this case, the studies for the selected parts of the material are carried out for cases of the boundary conditions of rigid fixation or hinged support at the "ends."

With the considered approach, *the theoretical ultimate strength corresponds to the value of the critical load calculated within the framework of the applied theory of stability of the corresponding mechanics of thin-walled systems.* The parts of the material that are highlighted or isolated by the above method are marked with a relatively "dense" shading on the simplest design schemes presented in Fig. 8.1c–e and Fig. 8.2c–e.

Apparently, the first application of the beam approach was implemented in the article by Obreimoff [130] of 1930. Thus, Obreimoff can be considered to be the initiator of *the beam approximation.* In subsequent years, in the articles of Entov and Salganik [131] of 1965 and Mikhailov [132] of 1966, the terminology "beam approximation, beam approach" began to be used in the titles of articles.

The review article [1] provides brief information on the further development and application of the beam approach to the study of specific classes of problems. This article also indicates the leading scientists who took part in the development of the discussed scientific direction based on **the First General Approach** (beam approximation).

Of course, the beam approach simplifies essentially the study of these problems and makes it possible to obtain the specific numerical results in a complete form. In this regard, the beam approach *has become widespread in some or most cases without proper justification.* Obviously, the analysis of the reliability and accuracy

Fig. 8.3

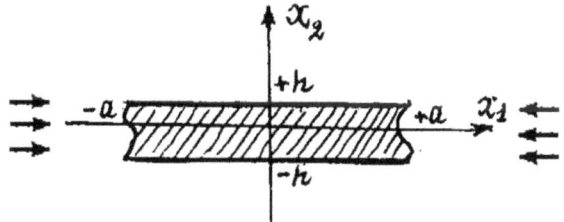

of the results obtained using the beam approximation requires a qualified discussion. A brief analysis of the beam approximation is presented below, following mainly the article [1] Fig. 8.3.

Following the design schemes of Figs. 8.1 and 8.2 with the beam approximation, the study is carried out for the selected or isolated part of the material, which in Figs. 8.1c–e and 8.2c, d is shown by "thick" shading, under certain boundary conditions on the "mentally highlighted" ends.

Following the above approximation, the selected part of the material is shown in Fig. 8.3. As already noted in the description of the **First General Approach**, the theoretical ultimate strength corresponds to the value of the critical load, calculated within the framework of the corresponding applied theory of stability of thin-walled systems. In this case, the value of the critical load is determined, as a rule, by the first eigenvalue of the problem.

It should be noted that the eigenvalues, as they were, characterize "integrally" this eigenvalue problem for the entire domain, and not the behavior of the sought function at the individual points. In connection with the foregoing, the theoretical ultimate compressive strength obtained with this approach is also an "integral" characteristic for this entire problem and, it would seem, should not strongly depend on the behavior of the desired function at the individual points.

Let us also dwell on the discussion of the boundary conditions at the "mentally selected" ends, i.e., when $x_1 = \pm a$ in Fig. 8.3. For this purpose, consider a small neighborhood near the right tip in relation, for the sake of clarity, to the design schemes in Figs. 8.1d and 8.2d, corresponding to the near-the-surface fracture. At that, the small vicinity of the right crack tip corresponding to the point $+a$ is indicated in Fig. 8.4 by the densest shading.

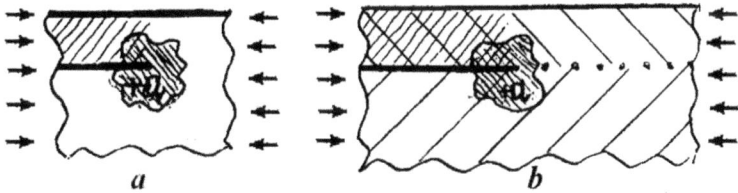

a   b

Fig. 8.4

Note that Fig. 8.4 a refers to the homogeneous material and corresponds to the design scheme in Figs. 8.1e, and 8.4b refers to the layered composite and corresponds to the design scheme in Fig. 8.1d. The designations in Figs. 8.1e and 8.4a are consistent with each other as also the designations in Fig. 8.2e

It seems obvious from Fig. 8.4 that in a small neighborhood near the tip of a crack (in the study of the stability problem) *the purely three-dimensional stress–strain state arises in the case of the spatial problem and the purely two-dimensional stress–strain state arises in the case of plane problems.* At that, within the framework of the beam approximation at point "$+a$" in Fig. 8.4 or with $x_1 = \pm a$ Fig. 8.3 the boundary conditions are formulated corresponding to the applied theory of thin-walled systems and obtained using the well-known hypotheses. In this regard, in the above two cases (the first is a rigorous statement, and the second is the application of applied theories of thin-walled systems), the features of various types appear near the tip of the crack.

Taking into account the above information and considerations, formulate below the results of a brief analysis of the beam approximation in the form of the following conclusions.

Conclusion 8.1 For any value of the thin-wall parameter $h \cdot a^{-1}$ (Fig. 8.3) in the beam approximation, *the unavoidable error* arises due to the formulation of the boundary conditions at the ends (for $x_1 = \pm a$ in Fig. 8.3) within the framework of the applied theories of thin-walled systems, which leads to a change in the order of the singularity at the crack tip (near the point "$+a$ " in Fig. 8.4).

Conclusion 8.2 *It is impossible* within the framework of the beam approximation, to rigorously formulate the boundary conditions at the ends (for $x_1 = \pm a$ in Fig. 8.3) within the framework of the applied theories of thin-walled systems. These boundary conditions should, as it were, correspond to the "elastic fixing."

Conclusion 8.3 In connection with Conclusion 8.2, the studies with the beam approximation, as a rule, are carried out for the boundary conditions at the ends in the form of rigid fixing or hinged support. The results obtained for the above two types of boundary conditions differ *by several times.*

Conclusion 8.4 In connection with Conclusion 8.3, the problem arises of estimating the results obtained within the framework of **the First General Approach** (beam approximation) from the point of view of the results obtained within the framework of more consistent and rigorous approaches.

Conclusion 8.5 The beam approach *cannot provide* studies within the design scheme presented in Figs. 8.1d and 8.2d (for a large number of cracks modeled by the periodic infinite row of cracks), since the interaction of cracks cannot be taken into account when rigidly fixing or hinging at the ends (Fig. 8.3).

Conclusion 8.6 Thus, **the First General Approach (*beam approximation*)** is a purely approximate approach, and the limit of its applicability depending on the thinness parameter $h \cdot a^{-1}$ (Fig. 8.3) can be determined for each specific class of problems.

We will restrict ourselves to the above information for a brief analysis of the beam approximation. Additional information can be obtained from the review article [1].

8.2.3.2   The Second General Approach. Application of the TLTDBS

This approach consists in applying the three-dimensional linearized theory of deformable bodies stability (TLTDBS), the main mathematical apparatus of which is described in a reduced form in Chap. 2 *(Part I) of this monograph, to the study of the stability of the local state of equilibrium of the material adjacent to cracks. Here, the strict equations of stability and the corresponding boundary conditions are considered in a three-dimensional formulation, which eliminates the need to invoke additional hypotheses. The theoretical ultimate strength is defined as the critical stress value corresponding to the local loss of stability of the equilibrium state near the cracks and calculated within the TLTDBS.*

The formulated approach refers to the homogeneous materials under compression along cracks concerning the design models shown in Fig. 8.1a–e and the layered composites with layers of constant thickness under compression along cracks located in the interface planes, as applied to the design schemes shown in Fig. 8.2a–e. In these cases (the design schemes in Figs. 8.1 and 8.2), the material is assumed to be isotropic or orthotropic (the symmetry planes of the material properties coincide with the horizontal planes) for the arbitrary models (elastic, plastic, or viscoelastic media) of deformable bodies. Under the noted conditions, the subcritical state will be homogeneous and the conditions (8.1) will be satisfied.

This **Second General Approach** *is the most rigorous, consistent, and accurate within the framework of the mechanics of deformable bodies. In this regard, the results obtained using the* **Second General Approach** *can be used to evaluate the results obtained in the framework of the various approximate approaches, including the* **First General Approach (Beam approximation).**

The Second General Approach *as applied to the mechanics of fracture under compression along the plane cracks in parallel planes in the material was proposed, developed, and implemented* **for the first time in the world** *in the Department of Dynamics and Stability of Continua at the S. P. Timoshenko Institute of Mechanics of the NASU.*

Apparently, the first publications in this direction were the articles [5, 89] of 1981 by the author of this monograph. In these articles, using the above **General Concept** and the **Second General Approach**, the fracture criteria were formulated in [89] for the plane problems and in [5] for the spatial problems for the hyperelastic materials with an arbitrary structure of the elastic potential (with the considered symmetry of properties) in the unified general form for **Theories 1, 2, and 3** (following the terminology of Chap. 2 of this monograph (Part I)). In the publications [35–37] for 1982, the Second General Approach was extended to the *plastic* materials taking into account the generalized concept of continuing loading, which is summarized in Sect. 8.2.3.2 (Part I) of this monograph.

Thus, it can be assumed that in the articles [5, 35–37, 89] of 1981–1982 the foundations of the **Second General Approach** in the unified general form were proposed for the compressible and incompressible isotropic and orthotropic elastic materials with an arbitrary structure of the elastic potential (for the brittle fracture) and plastic materials with constitutive relations of the sufficient general form (for the plastic fracture) as applied to **Problem 6** (Fracture under compression along parallel cracks).

At that, the research is carried out within the framework of the TLTDBS apparatus, which is set out in Chap. 2 of this monograph (Part I).

Note that according to the design schemes in Figs. 8.1 and 8.2, the studies concerning the discussed **Problem 6** are carried out under the action of an external "dead" load, which is typical for almost all publications on fracture mechanics.

In this regard, *it is strictly proved for the design schemes in* Figs. 8.1 and 8.2 *concerning the elastic and plastic materials that the sufficient conditions of the applicability of the static method of studying the stability problems are satisfied* (Sect. 8.2.4.2, the first result, Part I of this monograph) and, thus, these problems are reduced to *the eigenvalues problems; i.e., the Euler's method is applied*. Nevertheless, the results obtained in this case correspond to the application of the dynamical method by the above proof of the fulfillment of the indicated sufficient conditions.

Thus, as applied to **Problem 6**, *the studies within the framework of the above-formulated* **Second General Approach** *fully correspond to the generally accepted and rigorous method of studying the phenomenon of loss of stability—the analysis of the behavior of small perturbations within the framework of linearized three-dimensional* **dynamic problems**.

The above-stated *Conclusion, proof, and approach take place* for the elastic and plastic models and *do not take place* for the models with rheological properties.

Note 8.2 On the application of general solutions of TLTDBS: Within the framework of the above Second General Approach in the study of **Problem 6** for the design schemes presented in Figs. 8.1 and 8.2, the following situation takes place.

As noted before the expressions (8.1), *the homogeneous subcritical state* arises in the cases discussed for the applied models of materials. Besides, when describing above the **Second General Approach**, it was noted that for the considered form of loading for the elastic and plastic models of materials, *the sufficient conditions for the applicability of the static research method are satisfied*. This information indicates that, taking into account the noted situation for the study of problems corresponding to the design schemes in Figs. 8.1 and 8.2, it is advisable to apply *the general solutions for the static* plane and spatial problems, which are outlined in the monographs [123, 124, 127, 133–138]. These general solutions are presented in the most compact form in the monograph [127] (vol. 1).

The brief information on the general solutions of the static spatial and plane TLTDBS problems for the compressible and incompressible materials, including the use of complex potentials, is presented in Sect. 2.6 (Part I) of this monograph.

It is advisable to research the Lagrangian coordinates x_j ($j = 1, 2, 3$), which are introduced in the first reference state (for the elastic bodies—in the natural, undeformed state) and which in this state coincide with the Cartesian coordinates, as well as in the Cartesian curvilinear coordinates, which are linked by the known relations to the Cartesian coordinates.

Note 8.3 It should be noted that the three-dimensional linearized theory of elastic stability at the finite subcritical deformations was also used (within the plane problem) to study the stability of an infinite body under compression along one plane crack when the elastic potential of *a concrete particular structure* was set *from the very beginning of the study.* Apparently, for the first time, the results of the above studies were published in the articles [139, 140]. Thus, in 1979 [139], the case of a compressible isotropic elastic body with the harmonic elastic potential was investigated, and in 1980 [140], the case of an incompressible isotropic elastic body with an elastic potential of a concrete simplest structure was investigated.

In conclusion, it should be noted that the publications [139, 140] of 1979–1980 of the author and publications [5, 35–37, 89] of 1981–1982. The author of this monograph appeared *independently and in a form* that significantly differs *in the generality* of the problem statement. Besides, the publications [5, 35–37, 89] from the very beginning were focused on the study of non-classical problems of fracture mechanics.

Note 8.4 On delamination of layered composites: Usually, in a broad sense, the phenomenon of delamination of the layered composites is understood as the *exfoliation* or *delamination* along the interface in the sufficiently small or rather large areas in the case of compression along the interfaces, when the specified exfoliation or delamination is initiated, including by the presence of cracks at the interface.

The delamination phenomenon can be investigated within the framework of the above **Second General Approach** in the research of **Problem 6** (*Fracture under compression along parallel cracks*) *if the following condition is met.*
For the occurrence of these exfoliations or delaminations, *it is necessary* that the considered ultimate strength (value of the critical load) the modes of loss of stability that determine *the opening* of cracks correspond to this value.
As applied to design schemes for the laminated composites with layers of the constant thickness under compression along the layers, which are shown in Fig. 8.2a–e, the modes of loss of stability with *the crack opening* are shown in Fig. 8.5b, c, and e. At that, Fig. 8.2a–e and Fig. 8.5b, c, and e are consistent with each other. Note that in Fig. 8.5c the corresponding to the calculation scheme in Fig. 8.2c only concerning the "lower" crack, the exfoliation or delamination in the layered composites is shown, corresponding to the possible modes of loss of stability. The fact is that for the layered composites under compression along the layers, the modes of loss of stability are possible, which are close to the bending buckling forms. Such bending modes of loss of stability also called the first-order modes of loss of stability or shear modes are shown in Fig. 3.27 in Sect. 3.3.3.1 of this monograph.

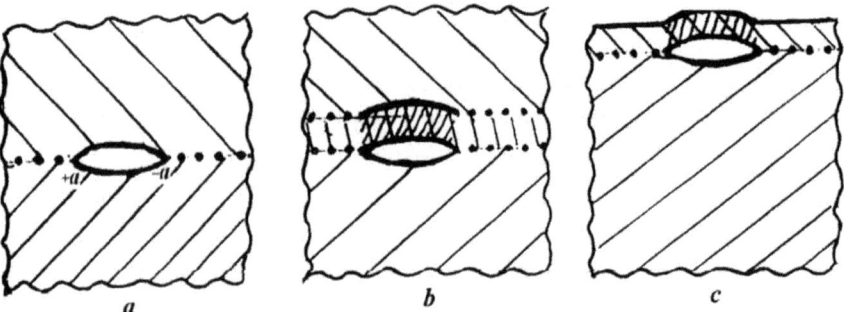

Fig. 8.5

With the loss of stability in modes close to the bending ones in the above Fig. 3.27, *there is no* noticeable delamination or exfoliation characteristic of the "delamination" phenomenon at the considered stage of the study. Seemingly, the discussed phenomenon occurs at a subsequent stage of deformation, corresponding to the nonlinear post-critical behavior.

We restrict ourselves to the above information when discussing **the Second General Approach**, which, as noted above, was developed and implemented in the Department of Dynamics and Stability of Continua of the S. P. Timoshenko Institute of Mechanics of the NASU.

In subsequent Sects. 8.3, 8.4, 8.5 and 8.6 of this chapter, brief information is presented about the results of the study of some classes of problems in **Problem 6** (*Fracture under compression along parallel cracks*) within the framework of the **Second General Approach**, which were obtained in the Department of Dynamics and Stability of Continua Media, starting with 1981–1982.

The more detailed information on the discussed results is presented on pages 40–80 of the review article [1] of 2014, At that, the most complete information on these results is presented in the monographs [127] (vol. 2, pp. 145–91) of 2008 and [128] of 2017.

Following the style of presentation of the survey article [1], when presenting results for the particular classes of problems in Sects. 8.3, 8.4, 8.5, and 8.6, the scientists are indicated who received the main results for each particular class of problems.

8.3   Results for Homogeneous Materials with Cracks Under Brittle and Plastic Fracture: The Second General Approach

The research was carried out for the plane and spatial problems. The main design schemes are presented in Fig. 8.1a–e *for the plane problems* concerning which the materials are indicated (isotropic and orthotropic; for the orthotropic materials, one of the planes of symmetry of properties coincides with the planes in which the cracks are located).

For *the spatial problems*, the basic design schemes in Fig. 8.1a–e remain valid for the materials under consideration (isotropic and transversely isotropic).

For the transversely isotropic materials, an *additional* condition is introduced—the isotropy planes coincide with the planes in which the cracks are located.

The results were obtained for two typical situations:

The first situation—the analysis is carried out for the cracks located in the same plane, and the interaction of cracks located in the neighboring parallel planes *is not taken into account*;
The second situation—the analysis is carried out for the design schemes presented in *taken into account*.

Figure 8.1a–e, and the interaction of cracks located in neighboring parallel planes *is taken into account*.

It is worth noting that the first situation is a special case of the second situation. In this case, the transition to the study of the problems corresponding to *the first situation* can be considered justified if *the distances* between the neighboring planes in which the cracks are located are *significantly (by an order of value or more) larger* than the maximum dimensions of the cracks in the plane.

8.3.1 Results for Brittle and Plastic Fracture of Homogeneous Materials with Cracks Located in the Same Plane. The Second General Approach. Exact Solutions

The main results were obtained by the author of this book.

The research was carried out for the anti-plane, plane, and spatial problems. The main publications are presented in the Introduction (Sect. 8.1) to this chapter. For the anti-plane and plane problems, the studies were carried out using the complex potentials of the TLTDBS, information about which is given in Chap. 2 of this monograph (part I).

For the spatial problems, the studies were carried out using the general solutions of the TLTDBS, information about which is also given in Chap. 2 of this monograph (Part I). Here, the discussed problems were reduced to the corresponding mixed problems for the harmonic potential in the half-space.

It should be noted that when obtaining the exact solutions of the discussed anti-plane, plane, and spatial problems of fracture mechanics under compression along the plane cracks of various shapes located in the same plane, the various modes of loss of stability were analyzed, corresponding to different symmetry conditions concerning the plane in which the cracks are located.

As a result of the exact solution for the plane problem, *it is strictly proved*:

the theoretical ultimate strengths for the symmetric and flexural buckling are the same;
the theoretical ultimate strength is determined by the critical value of the loading parameter corresponding to the surface instability of the half-space within the plane deformation;
the theoretical ultimate strength does not depend on the number of cracks located in one plane.

The last conclusion (independence of the result from the number of cracks in one plane) *cannot be obtained* from the physical considerations. A discussion of this situation is presented in the monograph [127] (vol. 2) on page 167 in Note 8.4.

As a result of the exact solution for the spatial problem, *it is rigorously proved that for the spatial problem under axisymmetric loading, the theoretical ultimate strength* **is not greater than** *the critical value of the loading parameter corresponding to the surface instability of the half-space in the framework of the spatial problem.*

Above, concerning the results for the spatial problem, the estimate "no more" was used due to the fact that, in addition to the above results, which were obtained from the analysis of the exact solutions, for *the flexural* modes of loss of stability, *one more* mixed eigenvalue problem was also obtained for two harmonic functions for the half-space, which *in the general case of the constitutive equations remained unexplored*.

It is shown for specific models of the theory of elasticity and plasticity in the monographs [123, 124, 125, 127 vol. 4, book 1, 17] that the study of the above eigenvalue mixed problem for two harmonic functions for a half-space *does not introduce* new information on the value of the theoretical ultimate strength ...

Additional information on the results discussed in this section can be obtained from the above monographs.

8.3.2 Results for Brittle and Plastic Fracture of Homogeneous Materials with Cracks Located in Parallel Planes: The Second General Approach

The main results were obtained by V. M. Nazarenko.

The main publications related to the considered scientific area are included in the general list, which is given in the Introduction (Sect. 8.1) in this chapter. The studies were carried out for the plane and spatial problems concerning the design schemes presented in Fig. 8.1a–e. Moreover, in the case of the spatial problems, the cracks in the form of circular disks under axisymmetric loading were considered.

The developed research methods are based on the application of general TLTDBS solutions, brief information about which is given in Chap. 2 of this monograph (Part I), *with the full consideration of the interaction of cracks* located in the parallel planes.

These methods are based on the use of the integral transforms (Fourier transform for the plane problems and Hankel transform for the spatial problems), the reduction of problems to the dual integral equations, and the subsequent reduction to the integral equations of the following type:

> in the case of the plane problems, to the integral equations of the first kind with the integrable singularity of the logarithmic type;
> in the case of the spatial problems—to the Fredholm integral equations of the second kind.

It should be noted that the above methods relate to the study of the corresponding *eigenvalue problems* of mathematical physics. When developing the above methods for **Problem 6** of fracture mechanics, the well-known approaches [165, 166] related to the corresponding *boundary-value* mixed problems of statics of the classical linear theory of elasticity were essentially used.

At the final stage of the development of methods for studying the problems of Sect. 8.3.2, the discussed problems, using the above methods for the classical linear theory of elasticity, are reduced *to the eigenvalue problems for the one-dimensional integral equations.*

It has been rigorously proved that the obtained one-dimensional integral equations have the continuous kernels in the range from the zero values of loading parameters to the values of loading parameters corresponding to the theoretical ultimate strengths for the case of compression of a homogeneous material along the system of plane cracks located in the parallel planes, taking into account the interaction of such cracks.

The above-mentioned proof of the continuity of the kernels of one-dimensional integral equations makes it possible to effectively apply the various numerical and approximate methods to study the eigenvalues of one-dimensional integral equations, through which the theoretical strength limits for the design schemes in Fig. 8.1c–e concerning the analyzed fracture mechanisms are determined.

In the results considered in this section, the Bubnov–Galerkin method was used to determine the eigenvalues of the obtained one-dimensional integral equations. In this case, the power functions or orthonormalized on [0; 1] shifted Legendre polynomials were chosen as the coordinate functions, and the change in the values of the eigenvalues with an increase in the number of coordinate functions was analyzed.

In this way, in general terms, it is possible to describe the developed method for solving the problems of fracture mechanics of the homogeneous materials under compression along a system of the plane cracks located in the parallel planes, taking into account the interaction of such cracks.

The discussed solution method is an exact method from the point of view of satisfying the equations and boundary conditions. The accuracy of obtaining the specific results is determined only by the accuracy of solving the eigenvalue problem of the obtained one-dimensional integral equations.

At present, using the above method, a significant number of the plane and spatial problems in the brittle and plastic fracture for the various homogeneous materials were already studied concerning the design schemes presented in Fig. 8.1b-e, taking into account the interaction of cracks located in the different planes. The corresponding results are presented in sufficient detail in the monographs [124, 125 vol. 4, book 1, 126 vol. 5, 127 vol. 2, 128] of 1990–2017, and in a more reduced form, these results are presented in the review article [1] of 2014.

In connection with the above, the discussed results will not be considered below. The brief conclusions from the analysis of these results will be only given. The noted mutual influence of cracks is determined by the dependence of the final results on the following dimensionless parameter β

$$\beta = h \cdot a^{-1}, \tag{8.2}$$

where the geometric parameters a and h are shown in Fig. 8.1c.

It should be noted that parameter β (8.2) refers to the design schemes in Fig. 8.1c-e and characterizes the dimensions of the "shaded" part of the material.

Consider some of the conclusions related to Sect. 8.3.2.

1. *The proposed method is sufficiently effective since it allows to determine ε_T (the the theoretical value of the limiting shortening) with an accuracy of three signs using no more than four coordinate functions in the case of changing the parameter β (8.2) within the limits $1/8 \leq \beta < +\infty$.*
2. *With the high efficiency of the method, the results obtained for the considered minimum value $\beta(\beta = 1/8)$ are **an order of value or less** than the results for one isolated crack. So, for example, for the near-the-surface crack (Fig. 8.1e) at $\beta = 1/8$ the theoretical value of the limiting shortening is 30 times less of ε_T for one isolated crack.*
3. *With the closing of the planes in which the cracks are located, the significant decrease occurs in the theoretical ultimate strength σ_T in comparison with the case of one isolated crack (without taking into account the mutual influence). So, at $\beta \approx 1/8$, a decrease σ_T **by an order of value or more** can be observed.*
4. *With the values of the parameter $\beta > 4$, it is possible to ignore the interaction of cracks located in the parallel planes, with an accuracy of 5% when studying the spatial problems. In the case of plane problems, the noted conclusion takes place for somewhat large values of the parameter β.*

8.3.3 Results for Brittle and Plastic Fracture of Homogeneous Materials with Cracks Located in Parallel Planes. Second General Approach: The Combined Approach for Problems 4 and 6

The main results were obtained by V. L. Bogdanov.

The main publications related to the considered scientific area are included in the general list, which is given in the Introduction (Sect. 8.1) to this chapter.

It should be noted that the most complete presentation of the discussed results is given in the monograph [128] of 2017 and analysis of the results on the corresponding spatial problems is carried out in the review article [33] of 2015.

The combined approach to research refers to *the problems of the fracture mechanics* of the homogeneous materials under compression along the system of plane cracks located in the parallel planes, taking into account the interaction of these cracks for the design schemes in Fig. 8.1a–e (**Problem 6.** *Fracture under compression along parallel cracks*) and *to the problems of fracture mechanics* of the homogeneous materials with the system of plane cracks located in the parallel planes, taking into account the action of the initial (residual) stresses along such cracks (**Problem 4.** *Brittle fracture of materials with cracks, taking into account the action of the initial (residual) stresses along the cracks*).

Note that due to the application of the basic relations of the three-dimensional linearized mechanics of deformable bodies (TLMDB), *the first* of the above problems are the eigenvalue problems and *the second* of the above problems are the boundary value problems *for the same* equations, as well as for the same areas corresponding to the design schemes in Fig. 8.1a–e.

Thus, examining the corresponding boundary value problems with a continuous change in the loading parameters in the plane of cracks, one can approach the values of these loading parameters when the amplitude values (stresses and displacements) tend to "infinity" (all amplitude values or some of them). Therefore, the found in this way values of the loading parameters (in the plane of cracks) correspond to the eigenvalues of the corresponding eigenvalue problem.

The above situation, which characterizes the discussed joint research method on **Problems 6 and 4** of the non-classical fracture mechanics, is similar to the well-known situation in the theory of oscillations. This situation in the theory of oscillations consists in the following approach—to determine the natural frequencies of oscillations, the forced oscillations can be investigated with the continuous change in the frequency of the external load. In this case, with a significant change in the amplitude value and its tendency "to infinity," the frequency of natural oscillations is determined.

It is advisable to note that the discussed combined approach for studying the problems of **Problems 4 and 6** refers not only to the design schemes of **Problem 6**, which are presented in Fig. 8.1a–e but also to the design schemes of **Problem 4** formulated in Sect. 6.4.3, for which *the specific results were obtained* (both exact solutions and using computer methods).

Note that 11 corresponding design schemes are formulated in Sect. 6.4.3.

When implementing the discussed combined approach, all the mathematical aspects of the corresponding methods are applied, which are indicated in the first part of Sect. 8.3.2, taking into account the situation that in this section the eigenvalue problems are investigated, whereas in Sect. 8.3.3, mainly the boundary value problems are investigated.

When applying the combined approach, it is convenient to choose values—indicators of the stress–strain state concerning the problems of **Problem 4** (*Brittle fracture* *of materials with cracks, taking into account the action of the initial (residual) stresses along the cracks).*

Note that the above indicators, in some sense, are analogous to the amplitude with the above analogy with the theory of oscillations. In the monograph [128] of 2017 and in the review article [33] of 2015, as well as in the indicated references to [33, 128] articles, the following values are used as the indicators discussed

$$K_I/K_I^\infty, K_{II}/K_I^\infty, K_{III}/K_I^\infty. \qquad (8.3)$$

In (8.3), the following designations are introduced:

K_I, K_{II} and K_{III}—the stress intensity factors in problems of **Problem 4** (as noted above, 11 calculation schemes are indicated in Sect. 6.4.3, according to which the specific results were obtained within the framework of **Problem 4**);
K_I^∞, K_{II}^∞ and K_{III}^∞—the stress intensity factors in problems of **Problem 4** for one crack in an "infinite" material under the action of the considered initial (residual) stresses.

As shown in the monographs [123, 124, 125 vol. 2, 127 vol. 2, 17] for cracks, on the sides of which the stresses are set, the values K_I^∞, K_{II}^∞ and K_{III}^∞ *do not depend* on the initial stresses. Thus, the normalization in indicators (8.3) can also be carried out using the quantities K_{II}^∞ and K_{III}^∞.

Thus, the discussed unified approach is *that as a result of solving many boundary value problems for the considered case of* **Problem 4**, **the dependence of indicators** (8.3) *of the stress–strain state for the same case of* **Problem 4** *on the initial (residual) stresses is determined. From the obtained dependence, the values of the initial (residual) stresses are determined, when approaching which the values of some indicators (8.3) tend to "infinity".*

The values of the initial (residual) stresses obtained in this way are the eigenvalues of the corresponding problem of **Problem 6**. *Following the* **General Concept** *formulated in Sect. 8.2.2 for* **Problem 6**, *the above eigenvalues (the critical loads) are the theoretical ultimate strengths for* **Problem 6**, *which (the theoretical ultimate strengths) are obtained from solving a sequence of the boundary value problems corresponding to* **Problem 4**.

In several cases, for example, for the highly elastic materials, when presenting the graphical dependence of indicators (8.3) on the initial (residual) stresses for the problems in **Problem 4**, it is convenient to replace the initial (residual) stresses with the corresponding elongation coefficients using the known relations.

Focusing on the study of the problems of **Problem 6**, the design schemes for which are presented in Fig. 8.1c–e, it is convenient for the aforementioned graphical representation to choose the following elongation coefficients:

in the case of a plane problem.

λ_1—the elongation coefficient along the axis $0x_1$ (Fig. 8.1a–e);

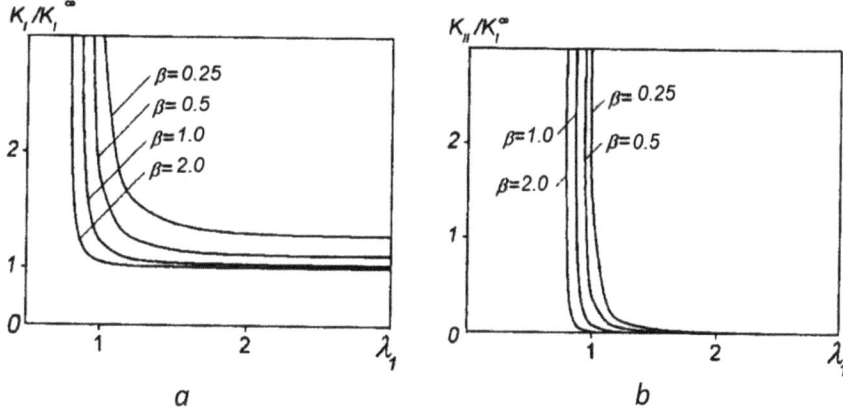

Fig. 8.6

in the case of the spatial problem under axisymmetric loading (Fig. 8.1b–e for the meridional section) for cracks in the form of circular discs.

$\lambda_1 = \lambda_2 = \lambda_r$—the elongation coefficient in the radial direction.

The above values were used in the monograph [17] of 2017 and the review article [33] of 2015.

In Fig. 8.6, which corresponds to Fig. 21 of the review article [33], the dependences are presented for the indicators (8.3) (Fig. 8.6a—for K_I/K_I^∞, Fig. 8.6 b—for K_{II}/K_I^∞) on the elongation coefficient $\lambda_1 = \lambda_2 = \lambda_r$ for the design model 8 of **Problem 4**, which is indicated in Sect. 6.4.3, concerning the highly elastic incompressible material with the Treloar elastic potential.

Note that the results in Fig. 8.6a and b are given for different values of the parameter β (8.2) following the accepted designations.

Also note that the design model 8 of **Problem 4**, indicated in Sect. 6.4.3, corresponds to the axisymmetric problems for the crack in the form of a circular disk near the half-space boundary (the results in Fig. 8.6 correspond to the crack mode I in the case of a uniform normal load on the crack faces). Therefore, the above design model 8 of **Problem 4** within the framework of **Problem 6** corresponds to the design model in Fig. 8.1e.

The curves shown in Fig. 8.6 have *the vertical asymptotes*, which are different for different values of the parameter β (8.2). Following the combined method, the indicated vertical asymptotes correspond to the theoretical ultimate strengths for the design model in Fig. 8.1e concerning **Problem 6** for the different values of the parameter β (8.2).

The vertical asymptotes in Fig. 8.6 correspond to the eigenvalues of the eigenvalue problems, which are investigated in **Problem 6**, and the eigenvalues correspond to the critical loads, which, by the **General Concept** (Sect. 8.2.2) for **Problem 6**, determine the values of the theoretical ultimate strength.

Additional information on the results for problems on **Problem 6**, obtained by the combined method, is presented in the monograph [128] of 2017 and the review article [33] of 2015 and the lists of references to which indicate the main publications in periodicals related to the considered scientific direction.

Note 8.5 In this section, a more detailed discussion of the results is carried out, in comparison with Sects. 8.3.1 and 8.3.2, since the generalizing publications on this point (monograph and review article) have appeared only in recent years.

8.4 Results for Layered Composites with Cracks at Interface Under Brittle and Plastic Fracture. The Second General Approach

The studies are carried out for the plane and spatial problems following the statement of the problems presented for the design schemes in Fig. 8.2a–e and Sect. 8.2.1 in the part with the subtitle "Design schemes for layered composite materials."

They are carried out within the framework of the **Second General Approach**, which is summarized in Sect. 8.2.3 in the part with the subtitle "The Second General Approach. Application of the three-dimensional linearized theory of deformable bodies stability." The main relations and approaches of this theory are indicated in Chapter 2 (Part I) of this monograph.

8.4.1 Introduction

The studies are carried out within the framework of the piecewise homogeneous medium model when the models of the homogeneous material (isotropic, transversely isotropic, or orthotropic elastic or plastic bodies) are used for the materials of the filler and binder layers, and certain continuity conditions are formulated at the interface *taking into account the presence of cracks*. The discussion of the model of a piecewise homogeneous medium in a brief form is carried out in the final part of Sect. 1.4.1 (Part I) of this monograph.

It is advisable to note that the statement of different problems determined by the relationships between the parameters h and a in Fig. 8.2c–e is possible. In this regard, it is possible to classify these problems.

The classification of problem statements for the composite materials with cracks was considered in the monographs [124, 127]. Concerning the mechanics of fracture under compression of the layered composites along the interfaces, in which the plane cracks are located (Fig. 8.2c–e), the classification of the formulations is considered in the publications [16, 65], which we will follow below, highlighting the *microcracks* and *macrocracks*.

The microcracks in the interface planes of the layered composite materials (for design schemes in Fig. 8.2a, c–e) will be called the cracks for which the following conditions are satisfied

$$h \cdot a^{-1} >> 1, \text{ including } h \cdot a^{-1} \to \infty. \tag{8.4}$$

In this case, we come to the problem for two interconnected half-spaces of different materials, in the interface of which the plane cracks exist. The design scheme for the analyzed situation in the case of one crack for the plane deformation in the plane $x_1 0 x_2$ is shown in Fig. 8.2b.

The macrocracks in the interface planes of the layered composite materials (for the design scheme in Fig. 8.2a) will be called the cracks for which condition (8.4) is not satisfied.

In this case, the macrocracks located in neighboring interface planes *interact with each other*. Thus, we come to the study of situations for which the characteristic simplest design schemes are presented in Fig. 8.2c–e.

It should be noted that in the publications [16, 65] also *the structural cracks* are identified, which it makes sense to consider for the layered composites composed of the filler in the form of layers of the significantly different thicknesses and a binder (matrix) in the form of layers of the significantly different thicknesses.

This monograph analyzes the results obtained within the framework of the **Second General Approach** for the layered composites, which are composed of the filler layers (reinforcing elements) of the same thickness h_a and binder (matrix) layers of the same thickness h_m, alternating along the vertical axis. In connection with the foregoing, for the discussed layered composites, it makes no sense to distinguish the structural cracks.

In this section, a summary of all the results obtained (within the framework of the **Second General Approach**, Sect. 8.2.3) for the brittle and plastic fracture of layered composites *under compression* along plane cracks located in the interface planes is grouped in the following sections:

Section 8.4.2—results for the layered composites with *microcracks* at the interfaces;
Section 8.4.3—results for the layered composites with *macrocracks* at the interfaces.

8.4.2   Results for Brittle and Plastic Fracture of Layered Composites with Microcracks at Interfaces: The Second General Approach

The main results were obtained by I. A. Guz and A. N. Guz.

The main publications related to the discussed scientific area are included in the general list, which is given in the Introduction (Sect. 8.1) to this chapter.

It should be noted that the most complete presentation of these results is presented in the monograph [127] (vol. 2). The results discussed in a fairly complete volume are considered in the review article [1] of 2014.

Apparently, the first publications in this scientific direction were the articles [16, 17, 66, 76] of 1992–1993 for the brittle fracture in the case of one and two microcracks at the interface of the layered composite (the design scheme for this situation is shown in Fig. 8.2b) under compression along the interface using the numerical research method proposed in [141].

The above results were obtained within the framework of **theory 3** of TLTDBS (in the terminology of Sect. 2.2 of Part I of this monograph) using the model of a linear elastic isotropic body to study the *brittle* fracture in the situation under discussion. As a result of research, in the above publications for the value ε_T (the theoretical value of the limiting shortening for a layered composite *with one microcrack* at the interface), the expression was obtained

$$\varepsilon_T = \min\{\varepsilon_T^a, \varepsilon_T^m\}, \tag{8.5}$$

where ε_T^a and ε_T^m are the theoretical values of the limiting shortening, respectively, of the material of the filler (reinforcing elements) and the binder (matrix), which correspond to the critical values of the shortening as applied to the near-the-surface instability of the half-space within the framework of plane deformation Fig. 8.7.

For the *brittle and plastic* fracture of the isotropic and orthotropic compressible and incompressible filler (reinforcing elements) and binder (matrix) materials *in a unified general form*, the results were obtained for the case of *a finite number of microcracks* at the interface of a layered composite under compression along the interface. The design scheme for the discussed case is shown in Fig. 8.7 for the plane deformation.

The above results were obtained in the articles [42–45, 74] of 2000 and 2001 in the unified general form for **theories 1–3** of the TLTDBS (according to the terminology of Sect. 2.2 of Part I of this monograph.

The following designations are introduced in Fig. 8.7:

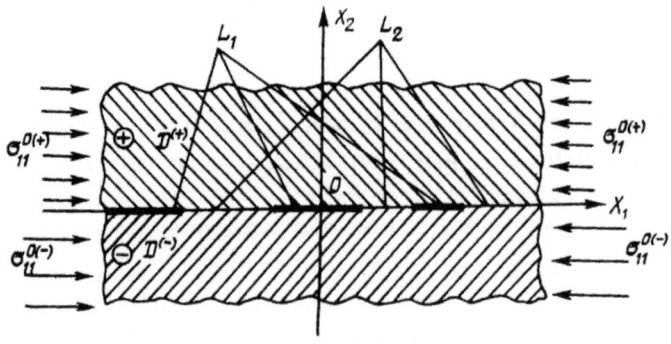

Fig. 8.7

Fig. 8.8

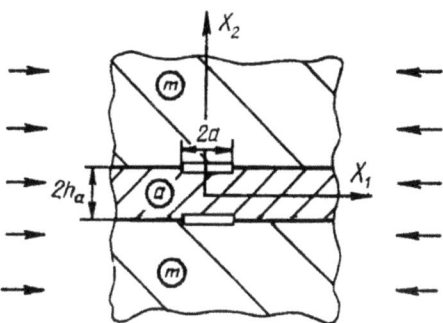

$D^{(\pm)}$—the half-planes of various materials;
L_1—a plurality of segments at the interface corresponding to cracks;
L_2—a plurality of segments at the interface corresponding to the complete connection of two materials.

Besides, Fig. 8.7 shows the compressive stresses $\sigma_{11}^{0(+)} \neq \sigma_{11}^{0(-)}$ because, following Note 8.3 (Part II of this monograph), in the study of loss of stability in the internal structure of the composite and the case of near-the-surface loss of stability of the composite, *the same shortening of the filler and matrix along the direction of compression is provided*, which is, apparently, the only possible condition that allows one to analyze the phenomena inside the composite.

The method for solving the problem, the design scheme for which is shown in Fig. 8.7, is to reduce to the homogeneous conjugation problem (or the Riemann–Hilbert problem) for two holomorphic functions defined in the entire plane, which corresponds to the above-formulated eigenvalue problem TLTDBS. This method is a development (for the eigenvalue problems) of Muskhelishvili's method [148], developed for the boundary value problems of statics of the classical linear theory of elasticity.

To realize the above method, the representation of stresses and displacements *for each of the half-planes in terms of one function of complex variables, which is analytical and defined in the entire plane*, was used in the articles [42–45, 74] for the plane static TLTDBS problems, The representations of this type have the form (2.129) for the case of unequal roots and the form (2.130) for the case of equal roots (Part I of this monograph). Using the above concepts for the plane static TLTUDT problem as applied to the design scheme in Fig. 8.7, a homogeneous conjugation problem is obtained for two functions holomorphic in the whole plane for the following three cases of layered composite arrangement:

the material $D^{(+)}$ has *unequal* roots, and the material $D^{(-)}$ has *unequal* roots;
the material $D^{(+)}$ has *equal* roots, and the material $D^{(-)}$ has the *equal* roots;
the material $D^{(+)}$ has *unequal* roots, and the material $D^{(-)}$ has *equal* roots.

Note that the three above-mentioned cases of the layered composites arrangement exhaust *all possible cases* of the layered composites arrangement as applied to the study of microcracks.

In publications [42–45, 74] of 2000 and 2001, **the exact solution** of the homogeneous problem of the conjugation of two functions holomorphic in the entire plane is obtained in the general formulation stated above for the above three arrangements of the layered composite. From this **exact solution**, *regardless of the number of cracks* (Fig. 8.7) for ε_T (**the theoretical value of the limiting shortening**) the expression (8.5) is obtained, where ε_T^a and ε_T^m correspond to the sufficient general models of the theory of elasticity and plasticity in the general statement of TLTDBS problems in this section.

The above results **on the exact determination** ε_T for the layered composite with the microcracks at the interface (Fig. 8.7) (ε_T is **the theoretical value of the limiting shortening**) are described in sufficient detail in the monograph [127] (vol. 2, Chap. 8, Sect. 8.2).

It is advisable to note that the above **exact solution** *has no analog in the world science* in the publications on the mechanics of fracture of layered composites under compression.

Note 8.6 The assumption *about the independence of the results from the number of cracks* concerning the situation in Fig. 8.7 *cannot be obtained* a priori for reasons of a physical nature due to the following fact.

The point is that the statement of problems for the situation in Fig. 8.7 includes a dimensionless geometric parameter corresponding to the ratio of the length of one of the bridges between the cracks to the length of one of the cracks. In this connection, one would expect from considerations of physical nature the dependence of the final result on the indicated dimensionless geometric parameter.

8.4.3 Results for the Brittle Fracture of Layered Composites with Macrocracks at the Interfaces: The Second General Approach

The main results were obtained by I. A. Guz.

The main publications related to the discussed scientific area are included in the general list, which is given in the Introduction (Sect. 8.1) to this chapter. The results discussed in Sect. 8.4.3 are presented in the publications [19, 67–98, 103–107, 481] since 1994. The corresponding analysis of these results is presented in the review article [1] of 2014, which we will mainly follow below.

These publications present the research results for the macrocracks at the interface within the design schemes presented in Fig. 8.2c-e for the typical simple cases, as well as for the related design schemes. These studies were carried out mainly for the plane problem in the plane $x_1 0 x_2$ as applied to the *brittle* fracture. At that, for the relatively rigid structural layered composites, **theory 3** TLTDBS (in the terminology of Sect. 2.2 of Part I of this monograph) and the model of the linear elastic isotropic or orthotropic body was applied, and for the highly elastic layered composites, **theory 1** of TLTDBS and the corresponding elastic potentials of the simplest structure was applied.

It is advisable to note that the results of this section were *obtained exclusively using the numerical research methods*—the finite differences and finite element methods. To obtain the results discussed, it is not possible to use the analytical research methods by analogy with the approaches set out in the previous section.

It should also be noted that the specific results outlined in the publications mentioned at the beginning of this section and which are presented below, were obtained mainly for the layered composites formed from the longitudinal-transverse laying of a unidirectional fibrous composite. This monolayer of the unidirectional fibrous composite in the continuous approximation is considered as a homogeneous orthotropic material with the averaged elastic constants, which are determined from the corresponding expressions of the well-known monographs. Thus, during the formation of a layered composite by the longitudinal-transverse laying of layers, for example, as in Fig. 8.8, in the filler layers (reinforcing elements, all values of which are marked with index "a ") the fibers are directed along the axis $0x_1$ and in the layers of the binder (matrix, all the values of which are marked with index "m ") the fibers are directed perpendicular to the plane of Fig. 8.8. In this case, when loaded along the axis $0x_1$ (Fig. 8.8), the filler is the more rigid material and the binder is the less rigid material. Figure 8.9.

Figure 8.8 shows the design scheme for the layered composite with layers of the constant thickness as applied to the plane problem in the plane $x_1 0x_2$ (plane deformation) when the binder layers are much thicker than the filler layers (for $h_m \gg h_a$, all designations are shown in Fig. 8.8). In this case, the layers of the binder are approximately replaced by two half-planes.

The study was carried out for two cracks located in the neighboring planes of separation, taking into account the interaction between these cracks, which, in the

Fig. 8.9

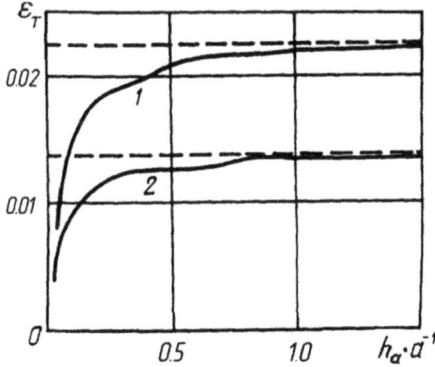

above statement, following the classification [16, 65], can be called the *structural cracks*.

The concrete results for the layered composite concerning the design scheme in Fig. 8.8 for the brittle fracture, which are obtained using the finite difference method, are presented in the article [67] of 1994. At that, the number of grid nodes was changed to ensure the reliability of the results obtained.

The studies in [67] were carried out for the various materials in a monolayer (in a unidirectional fiber composite). In Fig. 8.9, as an example, the results are shown for only two materials in a monolayer (in the unidirectional fiber composite). The first material, the results for which in Fig. 8.9 marked with "1", formed by the carbon fibers (Tornel - 300) in the form of a filler and the epoxy resin in the form of a binder. The second material, the results for which are marked with the number "2", is formed by the boron fibers (AVCO firm) in the form of filler and epoxy resin in the form of a binder.

Additional information about the material for a monolayer, including materials "1" and "2", as well as other materials considered, and on the implementation of numerical analysis that ensures the reliability of the results obtained, can be obtained from [67].

Figure 8.9 shows, according to [67], the dependence of the quantity ε_T (*the theoretical value of the ultimate shortening* for the layered composite with two cracks (Fig. 8.8)) on the parameter $h_a \cdot a^{-1}$ characterizing the relative crack size (Fig. 8.8), for *the first and second* above-mentioned layered composite. In Fig. 8.9, the solid lines show the results for two cracks (Fig. 8.8) taking into account their mutual influence and the dashed lines show the results for one microcrack corresponding to expression (8.5), which were first obtained by the numerical methods in the publications [16, 17, 66, 76], since 1992. It follows from the results in Fig. 8.9 that when changing the parameter $h_a \cdot a^{-1}$, it can get a decrease in the quantity ε_T by hundreds of percent (Figs. 8.10 and 8.11).

The publications indicated at the beginning of Sect. 8.4.3 present the results for the various problems concerning the *macrocracks* or *structural cracks* in the layered composites with the brittle fracture under compression along the interfaces.

Fig. 8.10

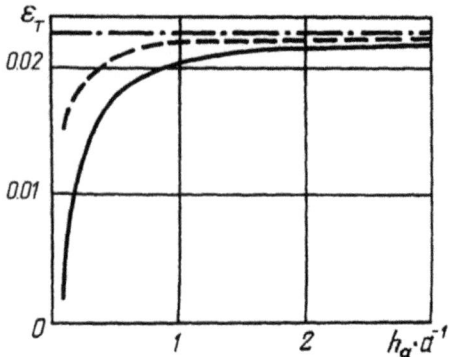

Fig. 8.11

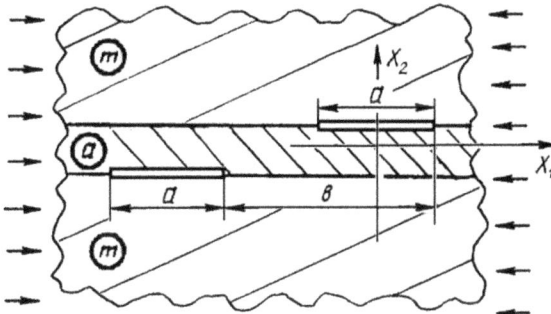

So, in the article [68] of 1995, the results for the layered composite with a periodic system of cracks are presented for the design scheme close to Fig. 8.2d under compression along cracks.

Figure 8.10, following [67, 68], shows the dependence ε_T (*the theoretical value of the limiting shortening* along the axis $0x_1$ (Fig. 8.8, Fig. 8.2d)) on the parameter $h_a \cdot a^{-1}$ characterizing the relative sizes of the cracks. The results in Fig. 8.10 are given for the composite, which concerning Fig. 8.9 named "*the first*" material. The solid curve corresponds to the layered composite with a periodic system of cracks [68]. The dashed line corresponds to the layered composite with two cracks [67]. The dash-dotted line corresponds to the layered composite with one microcrack [16, 17, 66, 76].

From the analysis of the results presented in Fig. 8.10 and other results presented in [67, 68], the following conclusion was made in [68].

When $h_a \cdot a^{-1} > 2$, it is possible to disregard the mutual influence of cracks in the parallel planes of the interface concerning two cracks (Fig. 8.8) [67] and to a periodic row of cracks [68], and to estimate the quantity ε_T (the theoretical value of the limiting shortening) one can apply expression (8.5), corresponding to one microcrack.

The above conclusion follows, for example, from Fig. 8.10, since for $h_a \cdot a^{-1} > 2$, the solid, dashed and dash-dotted lines practically coincide.

The publications indicated at the beginning of this section also provide an analysis of other regularities corresponding to the discussed problems of the mechanics of brittle fracture of layered composites under compression along the plane cracks located in the parallel interface planes.

In subsequent years, the results were obtained on the mechanics of brittle fracture of layered composites under compression along the plane cracks located in the parallel interface planes as applied to more complex design schemes as compared to Figs. 8.2 and 8.8. These results, which are presented, for example, in the publications [84–88] of 2006 and 2008, refer to the macrocracks that are not located one above the other (Figs. 8.2 and 8.8) but are displaced in the parallel planes relative to each other. The simplest design scheme of this type is shown in Fig. 8.11.

We will restrict ourselves to the information outlined in Sect. 8.4 when discussing the results on the mechanics of *brittle and plastic* fracture of the layered composites under compression along the plane cracks that are located in the parallel interface planes. All the considered results were obtained within the framework of the **Second General Approach**, which is summarized in the second part of Sect. 8.3.3.

8.5 Results for Brittle Fracture of Homogeneous Materials with Cracks Located in the Close Arranged Parallel Planes. Passage to the Limit. The Second General Approach

The main results were obtained by V. M. Nazarenko and M. V. Dovzhik.

They are presented in publications [20–23, 41, 142] of 2011–2013, and a sufficiently detailed analysis of these results is presented in the review article [1] of 2014. In these publications, a research method was developed that is some development of the method of Sect. 8.3.2, which made it possible to obtain the results for the cracks drawing together up to the passage to the limiting and located in the parallel planes.

This method was developed using **theories 1 and 3** of the TLTDBS (in the terminology of Sect. 2.2 of Part I of this monograph) within the framework of the **Second General Approach** of **Problem 6** (*Fracture under compression along parallel cracks*), which is summarized in the second part of Sect. 8.2.3.

8.5.1 Short Description of Developed Research Method

The discussed results [20–23, 41, 142] refer to the design schemes in Fig. 8.1c–e and are obtained concerning the spatial problems for identical cracks in the form of a circular disk (penny-shaped cracks) under axisymmetric external loading. Taking into account the information in the introductory part of Sect. 8.3.2, the above results can be considered the continuation of the studies in Sect. 8.3.2.

It was indicated in Sect. 8.3.2 that the discussed spatial problems in the general case for the brittle and plastic fracture are reduced to the problems on the eigenvalues of one-dimensional Fredholm integral equations of the second kind. The expressions for determining the kernels of these one-dimensional integral equations in the expanded form are presented in the monographs [124, 127 vol. 2, 125 vol. 4, book 1, 128].

Note that the kernels have a complex structure and are represented in the form of the definite integrals of the various special functions, including the functions of a circular cylinder. The obtained one-dimensional integral equations are only the final result of a long chain of the complex and cumbersome mathematical transformations corresponding to the solution method in Sect. 8.3.2.

For the numerical determination of the eigenvalues of the obtained homogeneous one-dimensional Fredholm integral equations of the second kind, the Bubnov–Galerkin method was used. In this case, the sought function $g(\xi)$ was represented as a series in power functions

$$g(\xi) = \sum_{i=1}^{N} a_i \cdot \xi^i, \tag{8.6}$$

where a_i are unknown coefficients.

Due to the complex structure of the kernels of one-dimensional integral equations in the current situation (the application of the Bubnov–Galerkin method and power functions as the coordinate functions), *it was not possible to perform in the analytical form by integrating* the expressions corresponding to these kernels. In connection with the above, the numerical integration and the Gauss quadrature formula were additionally applied, which led to cumbersome transformations and a large amount of computation.

The above situation led to the fact that *the specific results were not obtained for a relatively large number of the coordinate functions in* (8.6). So, even to prove the reliability of the results, the calculations were carried out at N_{max}, in other cases, they were limited to N_{max}. With this approach, the specific results in Sect. 8.3.2 were obtained by changing the parameter $\beta = h \cdot a^{-1}$ in the following interval

$$0.063 \leq \beta \leq \infty. \tag{8.7}$$

In publications [20–23, 41, 142], an approach was proposed that made it possible to obtain the *concrete results* in almost the entire following interval:

$$0 \leq \beta \leq \infty. \tag{8.8}$$

Since in Sect. 8.3.2 the concrete results were already obtained in the interval (8.7), it was sufficient to obtain in [20–23, 41, 142] the results for the following interval:

$$0 \leq \beta \leq 0.1, \tag{8.9}$$

However, in these publications, the results are presented for a wider interval in comparison with (8.9).

It is advisable to note that the approach and results of [20–23, 41, 142] are *fundamentally new*, which actually made it possible to obtain the passage to the limit at $\beta \to 0$. The approach of the above publications is based on the use of the package Wolfram Mathematica 7, which is a package of symbolic mathematics. Mathematica has a high speed and predetermined guaranteed accuracy, which allows it to work efficiently on the PC.

The main feature of this *new approach* is that in the previously indicated situation (using the Bubnov–Galerkin method and power functions as the coordinate functions) when determining the eigenvalues of one-dimensional Fredholm integral equations of the second kind, *it was possible to avoid the procedure* of numerical integration and the application of the Gauss quadrature formula.

Using the Mathematica symbolic computation package, **the discussed integrals were computed analytically**.

In this regard, the cumbersome transformations and a large number of computations, typical for the approach from Sect. 8.3.2, were excluded. The aforementioned allowed, in the further numerical calculations, *to increase the accuracy of calculations by eliminating the error of numerical integration.* To speed up the calculation of integrals, an algorithm based on the use of recurrence relations was used.

We restrict ourselves to the above information in a brief description of the approach described in the publications [20–23, 41, 142], which made it possible to obtain the results practically in the entire interval (8.8).

The concrete results were obtained in [20–23, 41, 142] for the highly elastic compressible and incompressible materials with the elastic potentials of the simplest structure for the design schemes in Fig. 8.1c–e. When obtaining these results, **theory 1** of the TLTDBS was applied (in the terminology of Sect. 2.2 of Part I of this monograph) and the research was carried out within the framework of the **Second General Approach of Problem 6**, which is set out in the second part of Sect. 8.2.3.

In this way, the results were obtained in [20–23, 41, 142] for three situations:

two cracks in the form of identical circular disks (the design scheme in Fig. 8.1c);
the periodic series of cracks in the form of identical circular disks (the design scheme in Fig. 8.1d);
one crack in the form of a circular disk, located parallel to the boundary of the half-space (near-the-surface crack) (the design scheme in Fig. 8.1e).

Moreover, the concrete results were obtained practically in the entire interval (8.8).

It is advisable to note that the specific results obtained in [20–23, 41, 142] for the above three design schemes and the above materials for the brittle fracture are *practically exact* since they were obtained for the entire interval (8.8), including the case $\beta = h \cdot a^{-1} \to 0$.

In connection with the above, the results of [20–23, 41, 142] can be applied to evaluate the corresponding results obtained using the **First General Approach** (*the beam approximation*), according to Conclusion 8.4, formulated as a result of a brief analysis of the beam approximation at the end the first part of Sect. 8.2.3, directly related to the **First General Approach**. However, it is necessary to take into account that *the analysis of the accuracy* of the beam approximation applied to a large number of cracks modeled by an infinite periodic series of cracks is *pointless* according to Conclusion 8.5, also formulated at the end of the first part of Sect. 8.2.3, which relates directly to the **First General Approach**.

Below in this section, as an example of the results [20–23, 41, 142], the results relating to one near-the-surface crack are considered.

8.5.2 Near-the-Surface Crack

The design scheme is shown in Fig. 8.1e, the study is carried out for the spatial problem under axisymmetric loading applied to one near-the-surface crack in the form of a circular disk, which is located parallel to the boundary of the half-space. Figure 8.1e shows the cross section of the plane $x_3 = 0$, where h is the distance from the crack to the surface of the half-space, a is the radius of the circular crack.

The concrete results are presented: in the article [142] for the compressible material with the elastic potential of the harmonic type and for an incompressible material with the Bartenev–Khazanovich elastic potential and in the article [41] for the incompressible material with the Treloar elastic potential (material of the NeoHookean type).

Below in this section, as an example, the concrete results are given following [41] for the incompressible highly elastic material with the Treloar elastic potential. In this case, the following notation is used:

ε_T—the theoretical value of the ultimate radial shortening;
σ_T^I—the theoretical value of the ultimate strength in the **First General Approach** (the beam approximation, the first part of Sect. 8.2.3);
σ_T—the theoretical value of the ultimate strength in the **Second General Approach** (the use of TLTDBS, the second part of Sect. 8.2.3);
$\beta = h \cdot a^{-1}$—the parameter according to (8.2);
c_{10}—the constant which corresponds to the expression for the Treloar elastic potential (e.g.,, the monograph [137], p. 358, expression (3.431)).

Figure 8.12 shows the dependence of ε_T on β: in Fig. 8.12a—for the interval $0.1 \le \beta \le 3.0$ and in Fig. 8.12b—for the interval $0.01 \le \beta \le 0.10$.

Following the above notations, the following expression holds [41] for the **First General Approach**.

$$\sigma_T^I = A^I c_{10} \beta^2, \beta = h \cdot a^{-1} \tag{8.10}$$

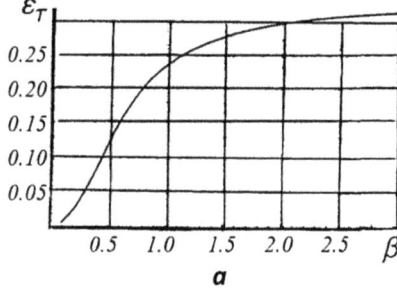 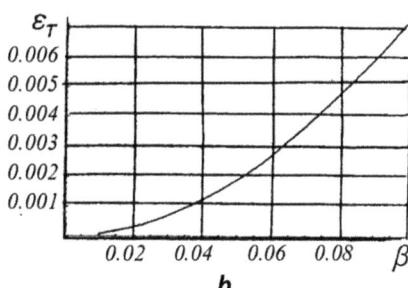

Fig. 8.12

In this case, the coefficient A^I for the rigid clamping of the circular plate, which is the most popular in studies using this approach, is determined as follows

$$A^I = 9.7866. \tag{8.11}$$

Taking into account the structure of expression (8.10) corresponding to the **First General Approach**, for comparison, the expression for the determination of σ_T corresponding to the **Second General Approach** is also presented in a similar form

$$\sigma_T = A c_{10} \beta^2, \ \beta = h \cdot a^{-1}; \ A = A(\beta). \tag{8.12}$$

Of course, to determine the quantity $A(\beta)$ corresponding to each value of β in (8.12), it is necessary to perform the significant work of a mathematical and computational nature, which is briefly described in Sects. 8.3.2 and 8.5.1. Table 8.1.

Taking into account the designations (8.10) and (8.12), Table 8.1 shows the values of ε_T and A when changing the parameter β in the interval $1 \cdot 10^{-9} \leq \beta \leq 9 \cdot 10^{-2}$.

To obtain specific results, which are presented in Fig. 8.12 and Table 8.1, in [41], the maximum number of coordinate functions N_{max} was used in an expression (8.6). At that, the practical convergence was achieved both by increasing the number of coordinate functions and by increasing the accuracy of calculations, using a larger number of significant digits after the decimal point.

From the analysis of the specific results presented in Fig. 8.12 and Table 8.1, the following significant conclusions can be drawn.

1. The results are obtained in a very wide range of variation of the parameter β $(1 \cdot 10^{-9} \leq \beta \leq 3)$, *including practically also the passage to the limit (at*

Table 8.1

β	ε_T	A
$9 \cdot 10^{-2}$	$5.96120 \cdot 10^{-3}$	8.83140
$8 \cdot 10^{-2}$	$4.76105 \cdot 10^{-3}$	8.92798
$7 \cdot 10^{-2}$	$3.68511 \cdot 10^{-3}$	9.02475
$6 \cdot 10^{-2}$	$2.73747 \cdot 10^{-3}$	9.12490
$5 \cdot 10^{-2}$	$1.92242 \cdot 10^{-3}$	9.22762
$4 \cdot 10^{-2}$	$1.24441 \cdot 10^{-3}$	9.33309
$3 \cdot 10^{-2}$	$7.08118 \cdot 10^{-4}$	9.44157
$2 \cdot 10^{-2}$	$3.18445 \cdot 10^{-4}$	9.55334
$1 \cdot 10^{-2}$	$8.05725 \cdot 10^{-5}$	9.66870
$1 \cdot 10^{-3}$	$8.14575 \cdot 10^{-7}$	9.77490
$1 \cdot 10^{-4}$	$8.15068 \cdot 10^{-9}$	9.78076
$1 \cdot 10^{-5}$	$8.14758 \cdot 10^{-11}$	9.77710
$1 \cdot 10^{-6}$	$8.14606 \cdot 10^{-13}$	9.77527
$1 \cdot 10^{-9}$	$8.14575 \cdot 10^{-19}$	9.77490

$\beta \rightarrow 0$) *to the results* of the **First General Approach** (*beam approximation,* the first part of Sect. 8.2.3).

2. For the case of a near-the-surface crack (design scheme in Fig. 8.1e), *at sufficiently small values of the parameter β (at $\beta \rightarrow 0$), it is possible to obtain* **the decrease in the quantity ε_T by two or more orders of value** *in comparison with the case of one isolated crack without taking into account the effect of the boundary surface of the material* (results of Sect. 8.3.1, which in Fig. 8.12a corresponds to the value ε_T at $\beta \geq 3$).

3. For the case of a near-the-surface crack (the design scheme in Fig. 8.1e), the application of the boundary conditions of rigid fixing in the **First General Approach** (the first part of Sect. 8.2.3) *is justified*, since **when for $\beta = h \cdot a^{-1} <$** 0.01 according to Table 8.1, the value of the quantity A in (8.12), corresponding to the **Second General Approach** *practically coincides* with the value of the quantity A^I (8.11) in (8.10), corresponding to the **First General Approach**.

4. For the case of a near-the-surface crack (the design scheme in Fig. 8.1e), the limit of applicability of the **First General Approach** (the first part of Sect. 8.2.3) **is established** in terms of the results (Fig. 8.12 and Table 8.1) of the **Second General Approach** (the second part of Sect. 8.2.3). Thus, when $\beta = h \cdot a^{-1} <$ 0.01, *the results as applied to σ_T for the* **Second General Approach** and *as applied to σ_T^I for the* **First General Approach practically coincide.**

It should be noted that, as was already emphasized in the final part of Sect. 8.5.1, the specific results in [20–23, 41, 142] were obtained for the highly elastic compressible and incompressible materials with the elastic potentials of the simplest structure for the design schemes in Fig. 8.1c–e.

At that, when obtaining the specific results, **theory 1** of the TLTDBS was applied (in the terminology of Sect. 2.2 of Part I of this monograph) and the studies were carried out within the framework of the **Second General Approach of Problem 6** (the second part of Sect. 8.2.3).

In particular, the above Conclusions 8.1 and 8.4 are presented according to [41] for the incompressible material with the Treloar elastic potential. Naturally, concerning other materials, the corresponding conclusions can be obtained only after the formation of similar concrete results for these materials.

Note also that similar concrete results and conclusions were obtained: for two cracks [20, 22] and the periodic series of cracks in [21, 23].

A brief analysis of results from [20–23] is also presented in the review article [1] of 2014.

Note 8.7 As already noted, the concrete results obtained with the approach [20–23, 41, 142] *are practically exact.* Therefore, they can be used *to assess the accuracy and limits of applicability* of the **First General Approach of Problem 6** (the beam approximation and the first part of Sect. 8.2.3). It is advisable to note also that the above analysis can *reasonably* be carried out only for two cases:

for *two equal cracks* (design scheme in Fig. 8.1c);
for the *one near-the-surface crack* (design scheme in Fig. 8.1e).

The analysis for the case of *a periodic series of cracks* **is pointless** according to Conclusion 8.5, formulated at the end of the first part of Sect. 8.2.3, which relates directly to the **First General Approach**.

In [20–23, 41, 142], for the materials considered in them concerning two equal cracks and one near-the-surface crack, *the limit of applicability of the* **First General Approach in Problem 6** *was determined* in the form of the following condition

$$\beta = h \cdot a^{-1} < 0.01 \qquad (8.13)$$

based on the **criterion** of almost complete coincidence (*the sufficiently strong criterion*) of the results obtained with the **First General Approach**, with the results obtained with the **Second General Approach**.

According to condition (8.13), the **First General Approach** is applicable for the situation when the linear dimensions of the cracks **are 100 times greater than** the distance between two cracks or between the crack and the material surface. Apparently, for this situation, the question arises about *the realism* of such cracks. The crack sizes can be significantly reduced if, instead of the sufficiently strong criterion of coincidence of results (condition (8.13) is obtained from the sufficiently strong criterion of practical coincidence of results), the weaker criteria for the coincidence of results are applied (e.g., the coincidence of results with an accuracy of 5; 10; 20%, etc.).

We restrict ourselves to the above information and considerations when presenting in this section the results for the brittle fracture of homogeneous materials with the cracks located in the close parallel planes, including the passages to the limit.

Note that these results were obtained within the framework of the **Second General Approach of Problem 6** (*Fracture under compression along parallel cracks*), which is summarized in the second part of Sect. 8.2.3.

8.6 On Results for Viscoelastic Fracture

The main results were obtained by S. D. Akbarov.

These results for the viscoelastic fracture were obtained for the homogeneous materials under compression along the plane cracks located in the parallel planes, and most of the results were obtained for the layered composites and structural members under compression along the plane cracks located at the interface between layers.

These results are presented in the monograph [148], published in 2012, and in the separate articles [143–147], starting with 1998, and in the number of other publications indicated in the monograph [148] and in the review article [144], which was published in 2007. The list of references to this monograph includes, as an example, only a few publications related to the discussed scientific direction, since the subsequent research developed in Azerbaijan and Turkey with the involvement to a certain extent of the results on the mechanics of composites with the curved structures, which are briefly discussed in Chap. 7 **in Problem 5** *(Brittle fracture in the form of separation into the slender parts of the composite materials under tension or compression along the reinforcing elements).*

As already noted in the Introduction (Sect. 7.1) to Chap. 7 (Part III of this monograph), the main above results on **Problem 5** were obtained in the Department of Dynamics and Stability of Continua of the S. P. Timoshenko Institute of Mechanics of the NASU and published in several monographs specified in Sect. 7.1, including the monograph [149] in English.

The developed by S.D. Akbarov research methods in *a three-dimensional statement* of the stability of materials and structural elements for *the viscoelastic models* include the following positions.

1. The research is carried out with the use of the *statical* equations in stresses and boundary conditions TLTDBS, a summary of which is presented in Chapter 2 (part I of this monograph).
2. To describe the viscoelastic deformation of a material, mainly a binder (Matrix) in a composite, the linear theory of viscoelasticity of hereditary type is used with the use of Rabotnov's fractional exponential operators.
3. As a stability criterion under the viscoelastic deformation, *a generalization* (to the problems in the three-dimensional statement) of *the approximate Hoff's stability criterion* [150], originally proposed in 1954 for studying the stability of thin-walled systems (rods, plates, and shells) under creep, is used. Following the approach [150], the approximately small deviation from the initially rectilinear shape of the thin-walled systems is introduced, and then, the time variation of the introduced deviation is analyzed at the constant or increasing load. In the *three-dimensional statement*, the procedure for introducing the small deviations in the structure of the composite is carried out using *the methods of the mechanics of curved composites* described in monograph [149], a review of the results on which is presented in the review article [151].
4. To study the appropriately formulated problems, the Laplace transform in time is used. At that, the approximate Schapery's method [152] is used to go to the originals.
5. To obtain the concrete information, at the final stage of the study, mainly the numerical methods are used, the finite element method and the boundary element method.

Concerning the mechanics of fracture under compression along the parallel cracks (the subject of research in **Problem 6**, chapter 8 of this monograph) *in the viscoelastic deformation of the material*, the above methods of approximate research *in a three-dimensional statement* of the stability of materials and structural elements for the viscoelastic models are supplemented by the **General Concept** that determines the start of the process of fracture, **which is formulated in** Sect. 8.2.2.

The above results can be classified as an approximate approach in *the three-dimensional statement*, which makes it possible to study the stability of composites and structural elements under the viscoelastic deformation, as well as the problems of fracture mechanics under compression along the system of cracks as applied to the viscoelastic material models.

The discussed approximate approach is described in the publications of S.D. Akbarov and in the numerous publications of the author of the approach and his pupils, indicated in the monograph [148] of 2012 and the review article [144] of 2007, the broad classes of concrete problems are investigated.

It should be noted, however, that *the discussed approach, despite the three-dimensional statement* **is purely approximate and does not correspond** *to the rigorous method of studying the loss of stability that is generally accepted in mechanics. It differs in consistency and rigor from* **the results for the elastic and plastic models**, *which are obtained* **based on TLTDBS** *and which are analyzed in this monograph.*

The above opinion is substantiated in the following **Note**.

Note 8.8

1. Apparently, it can be considered that in the mechanics the generally accepted and rather rigorous method for studying the phenomenon of loss of stability is the analysis of the **behavior of small perturbations** within the framework of linearized three-dimensional or two-dimensional (in the case of mechanics of thin-walled structures) **dynamical problems**. The stability criterion is the non-increase of disturbances in time.

2. **For the elastic and plastic materials**, within the framework of strict TLTDBS with the involvement of a dynamical statement, the fulfillment of sufficient conditions for the applicability of the statical research method (Euler's method, method of eigenvalue problems) is proved. Thus, the concrete results obtained **for the elastic and plastic materials** by the **statical method** coincide with the results obtained based on **dynamical equations** (the generally accepted method for studying the phenomenon of stability in mechanics, item 1 of this note).

3. In this method of studying the stability problems **for the viscoelastic materials, the dynamical linearized equations of stability are not used**. The studies from the very beginning are carried out **for the statical equations of stability**. The introduction of deviations from the rectilinear form and their analysis in time following the **approximate Hoff's criterion** [150] does not coincide with the linearized **dynamical equations.**

The considerations of Note 8.8 indicate that the approach discussed in Sect. 8.6 for studying the stability of the viscoelastic materials **is purely approximate**. Of course, it is advisable to analyze the concrete results obtained with the discussed approach in the review article, since this monograph considers the results only for the elastic and plastic materials, obtained within the framework of the generally accepted in mechanics and a rather rigorous method of analyzing the phenomenon of stability.

We restrict ourselves to the information presented above in this chapter when analyzing **Problem 6** (Fracture under compression along parallel cracks). A more detailed discussion of the analyzed results is presented in the review article [1] of 2014.

References

1. Guz, A.N.: O postroenii osnov mekhaniki razrusheniia materialov pri szhatii vdol treshchin (obzor) (On the construction of the foundations of the fracture mechanics of materials in compression along cracks (review)). Prikladnaya Mekhanika **50**(1), 5–89 (2014)
2. Bogdanov, V.L.: Neosesimmetrichnaia zadacha o razrushenii poluprostranstva pri szhatii vdol pripoverkhnostnoi krugovoi treshchiny (Nonaxisymmetric problem of the fracture of a half-space compressed along a near-surface circular crack). Dokl. Akad. Nauk USSR Ser. B. **5**, 42–47 (1991)
3. Bogdanov, V.L., Nazarenko, V.M.: Szhatie kompozitnogo materiala vdol pripoverkhnostnoi makrotreshchiny (Compression of a composite material along a near-surface macrocrack). Mekhanika Kompozitnykh Materialov **30**(3), 352–358 (1994)
4. Guz, A.N.: O postroenii teorii prochnosti odnonapravlennykh armirovannykh materialov na szhatie (On construction of the strength theory of unidirectional reinforced materials on compression). Probl. Prochn. **3**(3), 37–40 (1971)
5. Guz, A.N.: Ob odnom kriterii razrusheniia tverdykh tel pri szhatii vdol treshchin. Prostranstvennaia zadacha (Criterion of fracture of solids upon compression along cracks. Spatial problem). Dokl. Akad. Nauk SSSR **261**(1), 42–45 (1981)
6. Guz, A.N.: Tochnoe reshenie ploskoi zadachi o razrushenii materiala pri szhatii vdol treshchin, lezhashchikh v odnoi ploskosti (The exact solution of the plane problem of material fracture under compression along coplanar cracks). Dokl. Akad. Nauk SSSR **310**(3), 563–566 (1990)
7. Guz, A.N., Knyukh, V.I., Nazarenko, V.M.: Rassloenie kompozita pri szhatii vdol dvukh parallelnykh makrotreshchin (Delamination of a composite upon compression along two parallel macrocracks). Fiziko-himicheskaya Mekhanika Materialov **23**(1), 2–78 (1987)
8. Guz, A.N., Nazarenko, V.M., Bogdanov, V.L.: Prostranstvennaia neosesimmetrichnaia zadacha o razrushenii poluprostranstva s pripoverkhnostnoi krugovoi treshchinoi (Spatial nonaxisymmetric problem of the fracture of a half-space with a near-surface circular crack). Dokl. Akad. Nauk SSSR **319**(4), 835–839 (1991)
9. Guz, A.N., Nazarenko, V.M.: Osesimmetrichnaia zadacha o razrushenii poluprostranstva s poverkhnostnoi diskoobraznoi treshchinoi (Axisymmetric problem of the fracture of a half-space with a near-surface penny-shaped crack). Dokl. Akad. Nauk SSSR **274**(1), 38–41 (1984)
10. Guz, A.N., Nazarenko, V.M.: Prostranstvennaia zadacha o plasticheskom pripoverkhnostnom razrushenii materiala pri szhatii vdol makrotreshchin (Spatial problem of plastic near-surface fracture of a material under compression along macrocracks). Dokl. Akad. Nauk SSSR **284**(4), 812–815 (1985)
11. Guz, A.N., Nazarenko, V.M.: K teorii pripoverkhnostnogo otslaivaniia kompozitnykh materialov pri szhatii vdol makrotreshchin (Theory of near-surface delamination of composite materials under compression along macrocracks). Mekhanika Kompozitnykh Materialov **5**, 826–833 (1985)
12. Guz, A.N., Nazarenko, V.M.: Razrushenie materialov pri szhatii vdol periodicheskoi sistemy treshchin v usloviiakh ploskoi deformatsii (Fracture of materials under compression along a periodic system of cracks under plane strain conditions). Prikladnaya Matematika Mekhanika **51**(2), 323–329 (1987)
13. A.N. Guz, V.M. Nazarenko, I.P. Starodubtsev, in Ploskaia zadacha razrusheniia materialov s dvumia parallelnymi treshchinami pri szhatii vdol treshchin (Plane problem of fracture of materials with two parallel cracks under compression along cracks), ed. by V.G. Zubchaninov. Problemy mekhaniki deformiruemogo tverdogo tela (Kalinin.Univ., Kalinin, 1986), pp.138–151
14. Guz, A.N., Nazarenko, V.M., Khoma, Yu.I.: Razrushenie kompozitnogo materiala pri szhatii vdol tsilindricheskoi treshchiny (Fracture of composite materialupon compression along a cylindrical crack). Dopovidi NAN Ukrainy **10**, 48–52 (1995)
15. Guz, I.A.: Ustoichivost kompozita pri szhatii vdol treshchiny na granitse razdela sloev (Stability of a composite under compression along a crack at an interface between layers). DAN **325**(3), 455–458 (1992)

16. Guz, I.A.: Ustoichivost kompozitnykh materialov s mezhsloinymi treshchinami (Stability of composite materials with interlayer cracks). Mekhanika Kompozitnykh Materialov **5**, 603–608 (1992)

17. Guz, I.A.: Ustoichivost kompozita pri szhatii vdol dvukh mikrotreshchin na granitse razdela sloev (Stability of a composite in compression along two microcracks at the interface between layers). DAN **328**(4), 437–439 (1993)

18. Guz, I.A.: Kompozity s mezhsloinymi treshchinami: ustoichivost pri szhatii vdol dvukh mikrotreshchin mezhdu ortotropnymi sloiami (Composites with interlaminar cracks: stability under compression along two microcracks between orthotropic layers). Mekhanika Kompozitnykh Materialov **6**, 791–798 (1993)

19. Guz, I.A.: Ustoichivost kompozitov pri szhatii vdol sistemy parallelnykh mezhsloevykh treshchin (Stability of composites in compression along a system of parallel interlayer cracks). Dopovidi NAN Ukrainy. **6**, 44–47 (1995)

20. Dovzhyk, M.V.: Razrushenie materiala s dvumya diskoobraznymi treshchinami pri szhatii vdol treschin dlya malykh rasstoyaniy mezhdu treschinami (Fracture of a material compressed along two closely spaced penny-shaped cracks). Prikladnaya Mekhanika **49**(1), 100–108 (2013)

21. Dovzhyk, M.V.: Razrushenie materiala s periodicheskoy sistemoy diskoobraznykh treshchin pri szhatii vdol treschin dlya malykh znacheniy rasstoyaniy mezhdu treschinami (Fracture of a material compressed along periodic closely spaced penny-shaped cracks). Dopovidi NAN Ukrainy. **10**, 100–105 (2013)

22. Dovzhyk, M.V., Nazarenko, V.M.: Razrushenie materiala pri szhatii vdol dvuh diskoobraznykh treshchin dlya malykh rasstoyaniy mezhdu treschinami (Fracture of a material compressed along two closely spaced penny-shaped cracks). Prikladnaya Mekhanika **48**(4), 78–85 (2012)

23. Dovzhyk, M.V., Nazarenko, V.M.: Razrushenie materiala pri szhatii vdol periodicheskoy sistemy blizko raspolozhennykh treshchin (Fracture of a material compressed along a periodic set of closely spaced cracks). Prikladnaya Mekhanika **48**(6), 106–113 (2012)

24. Nazarenko, V.M.: Prostranstvennaya zadacha o szhatii materiala vdol periodicheskoy sistemy parallelnykh krugovykh treschin (The spatial problem of the compression of a material along a periodic system of parallel circular cracks). Prikladnaya Matematika Mekhanika **52**(1), 145–152 (1988)

25. Nazarenko, V.M.: Neosesimmetrichnaya zadacha o szhatii materiala vdol pripoverkhnostnoy krugovoy treschiny (Nonaxisymmetric problem of material compression along a near-surface circular crack). Prikladnaya Mekhanika **25**(1), 124–127 (1989)

26. Nazarenko, V.M., Khoma, Yu.I.: O metode resheniya zadach razrusheniya beskonechnogo materiala s cilindricheskoy treshchinoy pri osevom szhatii (sluchay neravnykh korney) (On a method for solving problems of fracture of an infinite material with a cylindrical crack under axial compression (the case of unequal roots)). Dopovidi NAN Ukrainy. **7**, 62–67 (1994)

27. Nazarenko, V.M., Khoma, Yu.I.: Szhatie beskonechnogo kompozitnogo materiala vdol konechnoy cilindricheskoy treshchiny (Compression of an infinite composite material along a finite cylindrical crack). Mekhanika Kompozitnykh Materialov **31**(1), 27–34 (1995)

28. Bogdanov, V.L.: Nonaxisymmetric problem of the stress-strain state of an elastic half-space with a near-surface circular crack under action of loads along it. J. Math. Sci. **174**(3), 341–366 (2011)

29. Bogdanov, V.L., Guz, A.N., Nazarenko, V.M.: Fracture of semiinfinite material with a circular surface crack in compression along the crack plane. Int. Appl. Mech. **28**(11), 687–704 (1992)

30. Bogdanov, V.L., Guz, A.N., Nazarenko, V.M.: Nonaxisymmetric compressive failure of a circular crack parallel to a surface of halfspace. Theor. Appl. Fract. Mech. **22**, 239–247 (1995)

31. Bogdanov, V.L., Guz, A.N., Nazarenko, V.M.: Fracture of a body with a periodic set of coaxial cracks under forces directed along them: an axisymmetric problem. Int. Appl. Mech. **45**(2), 111–124 (2009)

32. Bogdanov, V.L., Guz, A.N., Nazarenko, V.M.: Stress-strain state of a material under forces acting along a periodic set of coaxial mode II penny-shaped cracks. Int. Appl. Mech. **46**(12), 1339–1350 (2010)
33. Bogdanov, V.L., Guz, A.N., Nazarenko, V.M.: Spatial problems of the fracture of materials loaded along cracks (Review). Int. Appl. Mech. **51**(5), 489–560 (2015)
34. Bogdanov, V.L., Nazarenko, V.M.: Study of the compressive failure of a semi-infinite elastic material with a harmonic potential. Int. Appl. Mech. **30**(10), 760–765 (1994)
35. Guz, A.N.: Fracture mechanics of solids in compression along cracks. Sov. Appl. Mech. **18**(3), 213–224 (1982)
36. Guz, A.N.: Mechanics of fracture of solids in compression along cracks (three-dimensional problem). Sov. Appl. Mech. **18**(4), 283–293 (1982)
37. Guz, A.N.: Fracture mechanics of composites in compression along cracks. Sov. Appl. Mech. **18**(6), 489–493 (1982)
38. Guz, A.N.: General case of the plane problem of the mechanics of fracture of solids in compression along cracks. Sov. Appl. Mech. **25**(6), 548–552 (1989)
39. Guz, A.N.: Construction of fracture mechanics for materials subjected to compression along cracks. Int. Appl. Mech. **28**(10), 633–639 (1992)
40. Guz, A.N.: Some modern problems of physical mechanics of fracture. In: Cherepanov, G.P., (ed.), FRACTURE. A Topical Encyclopedia of Current Knowledge, pp. 709–720. Krieger Publishing Company, Malabar, Florida (1998)
41. Guz, A.N., Dovzhik, M.V., Nazarenko, V.M.: Fracture of a material compressed along a crack located at a short distance from the free surface. Int. Appl. Mech. **47**(6), 627–635s (2011)
42. Guz, A.N., Guz, I.A.: Analytical solution of stability problem for two composite half-plane compressed along interfacial cracks. Compos. B **31**(5), 405–418 (2000)
43. Guz, A.N., Guz, I.A.: The stability of the interface between two bodies compressed along interface cracks. 1. Exact solution for the case of unequal roots. Int. Appl. Mech. **36**(4), 482–491 (2000)
44. Guz, A.N., Guz, I.A.: The stability of the interface between two bodies compressed along interface cracks. 2. Exact solution for the case of equal roots. Int. Appl. Mech. **36**(5), 615–622 (2000)
45. Guz, A.N., Guz, I.A.: The stability of the interface between two bodies compressed along interface cracks. 3. Exact solution for the case of equal and unequal roots. Int. Appl. Mech. **36**(6), 759–768 (2000)
46. Guz, A.N., Khoma, Yu.I.: Stability of an infinite solid with a circular cylindrical crack under compression using the Treloar potential. Theor. Appl. Fract. Mech. **39**(3), 276–280 (2002)
47. Guz, A.N., Khoma, Yu.I.: Integral formulation for a circular cylindrical cavity in infinite solid and finite length coaxial cylindrical crack compressed axially. Theor. Appl. Fract. Mech. **45**(2), 204–211 (2006)
48. Guz, A.N., Khoma, Yu.I., Nazarenko, V.M.: On fracture of an infinite elastic body in compression along a cylindrical defect. In: Proceedings of ICF 9. Advance in Fracture Research, vol. 4, pp. 2047–2054. Sydney, Australia (1997)
49. Guz, A.N., Knyukh, V.L., Nazarenko, V.M.: Three-dimensional axisymmetric problem of fracture in material with two discoidal cracks under compression along latter. Sov. Appl. Mech. **20**(11), 1003–1012 (1984)
50. Guz, A.N., Knyukh, V.L., Nazarenko, V.M.: Cleavage of composite materials in compression along internal and surface macrocracks. Sov. Appl. Mech. **22**(11), 1047–1051 (1986)
51. Guz, A.N., Knyukh, V.L., Nazarenko, V.M.: Fracture of ductile materials in compression along two parallel disk-shaped cracks. Sov. Appl. Mech. **24**(2), 112–117 (1988)
52. Guz, A.N., Knyukh, V.L., Nazarenko, V.M.: Compressive failure of material with two parallel cracks: small and large deformation. Theor. Appl. Fract. Mech. **11**(3), 213–223 (1989)
53. Guz, A.N., Nazarenko, V.M.: Symmetric failure of the halfspace with penny-shaped crack in compression. Theor. Appl. Fract. Mech. **3**(3), 233–245 (1985)
54. Guz, A.N., Nazarenko, V.M.: Fracture of a material in compression along a periodic system of parallel circular cracks. Sov. Appl. Mech. **23**(4), 371–377 (1987)

55. Guz, A.N., Nazarenko, V.M.: Fracture mechanics of material in compression along cracks (Review). Highly elastic materials. Sov. Appl. Mech. **25**(9), 851–876 (1989)

56. Guz, A.N., Nazarenko, V.M.: Fracture mechanics of materials under compression along cracks (survey). Structural materials. Sov. Appl. Mech. **25**(10), 959–972 (1989)

57. Guz, A.N., Nazarenko, V.M., Bogdanov, V.L.: Fracture under initial stresses acting along cracks: approach, concept and results. Theor. Appl. Fract. Mech. **48**, 285–303 (2007)

58. Guz, A.N., Nazarenko, V.M., Bogdanov, V.L.: Combined analysis of fracture under stress acting along cracks. Arch. Appl. Mech. **83**(9), 1273–1293 (2013)

59. Guz, A.N., Nazarenko, V.M., Khoma, Yu.I.: Failure of an infinite compressible composite containing a finite cylindrical crack in axial compression. Int. Appl. Mech. **31**(9), 695–703 (1995)

60. Guz, A.N., Nazarenko, V.M., Khoma, Yu.I.: Fracture of an infinite incompressible hyperelastic material under compression along a cylindrical crack. Int. Appl. Mech. **32**(5), 325–331 (1996)

61. Guz, A.N., Nazarenko, V.M., Nazarenko, S.M.: Fracture of composites under compression along periodically placed parallel circular stratifications. Sov. Appl. Mech. **25**(3), 215–221 (1989)

62. Guz, A.N., Nazarenko, V.M., Starodubtsev, I.P.: Planar problem of failure of structural materials in compression along two parallel cracks. Sov. Appl. Mech. **27**(4), 352–360 (1991)

63. Guz, A.N., Nazarenko, V.M., Starodubtsev, I.P.: On problems of fracture of materials in compression along two internal parallel cracks. Appl. Math. Mech. **18**(6), 517–528 (1997)

64. Guz, I.A.: Estimation of critical loading parameters for composites with imperfect layer contact. Int. Appl. Mech. **28**(5), 291–295 (1992)

65. Guz, I.A.: Computational schemes in three-dimensional stability theory (the piecewise-homogeneous model of a medium) for composites with cracks between layers. Int. Appl. Mech. **29**(4), 274–280 (1993)

66. Guz, I.A.: The strength of a composite formed by longitudinal–transverse stacking of orthotropic layers with a crack at the boundary. Int. Appl. Mech. **29**(11), 921–924 (1993)

67. Guz, I.A.: Investigation of the stability of a composite in compression along two parallel structural cracks at the layer interface. Int. Appl. Mech. **30**(11), 841–847 (1994)

68. Guz, I.A.: Problems of the stability of composite materials in compression along interlaminar cracks: periodic system of parallel macrocracks. Int. Appl. Mech. **31**(7), 551–557 (1995)

69. Guz, I.A.: Computer aided investigations of composites with various interlaminar cracks. ZAMM **76**(5), 189–190 (1996)

70. Guz, I.A.: Composites with interlaminar imperfections: Substantiation on the bounds for failure parameters in compression. Compos. B **29**(4), 343–350 (1998)

71. Guz, I.A.: On modelling of a failure mechanism for layered composites with interfacial cracks. ZAMM **78**(1), S429–S430 (1998)

72. Guz, I.A.: Compressive behaviour of metal matrix composites: Accuracy of homogenization. ZAMM **80**(2), S473–S474 (2000)

73. Guz, I.A.: The effect of the multi-axiality of compressive loading on the accuracy of a continuum model for layered materials. Int. J. Solids Struct. **42**, 439–453 (2005)

74. Guz, I.A., Guz, A.N.: Stability of two different half-planes in compression along interfacial cracks: analytical solutions. Int. Appl. Mech. **37**(7), 906–912 (2001)

75. Guz, I.A., Herrmann, K.P.: On the lower bounds for critical loads under large deformations in non-linear hyperelastic composites with imperfect interlaminar adhesion. Eur. J. Mech. A Solids **22**(6), 837–849 (2003)

76. Guz, I.A., Kokhanenko, Yu.V.: Stability of laminated composite material in compression along microcrack. Int. Appl. Mech. **29**(9), 702–708 (1993)

77. Knyukh, V.L.: Fracture of a material with two disk-shaped cracks in the case of axisymmetric deformation in compression along the cracks. Sov. Appl. Mech. **21**(3), 221–225 (1985)

78. Nazarenko, V.M.: Mutual effect of a circular surface crack and a free boundary in an axisymmetric problem of the fracture of an incompressible half space in compression along the crack plane. Sov. Appl. Mech. **21**(2), 133–137 (1985)

79. Nazarenko, V.M.: Plastic rupture of materials during compression along near-surface fractures. Sov. Appl. Mech. **21**(2), 133–137 (1986)
80. Nazarenko, V.M.: Two-dimensional problem of the fracture of materials in compression along surface cracks. Sov. Appl. Mech. **22**(10), 970–977 (1986)
81. Nazarenko, V.M.: Theory of fracture of materials in compression along near-surface cracks under plane-strain conditions. Sov. Appl. Mech. **22**(12), 1192–1199 (1986)
82. Nazarenko, V.M.: Fracture of plastic masses with translational strain-hardening in compression along near-surface cracks. Sov. Appl. Mech. **23**(1), 61–64 (1987)
83. Starodubtsev, I.P.: Fracture of a body in compression along two parallel cracks under plane-strain conditions. Sov. Appl. Mech. **24**(6), 604–607 (1988)
84. Winiarski, B., Guz, I.A.: The effect of cracks interaction on the critical strain in orthotropic heterogeneous material under compressive static loading. In: Proceedings of the 2006 ASME International Mechanical Engineering Congress & Exposition (IMECE 2006), p. 9. Chicago, USA, ASME, 5–10 Nov 2006
85. Winiarski, B., Guz, I.A.: Plane problem for layered composites with periodic array of interfacial cracks under compressive static loading. Int. J. Fract. **144**(2), 113–119 (2007)
86. Winiarski, B., Guz, I.A.: The effect of cracks interaction for transversely isotropic layered material under compressive loading. Finite Elem. Anal. Des. **44**(4), 197–213 (2008)
87. Winiarski, B., Guz, I.A.: The effect of fibre Volume fraction on the onset of fracture in laminar materials with an array of coplanar interface cracks. Compos. Sci. Technol. **68**(12), 2367–2375 (2008)
88. Winiarski, B., Guz, I.A.: The effect of cracks interaction in orthotropic layered materials under compressive loading. Phil. Trans. Royal Soc. A. **366**(1871), 1835–1839 (2008)
89. Guz, A.N.: Ob odnom kriterii razrusheniia tverdykh tel pri szhatii vdol treshchin. Ploskaia zadacha (Criterion of fracture of solids upon compression along cracks. Plane problem). Dokl. Akad. Nauk SSSR **259**(6), 1315–1318 (1981)
90. Guz, A.N.: On construction of mechanics of fracture of materials in compression along the cracks. In: Proceedings of ICF7: Advance in Fracture Research, vol. 6, pp. 3881–3892. Pergamon Press (1990)
91. Guz, A.N.: The study and analysis of non-classical problems of fracture and failure mechanics. In: Abstracts of IUTAM Symposium of nonlinear analysis of fracture, p. 19. Cambridge, 3–7 Sept 1995
92. Guz, A.N.: Non-classical problems of composite failure. In: Proceedings of ICCST/1, pp. 161–166. Durban, South. Africa, 18–20 June 1996
93. Guz, I.A.: On one mechanism of fracture of composites in compression along interlayer cracks. In: Proceedings of International Conference on Design and Manufacturing Using Composites, pp. 404–412. Montreal, Canada, 10–12 Aug 1994
94. Guz, I.A.: Stability of composites in compression along cracks. In: Proceedings of Enercomp 95, pp. 163–170. Technomic Publishing Co., Lancaster–Basel, Montreal, Canada, 8–10 May 1995
95. Guz, I.A.: Failure of layered composites with interface cracks. In: Proceedings of the 18th International Conference on Reinforced Plastics 95, pp. 175–182. Czech Republic, Karlovy Vary, 16–18 May 1995
96. Guz, I.A.: Stability and failure of layered composites with interface cracks. In: Proceedings of the International Conference on Computer Engineering and Science: Computational Mechanics, 95, pp. 2317–2322. Springer–Verlag, Hawaii, USA, July 30–3 Aug 1995
97. Guz, I.A.: Stability loss of composite materials with cracks between compressible elastic layers. In: Proceedings of the ECCM–7, Vol. 2, pp. 259–264. Woodhead Publishing Ltd., London, UK, 14–16 May 1996
98. Guz, I.A.: Composite structures in compression along parallel interfacial cracks. In: Proceedings of the ICCST/1, pp. 167–172. Durban, South Africa, 18–20 June 1996
99. Guz, I.A.: Analysis of a failure mechanism in compression of composites with various kinds of interface adhesion. In: Proceedings of EUROMAT 97, vol. 2, pp. 375–380. The Netherlands, 21–23 April 1997

100. Guz, I.A.: Modelling of fracture of composites in compression along layers. In: Proceedings of the 3rd International Conference, pp. 523–530. A. A. Balkema, Rotterdam, Dublin, Ireland, 3–5 Sept 1997

101. Guz, I.A.: Metal matrix composites in compression. Substantiation of the bounds. In: Proceedings of 5th International Conference on Automated Composites, pp. 387–393. Institute of Materials, London, Glasgow, UK, 4–5 Sept. 1997

102. Guz, I.A.: Composites with various interfacial defects. Bounds for critical parameters of instability in compression. In: Proceedings of DURACOSYS 97, pp. 7.51–7.54. Blacksburg, USA, 15–17 Sept 1997

103. Guz, I.A.: Instability in compression as a failure mechanism for layered composites with parallel interfacial cracks. In: Proceedings of the ICF 9: Advances in Fracture Research, vol. 2, pp. 1053–1060. Sydney, Australia (1997)

104. Guz, I.A.: On one fracture mechanism for composites with parallel interfacial cracks. In: Proceedings of the 4th International Conference on Deformation and Fracture of Composites, pp. 579–588. Institute of Materials, London, Manchester, UK, 24–26 Mar 1997

105. Guz, I.A.: On calculation of critical strains for periodical array of parallel interfacial cracks in layered materials. In: Proceedings of the 6th EPMESC Conference, pp. 375–380. Guang-Zhou, China, 4–7 Aug 1997

106. Guz, I.A.: On fracture of brittle matrix composites: compression along parallel interfacial cracks. In: Proceedings of the 5th International Symposium, pp. 391–400. Woodhead Publishing Ltd. Cambridge, Warsaw, Poland, 13–15 Oct 1997

107. Guz, I.A.: Numerical investigation on one mechanism of fracture for rock with parallel interlaminar cracks. In: Atluri, S.N., Yagawa, G. (eds.), Advances in Computer Engineering and Sciences, pp. 956–961. Tech. Science Press, Forsyth, USA (1997)

108. Guz, I.A.: Analysis of stability and failure in compression of composites with various kinds of interfacial defects. In: Proceedings of 6th Asia–Pacific Conference on Structural Engineering and Construction, vol. 2, pp. 1337–1342. Taipei, Taiwan, 14–16 Jan 1998

109. Guz, I.A.: On asymptotic accuracy of the theory of plastic fracture in compression for layered materials. In: Proceedings of EUROMECH Coll. 378: Nonlocal Aspects in Solid Mechanics, pp. 118–123. Mulhouse, France, 20–22 April 1998

110. Guz, I.A.: Composites with various kinds of interfacial adhesion: compression along layers. In: Proceedings of ECCM–8, vol. 4, pp. 677–683. Woodhead Publishing Ltd., Naples, Italy, 3–6 June 1998

111. Guz, I.A.: On continuum approximation in compressive fracture theory for metal matrix composites: Asymptotic accuracy. In: Proceedings of ICCST/2, pp. 501–506. Durban, South Africa, 9–11 June 1998

112. Guz, I.A.: Investigation of accuracy of continuum fracture theory for piecewise-homogeneous medium. In: Proceedings of ICNM-III, pp. 224–227. Shanghai University Press, Shanghai, China, 17–20 Aug 1998

113. Guz, I.A.: On two approaches to compressive fracture problems. In: Proceedings of 12th European Conference on Fracture, vol. 3, pp. 1447–1452. EMAS Publishing, Sheffield, UK, 14–16 Sept 1998

114. Guz, I.A.: Asymptotic analysis of fracture theory for layered rocks in compression. In: Modelling and Simulation Based Engineering, vol. 1, pp. 375–380. Tech. Science Press, Palmdale, USA (1998)

115. Guz, I.A.: On calculation of accuracy for continuum fracture theory of metal matrix composites in compression. In: Proceedings of ICAC 96, pp. 757–764. Hurghada, Egypt, 15–18 Dec. 1998

116. Guz, I.A.: On estimation of critical loads for rocks in compression: 3-D approach. In: Proceedings of ARCOM'99, vol. 2, pp. 847–852. Elsevier, Singapore, 15–17 Dec., 1999

117. Guz, I.A.: Bounds for critical parameters in the stability theory of piecewise-homogeneous media: Laminated rocks. In: Proceedings of SASAM 2000, pp. 479–484. Durban, South Africa, 11–13 Jan 2000

118. Guz, A.N.: O neklassicheskikh problemakh mekhaniki razrusheniia (On non-classical problems of fracture mechanics). Fiziko-himicheskaya Mekhanika Materialov **29**(3), 86–97 (1993)

119. Guz, A.N.: Study and Analysis of Non-classical Problems of Fracture and Failure Mechanics and Corresponding Mechanisms. Lecture presented at Institute of Mechanics, HANOI (1998)
120. Guz, A.N.: Description and study of some nonclassical problems of fracture mechanics and related mechanisms. Int. Appl. Mech. 36(12), 1537–1564 (2000)
121. Guz, A.N.: On study of Nonclassical Problems of Fracture and Failure Mechanics and Related Mechanisms, pp. 35–68. ANNALS of the European Academy of Sciences, Liège, Belgium (2006–2007)
122. Guz, A.N.: On study of nonclassical problems of fracture and failure mechanics and related mechanisms. Int. Appl. Mech. 45(1), 1–31 (2009)
123. Guz, A.N.: Mekhanika khrupkogo razrusheniia materialov s nachalnymi napriazheniiami (Mechanics of Brittle Fracture of Materials with Initial Stresses). Naukova Dumka, Kyiv (1983)
124. Guz, A.N.: Mekhanika razrusheniia kompozitnykh materialov pri szhatii (Fracture Mechanics of Composite Materials under Compression). Naukova Dumka, Kyiv (1990)
125. Guz, A.N. (ed.): Neklassicheskie problemy mekhaniki razrusheniya, v 4 tomah, 5 knigah (Non-Classical Problems of Fracture Mechanics, in 4 volumes, 5 books). Naukova Dumka, Kyiv (1990–1993)
126. Guz, A.N. (ed.): Mekhanika kompozitov, v 12 tomah (Mechanics of Composites, in 12 volumes), vol. 1–4, Naukova Dumka, vol. 5–12. A.S.K., Kyiv (1993–2003), Golovach, V.T. (ed.): T.1. Statika materialov. (Statics of Materials, vol. 1) (1993), Shulga, N.A. (ed.): T.2. Dinamika i ustoichivost materialov. (Dynamics and Stability of Materials, vol. 2) (1993), Khoroshun, L.P., (ed.): T.3. Statisticheskaia mekhanika i effektivnye svoistva materialov (Statistical Mechanics and Effective Material Properties, vol. 3) (1993), Guz, A.N., Akbarov, S.D., (eds.) T.4. Mekhanika materialov s iskrivlennymi strukturami (Mechanics of Materials with Curved Structures, vol. 4) (1995), Kaminsky, A.A., (ed.): T.5. Mekhanika razrusheniia. (Fracture Mechanics, vol. 5) (1996), Shulga, N.A., Tomashevsky, V.T., (eds.): T.6. Tekhnologicheskie napriazheniia i deformatsii v materialakh. (Technological Stresses and Deformations in Materials, vol. 6) (1997), Guz, A.N., Kosmodamiansky, A.S., Shevchenko, V.P., (eds.): T.7. Kontsentratsiia napriazhenii. (Stress Concentration, vol. 7) (1998), Grigorenko, Ya.M., (ed.): T.8. Statika elementov konstruktsii. (Structural Elements Statics, vol. 8) (1999), Kubenko, V.D., (ed.): T.9. Dinamika elementov konstruktsii. (Dynamics of Structural Elements, vol. 9) (1999), Babich, I.Yu. (ed.): T.10. Ustoichivost elementov konstruktsii. (Stability of Structural Elements, vol. 10) (2001), Grigorenko, Ya.M., Shevchenko, Yu.N. (eds.): T.11. Chislennye metody. (Numerical Methods, vol. 11) (2002), Guz, A.N., Khoroshun, L.P. (ed.): T.12. Prikladnye issledovaniia. (Applied Research, vol. 12) (2003)
127. Guz, A.N.: Osnovy mekhaniki razrusheniia kompozitov pri szhatii: V 2-kh tomakh (Fundamentals of the Fracture Mechanics of Composites under Compression: In 2 volumes). Litera, Kyiv (2008), T. 1. Razrushenie v strukture materiala. (Fracture in Structure of Materials), T. 2. Rodstvennye mekhanizmy razrusheniia. (Related Mechanisms of Fracture)
128. Bogdanov, V.L., Guz, A.N., Nazarenko, V.M.: Obieedinennyi podkhod v neklassicheskikh problemakh mekhaniki razrusheniia (A Unified Approach in Non-Classical Problems of Fracture Mechanics). LAP LAMBERT Academic Publishing, Saarbrücken, Deutschland (2017)
129. Guz, A.N., Bogdanov, V.L., Nazarenko, V.M.: Fracture of Materials under Compression along Cracks. Springer Nature Switzerland AG (2020)
130. Obreimoff, I.W.: The splitting strength of mica. Proc. Soc. Lond. A. 127A, 290–297 (1930)
131. Yentov, V.M., Salganik, R.L.: O balochnom priblizhenii v teorii treschin (Beam approximation in crack theory). Izvestiya Akad. Nauk SSSR, Mech. 5, 95–102 (1965)
132. Mikhailov, A.M.: Dinamicheskie zadachi teorii treschin v balochnom priblizhenii (Dynamic problems of crack theory in the beam approximation). J. Prikladnoy Mekhaniki Tekhnicheskoy Fiziki 5, 167–172 (1966)
133. Guz, A.N.: Ustoichivost trekhmernykh deformiruemykh tel (Stability of Three-Dimensional Deformable Bodies). Naukova Dumka, Kyiv (1971)

134. Guz, A.N.: Ustoichivost uprugikh tel pri konechnykh deformatsiiakh (Stability of Elastic Bodies under Finite Deformations). Naukova Dumka, Kyiv (1973)
135. Guz, A.N.: Osnovy teorii ustoichivosti gornykh vyrabotok (Fundamentals of the Theory of Stability of Mine Workings). Naukova Dumka, Kyiv (1977)
136. Guz, A.N.: Ustoichivost uprugikh tel pri vsestoronnem szhatii (Stability of Elastic Bodies under All-Round Compression). Naukova Dumka, Kyiv (1979)
137. Guz, A.N.: Osnovy trekhmernoi teorii ustoichivosti deformiruemykh tel (Fundamentals of the Three-Dimensional Theory of Stability of Deformable Bodies). Vyshcha Shkola, Kyiv (1986)
138. Guz, A.N.: Fundamentals of the Three-Dimensional Theory of Stability of Deformable Bodies. Springer, Berlin Hiedelberg New York (1999)
139. Wu, C.H.: Plane-strain buckling of a crack in harmonic solid subjected to crack-parallel compression. J. Appl. Mech. **46**, 597–604 (1979)
140. Wu, C.H.: Plane strain buckling of a crack in incompressible elastic solids. J. Elast. **10**(2), 161–177 (1980)
141. Kokhanenko, Yu.V.: Numerical solution of problems of the theory of elasticity and the three-dimensional stability of piecewise-homogeneous media. Sov. Appl. Mech. **22**(11), 1052–1058 (1986)
142. Dovzhyk, M.V.: Razrushenie poluprostranstva pri szhatii vdol pripoverhnostnoy diskoo-braznoy treschiny dlya malykh rasstoyaniy mezhdu svobodnoy poverhnostu i treschinoy (Fracture of a half-space compressed along a penny-shaped crack located at a short distance from the surface). Prikladnaya Mekhanika **48**(3), 79–88 (2012)
143. Akbarov, S.D.: On the three-dimensional stability loss problems of elements of constructions fabricated from the viscoelastic composite materials. Mech. Compos. Mater. **34**(6), 537–544 (1998)
144. Akbarov, S.D.: Three-dimensional stability loss problems of viscoelastic composite materials and structural members. Int. Appl. Mech. **43**(10), 1069–1089 (2007)
145. Akbarov, S.D., Cilli, A., Guz, A.N.: The theoretical strength limit in compression of viscoelastic layered composite materials. Compos. B Eng. **30**(5), 365–372 (1999)
146. Akbarov, S.D., Yahnioglu, N.: The method for investigation of the general theory of stability problems of structural elements fabricated from the viscoelastic composite materials. Compos. B Eng. **32**(5), 475–482 (2001)
147. Rzayev, O.G., Akbarov, S.D.: Local buckling of the elastic and viscoelastic coating around the penny-shaped interface crack. Int. J. Eng. Sci. **40**, 1435–1451 (2002)
148. Akbarov, S.D.: Stability Loss and Buckling Delamination. Springer, Berlln (2012)
149. Akbarov, S.D., Guz, A.N.: Mechanics of Curved Composites. Kluwer Academic Publisher, Dordrecht Boston London (2000)
150. Hoff, N.J.: Buckling and stability. J. R. Aeronaut. Soc. **58**(1), 1–11 (1954)
151. Akbarov, S.D., Guz, A.N.: Mechanics of curved composites and some related problems for structural members. Mech. Adv. Mater. Struc. **11**(6), 445–515 (2004)
152. Schapery, R.A.: Approximate methods of transform inversion for viscoelastic stress analyses. Proc. U.S. Nat. Congr. Appl. ASME. **4**, 1075–1085 (1966)

Chapter 9
Problem 7. Brittle Fracture of Materials with Cracks Under Action of Dynamic Loads (with Allowance for Contact Interaction of the Crack Edges)

In this chapter, in a very short form (in comparison with **Problems 4–6**, which have already been discussed in Part III of this monograph), the results on **Problem 7** are presented, which were obtained in the Department of Dynamics and Stability of Continua of the S. P. Timoshenko Institute of Mechanics of the NASU. The presentation of these results is made in the style announced in the Introduction (Part I) to this monograph (without excessively invoking the aspects of a mathematical nature).

9.1 Introduction

The main results on this problem, obtained in the Department of Dynamics and Stability of Continua, are presented in the monograph [1] (vol. 4, book 2) of 1993, the review articles [2–4], which are completely devoted to the analysis of results on **Problem 7**, and the review articles [5–9], which are devoted to the analysis of results on the various non-classical problems of fracture mechanics, including **Problem 7**. Initially, these results were presented in the publications [10–53] and several other publications and reports at the international conferences since 1990–1991.

On the considered scientific direction of the Department of Dynamics and Stability of Continua, three dissertations for the degree of Doctor of Physical and Mathematical Sciences (DSc) were prepared and defended: V. V. Zozulya, A. V. Menshikov, and V. A. Menshikov.

Note 9.1 The dissertation of V. A. Menshikov is devoted to the dynamical spatial problems of the brittle fracture mechanics for material with cracks, located at the interface between two materials, without taking into account the contact interaction of the crack edges. Nevertheless, the numerical methods developed in it can be used to construct a research algorithm taking into account the contact interaction of the

A. N. Guz, *Eight Non-Classical Problems of Fracture Mechanics*,
Advanced Structured Materials 159,
https://doi.org/10.1007/978-3-030-77501-8_9

crack edges, which gave them grounds to refer this dissertation to the considered scientific direction. An overview of the results on the dynamical spatial problems for cracks located at the interface is presented in the review article [54] for 2013. As an example, the article [55] is included in the list of references to this monograph on the subject of the survey [54].

It is advisable to note that, starting with the classical results on the diffraction of elastic waves on the plane cracks, all the corresponding publications did not take into account the contact interaction of the crack edges. In the subsequent years, the above research results were transferred to the dynamical brittle fracture mechanics.

It would seem that fracture mechanics has always been characterized by an in-depth study of the phenomena occurring near the crack and, especially, near the tip of the crack.

Unfortunately, this did not happen for studies on the dynamical brittle fracture mechanics.

In the dynamical mechanics of brittle fracture, the contact interaction of the crack edges *was also not taken into account until the publications of the Department of Dynamics and Stability of Continua* indicated at the beginning of this Introduction of Sect. 9.1. The above situation is confirmed by the fact that in the well-known mono-graphs [56] of 1985 and [57] of 1988, devoted entirely to the dynamical mechanics of brittle fracture (even judging by their name), the contact interaction of the crack edges *was not considered or analyzed.* Also, a similar situation takes place in the well-known monograph (Elastodynamic crack problems/Eds. G. G. Sih, Leyden: Noordhoff, 1977. 423 p.). In 1994, the review [58] of publications on the dynamical mechanics of brittle fracture was published without taking into account the contact interaction of the crack edges. At present, the publication of such articles continues, even without comments on the incomplete accounting of phenomena near the cracks. As an example, we can point to the article [59] of 2004 and several others.

Below, in this chapter, the substantiation of the formulation of the problems of dynamical brittle fracture mechanics, *taking into account the contact interaction of crack edges*, is presented in a brief form, and the statement of these problems, the developed research method, and, in the form of examples, several results on solving the specific problems *with taking into account the contact interaction of the crack edges* and analyzing the corresponding effects.

9.2   Substantiation of Statement of Problems. Method of Solving

To substantiate the statement of the problems, an analysis of the formulation of the simplest problem of the traditional dynamical mechanics of brittle fracture is carried out, which in a certain sense is a reference and is given in almost all monographical publications on the dynamical mechanics of brittle fracture.

9.2.1 Substantiation of the Discussed Problem Statement

Consider the case when the material is modeled by the linearly elastic isotropic body. The analysis will be carried out using the Cartesian coordinates $x_j (j = 1, 2, 3)$.

Consider the body as infinite, in which in the plane $x_2 = 0$ (Fig. 9.1), there is the infinite along the axis $0x_3$ (the axis $0x_3$ is directed perpendicular to the plane of Fig. 9.1) plane crack of constant width $2a$.

Assume also that a dynamic load is applied, which does not depend on the coordinate x_3. In this case, the plane problem can be considered in the plane $x_1 0x_2$ (Fig. 9.1) for an infinite body with the crack of width $2a$ along the axis $0x_1$ (Fig. 9.1). Let us analyze the formulation of problems when the plane harmonic longitudinal extension wave propagates along the axis $0x_2$ (Fig. 9.1). In this case, all quantities (stress and displacements) taking into account the phenomenon of the wave diffraction have a factor $(\exp i\omega t)$, including near the crack tip.

Consider the phenomena that occur in this case, for example, near the right tip of the crack (Fig. 9.2a). Under the action of the load, near the right tip of the crack, as in the whole body, three characteristic positions arise:

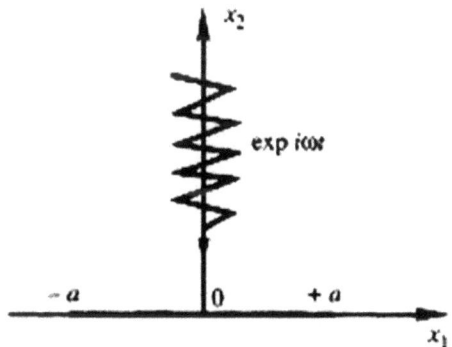

Fig. 9.1

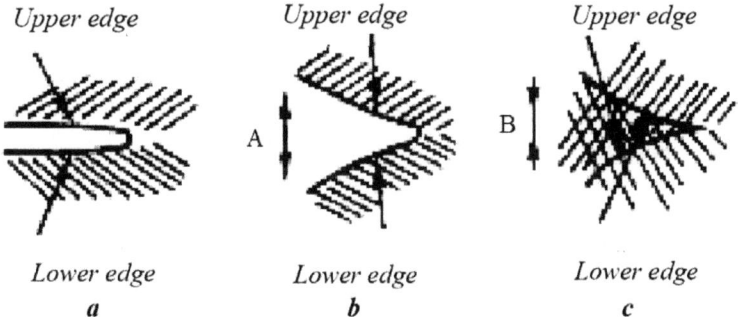

Fig. 9.2

1. position at rest (Fig. 9.2a);
2. position in the tension phase corresponding to the maximum crack opening (Fig. 9.2b);
3. position in the compression phase corresponding to the maximum crack closing (Fig. 9.2c).

It should be noted that position "3" occurs in the time of half a period after the occurrence of position "2". In Fig. 9.2c, A denotes the opening of the crack edges at a point near the right tip of the crack in the extension phase.

In all publications on the classical dynamical mechanics of fracture, **the boundary conditions are set on the crack edges that do not take into account the phenomena occurring in the compression phase, while the contact interaction is not taken into account and the crack edges allow as if the "free penetration" into each other.**

This situation, taken in the classical dynamical fracture mechanics in the compression phase, is schematically shown in Fig. 9.2c. In Fig. 9.2c, the crack edges seem to have changed places—the upper end passes into the position of the lower end, and the lower edge passes into the position of the upper edge. In Fig. 9.2c, through B, the value of "mutual penetration" of the crack edges (at the same point as in Fig. 9.2b) near the right tip of the crack in the compression phase is denoted. The situation shown in Fig. 9.2a–c occurs in all publications on the dynamical fracture mechanics when the boundary conditions are set corresponding to an unloaded crack. From the above and Fig. 9.2b, c, the equality follows

$$A = B. \tag{9.1}$$

Thus, in the classical dynamical fracture mechanics, the phenomena in the extension phase of the same order are *correctly* taken into account as those *incorrectly* taken into account in the compression phase. Therefore, the results of the classical dynamical fracture mechanics *include the unremovable error independent of the intensity of the applied load.*

It should also be noted that the approximate design scheme in the compression phase shown in Fig. 9.2c and used in fact in all publications on classical dynamic fracture mechanics indicates that the material is provided, as it were, with "additional inertia" when the contact interaction of the crack edges is not taken into account. In this regard, it should be expected that in the publications on the classical dynamical fracture mechanics, the overestimated values of the maximum values should be obtained, including also the information on the stress intensity factors.

General Conclusion In connection with the mentioned change in time of the sign of the displacements of the crack edges, which is characteristic of the physics of phenomena arising in the dynamical fracture mechanics, *it is necessary (with the correct and rigorous statement of the problems of the dynamical fracture mechanics) to take into account the contact interaction of the crack edges.*

The indicated contact interaction of the crack edges arises *apart from of the level of the acting load* and takes place, of course, when applying also the linear equations of the mechanics of deformable bodies, which describe the laws of wave propagation. Naturally, this phenomenon occurs when modeling the crack with the mathematical cut. Note that this simulation is used exclusively in all publications on dynamical fracture mechanics. The neglect to take into account the contact interaction of crack edges as applied to the dynamical fracture mechanics leads to *the non-removable errors* since the contact interaction is a necessary moment determined by the physics of phenomena.

In this connection, apparently, *all the previously obtained results on the dynamical fracture mechanics can be considered* **as inconsistent with the physics of the phenomenon under consideration**.

Of course, we are talking about the results obtained in the framework of the mechanics of a deformable body. Such a serious conclusion becomes obvious after considering above the statement of the simplest reference problem of the classical dynamical brittle fracture mechanics, which, as noted above, is given in all monographic publications on the dynamical brittle fracture mechanics.

In connection with the above **General Conclusion**, *it seems appropriate*, at least, according to the author of this monograph, *to consider* **all the problems** *of the dynamic mechanics of brittle fracture taking into account the contact interaction of crack edges, since it corresponds to* **the physics of phenomena**. *Only the concrete results obtained with allowance for the contact interaction of the crack edges* allow one to evaluate the effect of this phenomenon in certain classes of problems.

Proceeding from considerations of a physical nature, as noted above before the **General Conclusion**, one can expect that the classical dynamical mechanics of brittle fracture, which neglects tacitly the discussed physical phenomenon, should receive the overestimated values of the maximum values, including also information on the stress intensity factors.

9.2.2   On Research Method

Taking into account the above **General Conclusion** and comments to it, the construction of the dynamical mechanics of brittle fracture for materials with cracks, taking into account the contact interaction of the crack edges, was proposed as the topic of his doctoral dissertation when V. V. Zozulya entered the doctoral program of the S. P. Timoshenko Institute of Mechanics of the NASU in 1989. This moment can be considered the beginning of the development of **Problem 7** *(Brittle fracture of materials with cracks under the action of dynamic loads (with allowance for the contact interaction of crack ends))* in the Department of Dynamics and Stability of Continua.

On the topics discussed, the first articles of the Department of Dynamics and Stability of Continua [10–12, 14–19, 26] were published in 1990–1992, and the authors of these articles were V. V. Zozulya and, to a certain extent, the author of this monograph. As it turned out for all the past years, the above articles were the **first** *articles in the world scientific literature on the dynamical mechanics of brittle* *fracture of materials taking into account the contact interaction of crack edges.* In these articles, the statement of the indicated problems, the developed research method, and several results for the plane problems were presented. In more detail, all information with the corresponding aspects of a mathematical nature is presented in the monograph [1] (vol. 4, book 2) of 1993.

Naturally, the problems considered in the above statement are formulated as *nonlinear*, since the sizes of the contact zone are determined by the value of the displacements of the crack edges and other quantities. The resulting problem in the simplest case consists of *the linear equations and nonlinear boundary conditions.*

To solve such problems, the step-by-step methods are used, when at each step a problem with a fixed contact zone is obtained. This is a commonly accepted approach for solving the problems with a contact zone, the dimensions of which are determined from solving the same problem (nonlinear problem).

Specifically, in the above studies on the construction of the dynamical fracture mechanics taking into account the contact interaction of the crack edges, a solution method was used based on the reduction to the boundary integral equations with constraints in the form of inequalities on the crack edges. Later, the method of boundary elements with a step in time scheme and the corresponding iterative process were used.

The specific results were obtained for the plane cracks under the action of harmonical dynamical loads. It should be noted that this problem *should be attributed to the non-classical problems of fracture mechanics* because in its study *the change in the configuration of the body at the crack tip is taken into account—the contact interaction of the crack edges is taken into account.* Besides, the non-classical nature of this problem also lies in the fact that *a physical phenomenon is taken into account, which was not taken into account by other authors, and the need to consider which quite obviously follows from the considerations outlined in* Sect. 9.2.1.

9.3 Concrete Results

In this section, several specific results on the dynamic mechanics of brittle fracture of materials with cracks are briefly considered, which were obtained following the statement of problems and the research method of the previous section, *taking into account the contact interaction of the crack edges.*

The above specific results refer to **Problem 7** (*Brittle fracture of cracked materials under dynamic loads (with allowance for the contact interaction of crack ends)*), which is discussed in the present chapter. This brief information on the specific results will be presented separately for the two-dimensional problems in the spatial variables (plane, anti-plane, and other problems) and the three-dimensional problems in the spatial variables (spatial problems). These results, along with other related results, are set out in the publications listed in the Introduction of Sect. 9.1.

9.3.1 Two-Dimensional Problems

First of all, consider the specific results obtained within the framework of the classical dynamical mechanics of brittle fracture (**without taking into account** *the contact interaction of the crack edges*) and within the approaches of **Problem 7** (statement and method of Sect. 9.2, **taking into account** *the contact interaction of the crack edges*), as applied to the simplest reference problem, the design scheme for which is shown in Fig. 9.1. Additional information and designations are presented at the beginning of subsection 9.2.1.

The specific results concerning the discussed problem are presented in Fig. 9.3 in the form of dependence K_I^* (the dimensionless value of the mode 1 stress intensity factor from the dimensionless frequency $a\omega c_1^{-1}$ (the dimensionless wavenumber)), where the velocity of the longitudinal wave is denoted through c_1. In this case, the value K_I^* is determined by the following expression

$$K_I^* = \left(K_I^{\text{stat}}\right)^{-1} K_I^{\text{dyn}}. \tag{9.2}$$

In (9.2) and below in this chapter, the following designations are introduced:

K_I^{dyn}—the amplitude value (without a time factor of the type $\exp i\omega t$) of the mode 1 stress intensity factor of normal separation *under dynamic loading*;
K_I^{stat}—the mode 1 stress intensity factor *under the corresponding static loading*.

In Fig. 9.3, curve 1 corresponds to the dynamical mechanics of brittle fracture **without taking into account** *the contact interaction of the crack edges*, and curve 2 refers to the dynamical mechanics of brittle fracture **taking into account** *the contact interaction of the crack edges* (statement and method of Sect. 9.2, **Problem 7**).

From the analysis of the results in Fig. 9.3, two conclusions follow:

Conclusion 9.1 **Taking into account** the contact interaction of the crack edges, the maximum value of the quantity K_I^* (9.2) *decreases*.

Conclusion 9.2 **Taking into account** the contact interaction of the crack edges, the maximum value of the quantity K_I^* (9.2) is achieved at a different value of the

Fig. 9.3

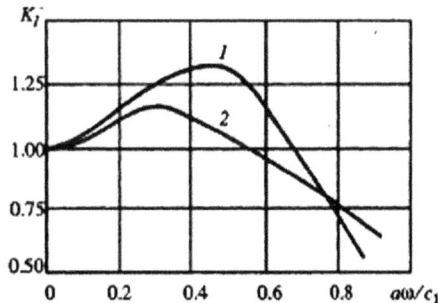

dimensionless frequency as compared with the classical dynamical mechanics of brittle fracture.

It is advisable to note that Conclusion 9.1 confirms the physical considerations (the last 8 lines before the **General Conclusion** of subsection 9.2.1 and the last few lines before subsection 9.2.2), which were formulated *before carrying out* the rigorous studies *taking into account the contact interaction of the crack edges.*

The results presented above in subsection 9.3.1 are presented in sufficient detail in Chap. 7 of the monograph [1] (vol. 4, book 2).

Below in this subsection, some problems of the dynamic mechanics of brittle fracture are briefly indicated, which were studied following the statement and method described in Sect. 9.2, *taking into account the contact interaction of the crack edges,* which determines their belonging to **Problem 7**.

The problem for two identical parallel plane cracks, displaced relative to each other when the harmonic longitudinal wave falls at an arbitrary angle (incidence angle) is analyzed. These studies were carried out for an infinite material (the reflected waves from the external boundary of the material are not taken into account). The results of a quantitative nature in the form of graphical dependencies are shown in 12 figures on pages 192–193 of the monograph [1] (vol. 4, book 2).

The problem for one crack (Fig. 9.1) under polyharmonic loading, corresponding to the incidence along the axis $0x_2$ of two longitudinal or two shear waves, the frequency in one of which differs twice from the other. The studies were carried out for an infinite material (reflected waves from the external boundary of the material are not taken into account). The quantitative results are shown in the form of graphical dependencies of the form shown in Fig. 9.3. At that, the curves of type 2 in Fig. 9.3 must be several instead of one curve of type 2 in Fig. 9.3. It should be noted that due to the nonlinear nature of problems in **Problem 7** (linear equations, but nonlinear boundary conditions), the interaction occurs under polyharmonic loading.

The results of the study of other two-dimensional (in the spatial variables) problems related to **Problem 7** (*taking into account the contact interaction of the crack edges*) can be found, using the list of publications at the beginning of the Introduction in Sect. 9.1.

9.3.2   Three-Dimensional (Spatial) Problems

As in the previous section concerning the two-dimensional (in the spatial variables), and this section concerning the three-dimensional (in the spatial variables) problems for the case of the simplest reference problems, the specific results are first of all considered which are obtained in the framework of the classical dynamical mechanics of brittle fracture (**without taking into account** *the contact interaction of the crack edges*) and within the framework of the approaches of **Problem 7** (statement and method of Sect. 9.2, **taking into account** *the contact interaction of the crack faces*) and **compare them**.

Fig. 9.4

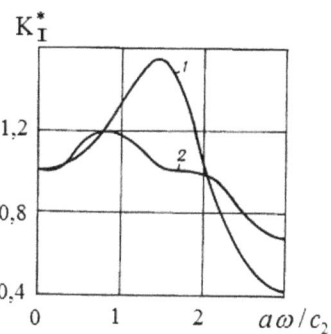

Example 9.1 So, the specific results are presented in Fig. 9.4 for the following simplest reference problem. An infinite material is considered (the reflected waves from the external boundaries of the area occupied by the material are not taken into account), in which in a plane $x_1 0 x_2$ the crack in the form of a circular disk (penny-shaped crack) of the radius a is located. A harmonic longitudinal wave with frequency ω is incident along the axis $0x_3$ (perpendicular to the crack). This situation is similar to the situation shown in Fig. 9.1 for the case of the two-dimensional problem (plane deformation in the plane $x_1 0 x_2$).

In Fig. 9.4 for the discussed spatial problem, the dependence of the quantity K_I^*—the dimensionless value of the mode I stress intensity factor, determined by expression (9.2), on the dimensionless frequency $a\omega c_2^{-1}$ (the dimensionless wavenumber), where through c_2 the transverse wave velocity is denoted.

It should be noted that the horizontal axis in Fig. 9.3 shows a dimensionless quantity $a\omega c_1^{-1}$, and Figs. 9.4 and 9.5 show a dimensionless quantity $a\omega c_2^{-1}$, which corresponds to the primary sources [1] (vol. 4, book 2), [35] and [30].

In Fig. 9.4, the curve 1 (as in Fig. 9.3) corresponds to the classical dynamical mechanics of brittle fracture (**without taking into account** *the contact interaction of the crack edges*), and curve 2 refers to the dynamical mechanics of brittle fracture **taking into account** *the contact interaction of the crack edges* (statement and method of Sect. 9.2, **Problem 7**).

Fig. 9.5

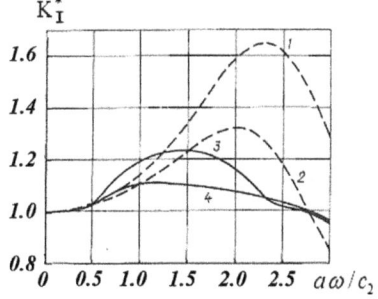

Conclusions 9.1 and 9.2 presented in subsection 9.3.1 for the two-dimensional problems, as well as the physical considerations also presented in subsection 9.3.1, follow from the analysis of the results in Fig. 9.4. But also Conclusion 9.3 follows from the analysis of the results in Figs. 9.3 and 9.4, which *is not the case* for the two-dimensional problems.

Conclusion 9.3 Unlike the two-dimensional problems (Fig. 9.3) in the three-dimensional problems (Fig. 9.4) along the horizontal axis from "0" to the value corresponding to *max* in curve 1, a small interval exists in which the results on curve 2 are higher than the results on curve 1.

Example 9.2 Figure 9.5 presents the specific results for the next more complex reference problem. An infinite material is considered (the reflected waves from the external boundaries of the area occupied by the material are not taken into account), in which in the plane $x_1 O x_2$ the crack in the form of an elliptical disk is located in the plane, where a and b are the major and minor semiaxes of the ellipse. A harmonic longitudinal wave with frequency ω is incident along the axis $0 x_3$ (perpendicular to the crack). The above situation is similar to the situation shown in Fig. 9.1 for the case of the two-dimensional problem (plane deformation in the plane $x_1 O x_2$) and the situation in Example 9.1 of this subsection.

In Fig. 9.5, for this spatial problem, the dependence of the quantity K_I^*—the dimensionless value of the mode I stress intensity factor, determined by the expression (9.2), on the dimensionless frequency $a \omega c_2^{-1}$ (the dimensionless wavenumber) is presented, where c_2 is the transverse wave velocity and a is the semimajor axis of the elliptical crack.

The results in Fig. 9.5 are presented as the dashed curves marked with numbers 1 and 2 and the solid curves marked with numbers 3 and 4. The results on curves 1 and 3 refer to the vertex of the semiminor axis of the ellipse, and the results on curves 2 and 4 3 refer to the vertex of the semimajor axis. The dashed curves 1 and 2 correspond to the results of the classical dynamical brittle fracture mechanics (**without taking into account** *the contact interaction of the crack edges*), the solid curves 3 and 4 correspond to the dynamical brittle fracture mechanics **taking into account** *the contact interaction of the crack edges* (statement and method of Sect. 9.2, **Problem 7**).

Conclusions 9.1 and 9.2, formulated in subsection 9.3.1, and Conclusion 9.3, formulated in Example 1 of this subsection, follow from the analysis of the results in Fig. 9.5. At that, from the analysis of the results of the article [30] and the results presented in Fig. 9.5, Conclusion 9.4 follows, which corresponds to studies **without and taking into account** the contact interaction of the crack edges.

Conclusion 9.4 At the normal incidence of a harmonic longitudinal wave on a plane crack in the form of an elliptical disk, the maximum quantity (9.2) K_I^* arises in the neighborhood of the continuation of the semiminor axis and the minimum of quantity K_I^* (9.2) appears in the neighborhood of the continuation of the semimajor axis of the elliptical cracks.

It is advisable to note that in the above two examples, in comparative detail from the point of view of the style of this monograph, brief information is given on the specific two spatial problems of the dynamical mechanics of brittle fracture (the incidence of a harmonic longitudinal wave on a plane crack in the form of a circular disk (the penny-shaped crack)—Example 9.1 or on a plane crack in the form of an elliptical disk—Example 9.2), which were investigated **taking into account** *the contact interaction of the crack edges* (statement and method of Sect. 9.2, **Problem 7**).

The list of references on the dynamical mechanics of brittle fracture of materials **taking into account** *the contact interaction of crack edges* (statement and method of Sect. 9.2, **Problem 7**), which is given in the Introduction (Sect. 9.1) to this chapter, contains other publications related to the spatial problems, which were published on the discussed topics of the Department of Dynamics and Stability of Continua of the S. P. Timoshenko Institute of Mechanics of the NASU. Due to the certain limited volume of Part III, it is not possible to provide brief information on the above publications. In this regard, below in this section, only a list of spatial problems is indicated that have been studied in the discussed statement of the dynamical mechanics of brittle fracture of materials, **taking into account** *the contact interaction of crack edges.*

The problem for a material with a penny-shaped crack when the combined harmonic longitudinal and transverse waves incident at an arbitrary angle to the crack plane. The studies are carried out for an infinite material (reflected waves from the external boundary of the area occupied by the material are not taken into account). The specific results are given in the form of graphical dependencies of the various quantities.

The problem for a material with a crack in the form of an elliptical disk at the incidence of a harmonic transverse wave propagating perpendicular to the plane of the crack and at an arbitrary angle to the major axis of the ellipse. The studies were carried out for an infinite material (the reflected waves from the external boundary of the area occupied by the material are not taken into account). The specific results are given in the form of graphical dependencies of the various quantities.

The problem for material with two identical cracks in the form of a circular disk (the penny-shaped crack), which are located in the same plane or in two parallel planes (these are two different problems), with the normal incidence of a harmonic longitudinal wave. The studies were carried out for an infinite material (the reflected waves from the external boundary of the area occupied by the material are not taken into account). Some specific results are given in the form of graphical dependencies of the various quantities. Obtaining some specific results and especially their analysis, at least, in the opinion of the author of this monograph, it is advisable to continue.

It should also be noted that the doctoral dissertation (Doctor of Sciences) of A. V. Menshikov, indicated in the Introduction of Sect. 9.1, is also devoted to the spatial problems of the dynamical mechanics of brittle fracture of materials, **taking into account** *the contact interaction of crack edges* (statement and method of Sect. 9.2, **Problem 7**).

The above brief information on **Problem 7** *(Brittle fracture of materials with cracks under the action of dynamic loads (with allowance for the contact interaction of the crack edges))* will be limited in this monograph.

It should be also noted in conclusion to this chapters, that *the first publications in the world scientific literature* on the study of problems of dynamical fracture mechanics **taking into account** *the contact interaction of crack edges* were the articles of 1990–1992. [10–12, 14–19, 26] of V. V. Zozulya and the author of this monograph.

References

1. Guz, A.N., (ed.): Neklassicheskie problemy mekhaniki razrusheniya, v 4 tomah, 5 knigah (Non-classical problems of fracture mechanics, in 4 volumes, 5 books). Naukova Dumka, Kyiv (1990−1993), Kaminsky, A.A. (ed.): T.1. Razrushenie viazkouprugikh tel s treshchinami (Fracture of Viscoelastic Bodies with Cracks, vol. 1) (1990), Guz, A.N. (ed.): T.2. Khrupkoe razrushenie materialov s nachalnymi napriazheniiami, (Brittle Fracture of Materials with Initial Stresses, vol. 2) (1991), Kaminsky, A.A., Gavrilov, D.N. (eds.): T.3. Dlitelnoe razrushenie polimernykh i kompozitnykh materialov s treshchinami (Long-Term Fracture of Polymer and Composite Materials with Cracks, vol. 3) (1992), Guz, A.N., Dyshel, M.Sh., Nazarenko, V.M., (eds.): T.4, kniga 1. Razrushenie i ustoichivost materialov s treshchinami (Fracture and Stability of Materials with Cracks, vol. 4, Book 1) (1992), Guz, A.N., Zozulya, V.V. (eds.): T.4, kniga 2. Khrupkoe razrushenie materialov pri dinamicheskikh nagruzkakh (Brittle Fracture of Materials under Dynamic Loads, vol. 4, Book 2) (1993)
2. Guz, A.N., Zozulya, V.V.: Problems of dynamic fracture mechanics without allowance for contact of the crack edges. Int. Appl. Mech. **31**(1), 1–31 (1995)
3. Guz, A.N., Zozulya, V.V.: Fracture dynamic with allowance for a crack edges contact interaction. Int. J. Nonlinear Sci. Numer. Simul. **2**(3), 173–233 (2001)
4. Guz, A.N., Zozulya, V.V.: Elastodynamic unilateral contact problem with friction for bodies with cracks. Int. Appl. Mech. **38**(8), 895–932 (2002)
5. Guz, A.N.: O neklassicheskikh problemakh mekhaniki razrusheniia (On non-classical problems of fracture mechanics). Fiziko-himicheskaya Mekhanika Materialov **29**(3), 86–97 (1993)
6. Guz, A.N.: Study and Analysis of Non-classical Problems of Fracture and Failure Mechanics and Corresponding Mechanisms. Lecture presented at Institute of Mechanics. HANOI (1998)
7. Guz, A.N.: Description and study of some nonclassical problems of fracture mechanics and related mechanisms. Int. Appl. Mech. **36**(12), 1537–1564 (2000)
8. Guz, A.N.: On Study of Nonclassical Problems of Fracture and Failure Mechanics and Related Mechanisms, pp. 35–68. ANNALS of the European Academy of Sciences, Liège, Belgium (2006–2007)
9. Guz, A.N.: On study of nonclassical problems of fracture and failure mechanics and related mechanisms. Int. Appl. Mech. **45**(1), 1–31 (2009)
10. Guz, A.N., Zozulya, V.V.: Dinamicheskaia zadacha dlia ploskosti s razrezom. Uchet vzaimod-eistviia beregov (Dynamic problem for a plane with a cut. Allowing for the interaction of edges). Dokl. Akad. Nauk SSSR **318**(2), 304–307 (1991)
11. Guz, A.N., Zozulya, V.V.: Dinamicheskaia kontaktnaia zadacha dlia ploskosti s dvumia razrezami (Dynamic contact problem for a plane with two cuts). Dokl. Akad. Nauk SSSR **321**(2), 278–280 (1991)
12. Guz, A.N., Zozulya, V.V.: Dinamicheskaia zadacha teorii uprugosti s ogranicheniiami v vide neravenstv (Dynamic problems of the theory of elasticity with constraints in the form of inequalities). Dokl. Akad. Nauk USSR **5**, 47–50 (1991)

13. Guz, A.N., Zozulya, V.V., Menshykov, A.V.: Kontaktnoe vzaimodeistvie beregov elliptich-eskoi treshchiny pod vozdeistviem normalnoi garmonicheskoi nagruzki (Contact interaction of edges of the elliptical crack under normal harmonic loading). In: Ivleva, D.D., Morozov, N.F., (eds.) Sbornik nauchnykh trudov «Problemy mekhaniki deformiruemykh tel i gornykh porod», pp.204–220. Fizmatgiz, Moscow, (2006)

14. Zozulya, V.V.: O dinamicheskikh zadachah teorii treschin s oblastyami kontakta, scepleniya i skolzheniya (On dynamic problems of the crack theory with contact, adhesion and slip areas). Dokl. Akad. Nauk USSR Ser. A. 1, 47–50 (1990)

15. Zozulya, V.V.: O razreshimosti dinamicheskih zadakh teorii treschin s oblastyami kontakta, scepleniya i skolzheniya (On solvability of dynamic problems of crack theory with contact, adhesion and slip areas). Dokl. Akad. Nauk USSR Ser. A. 3, 53–55 (1990)

16. Zozulya, V.V.: O deystvii garmonicheskoy nagruzki na treschinu v beskonechnom tele s uchetom vzaimodeystviya ee beregov (On effect of a harmonic load on a crack in an infinite body, allowing interaction of its edges). Dokl. Akad. Nauk USSR Ser. A. 4, 46–49 (1990)

17. Zozulya, V.V.: Integraly tipa Adamara v dinamicheskikh zadachah teorii treschin (Integrals of Hadamard type in dynamic problems of crack theory). Dokl. Akad. Nauk USSR Ser. A. 2, 19–22 (1991)

18. Zozulya, V.V.: Dinamicheskaya zadacha dlya ploskosti s dvumya treschinami uchet kontakta beregov (Dynamic problem for a plane with two cracks. Consideration of edge contact). Dokl. Akad. Nauk USSR 8, 75–80 (1991)

19. Zozulya, V.V.: O reshenii zadach dinamiki tel s treschinami metodom granichnykh integralnykh uravneniy (On solving of problems of dynamics of bodies with cracks using the method of boundary integral equations). Dokl. Akad. Nauk USSR Ser. A. 3, 38–43 (1992)

20. Menshykov, A.V.: Prostranstvennaya kontaktnaya zadacha dlya dvukh soosnykh krugovykh treschin pri normalnom garmonicheskom nagruzhenii (Spatial contact problem for two coaxial circular cracks under normal harmonic load). Dopovidi NAN Ukrainy 6, 44–49 (2005)

21. Menshykov, A.V.: Koefficienty intensivnosti napryazheniy dlya krugovoy treschiny pri garmonicheskom nagruzhenii i uchete kontakta beregov (Stress intensity factors for a circular crack under harmonic loading with concidering contact of edges). Problemy Mashinostroeniya 9(2), 43–47 (2006)

22. Menshykov, A.V., Guz, I.A.: Zavisimost kofficientov intensivnosti napryazheniy sdviga ot sily treniya pri garmonicheskom nagruzhenii krugovoy treschiny (The dependence of the shear stress intensity factors on the friction force under the harmonic loading of a circular crack). Problemy Mashinostroeniya 9(3), 65–71 (2006)

23. Menshykov, A.V., Menshykova, M.V.: Issledovanie kontaktnogo vzaimodeystviya beregov treschiny metodom Galerkina (Use of the Galerkin method for the investigation of cracks edges contact interaction). Teoreticheskaya Prikladnaya Mekhanika 41, 151–155 (2005)

24. Guz, A.N., Menshikov, A.V., Zozulya, V.V.: Surface contact of elliptical crack under normally incident tension-compression wave. Theor. Appl. Fract. Mech. 40(3), 285–291 (2003)

25. Guz, A.N., Menshikov, A.V., Zozulya, V.V., Guz, I.A.: Contact problem for the plane elliptical crack under normally incident shear wave. Comput. Model. Eng. Sci. 17(3), 205–214 (2007)

26. Guz, A.N., Zozulya, V.V.: Contact interaction between crack edges under dynamic load. Int. Appl. Mech. 28(7), 407–417 (1992)

27. Guz, A.N., Zozulya, V.V.: Investigation of the effect of frictional contact in III Mode crack under action of SH-wave harmonic load. Comp. Model. Eng. Sci. 22(2), 119–128 (2007)

28. Guz, A.N., Zozulya, V.V.: On dynamical fracture mechanics in the case of polyharmonic loading by P-waves. Int. Appl. Mech. 45(9), 1033–1036 (2009)

29. Guz, A.N., Zozulya, V.V.: On dynamical fracture mechanics in the case of polyharmonic loading by SH-waves. Int. Appl. Mech. 46(1), 113–116 (2010)

30. Guz, A.N., Zozulya, V.V., Menshikov, A.V.: Three-dimensional dynamic contact problem for an elliptic crack interacting with normally incident harmonic compression-expansion wave. Int. Appl. Mech. 39(12), 1425–1428 (2003)

31. Guz, A.N., Zozulya, V.V., Menshikov, A.V.: General spatial dynamic problem for an elliptic crack under the action of a normal shear wave with consideration for the contact interaction of the crack faces. Int. Appl. Mech. 40(2), 156–159 (2004)

32. Menshikov, A.V.: Elastodynamics contact problem for two penny-shaped cracks. In: Abstracts and Proceedings of the XXI ICTAM 04, p. 262. Warsaw, Poland (2004)
33. Menshikov, A.V., Guz, I.A.: Contact interaction of crack faces under oblique incidence of a harmonic wave. Int. J. Fract. **139**(1), 145–152 (2006)
34. Menshikov, A.V., Guz, I.A.: Effect of the contact interaction on the stress intensity factors for a crack under harmonic loading. Appl. Mech. Mater. **5–6**, 174–180 (2006)
35. Menshikov, A.V., Guz, I.A.: Effect of contact interaction of the crack faces for a crack under harmonic loading. Int. Appl. Mech. **43**(7), 809–815 (2007)
36. Menshikov, A.V., Menshikova, M.V., Wendland, W.L.: On use of the Galerkin method to solve the fracture mechanics problem for a linear crack under normal loading. Int. Appl. Mech. **41**(11), 1324–1328 (2005)
37. Menshikov, V.A., Menshikov, A.V., Guz, I.A.: Interfacial crack between elastic half-spaces under harmonic loading. Int. Appl. Mech. **43**(8), 865–873 (2007)
38. Zozulya, V.V.: Investigation of the contact of edges of cracks interacting with a plane longitudinal harmonic wave. Sov. Appl. Mech. **27**(12), 1191–1195 (1991)
39. Zozulya, V.V.: Contact interaction between the edges of a crack in an infinite plane under a harmonic load. Int. Appl. Mech. **28**(1), 61–64 (1992)
40. Zozulya, V.V.: Investigation of the effect of crack edge contact for loading by a harmonic wave. Int. Appl. Mech. **28**(2), 95–99 (1992)
41. Zozulya, V.V.: Harmonic loading of the edges of two collinear cracks in plane. Int. Appl. Mech. **28**(3), 170–172 (1992)
42. Zozulya, V.V.: Contact problem for a plane crack under normally incident antiplane shear wave. Int. Appl. Mech. **43**(5), 586–588 (2007)
43. Zozulya, V.V.: Stress intensity factor in a contact problem for a plane crack under an antiplane shear wave. Int. Appl. Mech. **43**(9), 1043–1047 (2007)
44. Zozulya, V.V., Fenchenko, N.V.: Influence of contact interaction between the sides of crack on characteristics of failure mechanics in action P- and SV waves. Int. Appl. Mech. **35**(2), 175–180 (1999)
45. Zozulya, V.V., Lukin, A.N.: Solution of three-dimensional problems of fracture mechanics by the method of integral boundary equations. Int. Appl. Mech. **34**(6), 544–551 (1998)
46. Zozulya, V.V., Menshikov, A.V.: Contact interaction of the faces of a rectangular crack under normally incident tension-compression waves. Int. Appl. Mech. **38**(3), 302–307 (2002)
47. Zozulya, V.V., Menshikov, A.V.: On one contact problem in fracture mechanics for a normally incident tension-compression wave. Int. Appl. Mech. **38**(7), 824–828 (2002)
48. Zozulya, V.V., Menshikov, A.V.: Contact interaction of the faces of a penny-shaped crack under a normally incident shear wave. Int. Appl. Mech. **38**(9), 1114–1118 (2002)
49. Zozulya, V.V., Menshikov, A.V.: Use of the constrained optimization algorithms in some problems of fracture mechanics. Optim. Eng. **4**(4), 365–384 (2003)
50. Zozulya, V.V., Menshikova, M.V.: Study of interactive algorithms for solution of dynamic contact problems for elastic cracked bodies. Int. Appl. Mech. **38**(5), 573–578 (2002)
51. Zozulya, V.V., Menshikova, M.V.: Dynamic contact problems for a plane with a finite crack. Int. Appl. Mech. **38**(12), 1459–1463 (2002)
52. Zozulya, V.V., Menshikov, V.A.: Contact interaction of the edges of a crack in a plane under harmonic loading. Int. Appl. Mech. **30**(12), 986–989 (1994)
53. Zozulya, V.V., Menshikov, V.A.: Solution of three-dimensional problems of the dynamic theory of elasticity for bodies with cracks using hypersingular integrals. Int. Appl. Mech. **36**(1), 74–81 (2000)
54. Guz, A.N., Guz, I.A., Menshykov, A.V., Menshikov, V.A.: Prostranstvennye zadachi dinamicheskoi mekhaniki razrusheniia materialov s treshchinami v granitse razdela (obzor) (Spatial problems in the dynamic fracture mechanics of materials with interface cracks (Review)). Prikladnaya Mekhanika **49**(1), 3–78 (2013)
55. Guz, A.N., Guz, I.A., Menshikov, A.V., Menshikov, V.A.: Penny-shaped crack at the interface between elastic half-space under the action of a shear wave. Int. Appl. Mech. **45**(5), 534–539 (2009)

56. Parton, V.Z., Borisovsky, V.G.: Dinamicheskaya mekhanika razrusheniya (Dynamic Fracture Mechanics). Mashinostroenie, Moscow (1985)
57. Parton, V.Z., Borisovsky, V.G.: Dinamika khrupkogo razrusheniya (Brittle Fracture Dynamics). Mashinostroenie, Moscow (1988)
58. Guz, A.N., Zozulya, V.V.: Problems of dynamic fracture mechanics without contact of the crack faces. Int. Appl. Mech. **30**(10), 735–759 (1994)
59. Mikhaskiv, V.V., Sladek, J., Sladek, V., Stepanyuk, A.I.: Stress concentration near an elliptic crack in the interface between elastic bodies under steady-state vibration. Int. Appl. Mech. **40**(6), 664–671 (2004)

Chapter 10
Problem 8. Fracture of Thin-Wall Bodies with Cracks Under Tension in the Case of Preliminary Loss of Stability

In this chapter, in a very short form (even in comparison with **Problem 7**, which has already been discussed in the considered Part III of this monograph), the results on **Problem 8** are presented, which were obtained in the Department of Dynamics and Stability of Continua of the S. P. Timoshenko Institute of Mechanics of NASU. The presentation of these results is written in the style announced in the Introduction (Part I) to this monograph (without excessively invoking aspects of a mathematical nature).

10.1 Introduction

This problem of fracture mechanics is non-classical because in this case *Condition 2* of the applicability of the classical fracture mechanics (loss of stability does not precede the fracture), which is formulated in Sect. 10.1 is not satisfied. This section also discusses the above condition. Indeed, in the classical fracture mechanics of materials with cracks under tension and shear, it is almost always tacitly assumed that the fracture begins from the configuration of the body that it had in the undeformed state. Therefore, it is assumed that in the process of deformation before the start of fracture, there is no sharp change in the configuration of the body; i.e., the loss of stability does not precede the fracture. *In fact, even under tension in the case of thin-walled structural members, the local loss of stability of the equilibrium state near the cracks may precede failure.*

The reference problems in this non-classical problem of fracture mechanics are the problems of stretching the thin-walled plates and shells perpendicular to the crack. In the case of the cylindrical shells, as a rule, the situation is investigated when the crack is located along the shell guiding and the shell is stretched along its axis. In this situation, as a result of the stress concentration near the crack, the local zones of

A. N. Guz, *Eight Non-Classical Problems of Fracture Mechanics*,
Advanced Structured Materials 159,
https://doi.org/10.1007/978-3-030-77501-8_10

compressive stresses arise, which can lead to a local loss of stability near the cracks before the start of the fracture process.

The various authors in the study of the above situation use the various approximate design schemes when analyzing the local loss of stability zones near cracks and holes.

In the works of the author of this monograph and his pupils, the strict equations of mechanics of the thin-walled systems were used to analyze this problem, followed by the use of the variational and numerical methods. At that, considerable attention was paid to the experimental studies and the use of their results in the analysis of this problem.

The main results on this problem, obtained in the Department of Dynamics and Stability of Continua, are presented in the monographs [1] of 1981 and [2] (vol. 4, book 1) of 1992, in the review articles [3–5], which are fully devoted to the analysis of the results for **Problem 8**, and in review articles [6–10], which are devoted to the analysis of results on the various non-classical problems of fracture mechanics, including **Problem 8**. Initially, the above results were presented in the publications [11–40] and in several other publications since 1975. The first report [41] at the international congress was presented already in 1976.

In this scientific direction of the Department of Dynamics and Stability of Continua, two dissertations were prepared and defended: G.G. Kuliev—for the degree of the Doctor of Physical and Mathematical Sciences and M.Sh. Dyshel—for the degree of the Doctor of Technical Sciences.

Note 10.1 The doctoral dissertation of G. G. Kuliev, along with the results of research on **Problem 8** (*Fracture of thin-walled bodies with cracks under tension in the case of preliminary loss of stability*), also included the results of studies on the stability of the equilibrium state of the rock mass near the mine workings, which are abbreviated as the theory of stability of the mining workings. The indicated studies of G. G. Kuliev on the theory of stability of the mine workings were carried out based on the approach of the monograph [42] of 1977 by the author of this monograph.

Note that a modern review of the results on the theory of stability of the mine workings is presented in a review article [43] of 2003. In subsequent years, G. G. Kuliev carries out the scientific research in Azerbaijan, where, to a large extent, the new results were published. The list of references to this monograph includes the articles by G. G. Kuliev related to the Kyiv period of his activity.

10.2   Statement of Problems

In the Introduction (Sect. 10.1), the qualitative considerations were formed that, when the plates and shells with cracks are stretched, as a result of the stress concentration near the cracks, the local zones of compressive stresses arise, which can lead to a local loss of stability before fracture.

Fig. 10.1

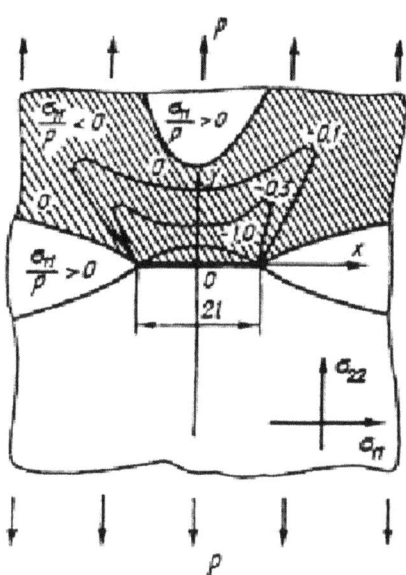

Note that the above consideration is *only of a qualitative nature*. Below, using the simplest reference problem as an example, the *quantitative* information on the discussed zones of compressive stresses is presented.

To illustrate the above, consider the well-known exact solution of Inglis–Muskhelishvili about the elastic equilibrium of an isotropic plate with a crack under uniaxial tension (Fig. 10.1) in the direction perpendicular to the crack.

At the top of Fig. 10.1 shows the distribution of stresses σ_{11} acting along the crack. The shaded areas correspond to the compressive stresses σ_{11} under the action of uniaxial tension "at infinity," and the curves with numbers $(-1; -0.3; -0.1)$ separate the areas with stresses σ_{11} of the indicated intensity. Since the compressive stresses acting along the crack, then the crack length is large in comparison with the plate thickness, the local loss of stability, characterized by the normal deflection w, may precede the failure. A similar situation can arise in the thin-walled shells under external loads, causing, as it were, the tensile stresses near cracks.

Taking into account the above information, it seems reasonable to consider that in **Problem 8**, in the general case, a relatively wide range of problems of constructing the fracture mechanics for the thin-walled plates and shells under the action of external loads providing the occurrence of *basically* the tensile stresses near cracks is considered, as applied to the case when the local loss of stability near cracks may precede the fracture.

To construct the fracture mechanics in this case, it is necessary to answer *two questions*:

The *first* question is to investigate the local loss of stability near the crack corresponding to the considered type of loading.

The *second* question is to investigate the fracture mechanism based on the shape of the thin-walled element and the distribution of the stress–strain state corresponding to the thin-walled element after the loss of stability.

The main publications to date, with the rare exceptions, are devoted to the study of the first of the above questions, including the theoretical and experimental approaches. Quite a few publications are devoted to the study of the second question, where, with the rare exceptions, the results of experimental studies are presented.

When analyzing the local loss of stability near the cracks and holes under tension of the plates and shells, which in the case of cracks corresponds to the first question in the construction of fracture mechanics in the framework of **Problem 8**, the overwhelming majority of authors propose or apply the various approximate design schemes and models.

As an example, the paper [44] of 1968 for the case of tension of a plate with a hole and paper [45] of 1978 for the case of a plate with a crack can be mentioned.

If the membrane is considered to be an approximate design scheme or a model when researching the plate, then it is also advisable to mention the article [46] back of 1963, which sets out *the exact solution* to the problem of the local loss of stability near the hole under membrane stretching.

In most cases, the studies are carried out within the framework of the fairly approximate design schemes and models.

Consider below for example the design scheme that was used in [45 for stretching the plate with a crack to analyze the local loss of stability near the crack.

Along the crack (Fig. 10.2a), a finite strip of a certain width and length is conventionally distinguished (in Fig. 10.2b, the selected strip is shaded). Subsequently, the stability of the thus selected strip under the action of loads corresponding to the stresses in the original plate with a crack is analyzed (Fig. 10.2c).

The approximation and disadvantages of this approach are obvious since it is associated with arbitrariness in choosing the width and length of the selected strip,

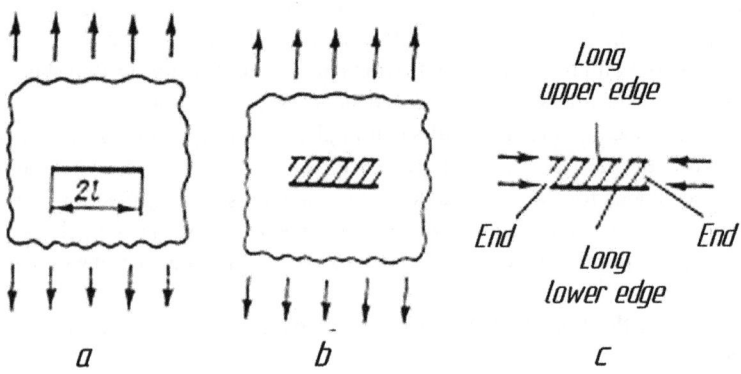

Fig. 10.2

the type of boundary conditions at the ends, and on the upper long side (Fig. 10.2c) of this strip, as well as the nature and intensity of the assigned external loads (Fig. 10.2c).

Conclusion It follows from the above design scheme of the article [45] shown in Fig. 10.2a which is given as an example that to obtain more rigorous results more reliable design schemes, more consistent methods, and additional research are needed.

10.3   Research Methods and Results Obtained

In the publications of the author of this monograph, his pupils and colleagues, the more rigorous methods were proposed and the research was carried out on the *first* question, and the targeted experimental studies on the *second* question were organized, the solutions of which jointly determine the construction of fracture mechanics within the framework of the **Problem 8**. In particular, the following relatively more rigorous approach was proposed for *solving the first question.*

To study the local loss of stability in the case of the plates and shells in a situation that is schematically shown in Fig. 10.1, *the linearized theory of stability of thin-walled plates or shallow shells is applied. The study is carried out for an "infinite" area (plane). In this case, the boundary conditions corresponding to a specific problem are set on the crack edges, and the attenuation conditions are set at "infinity" (with moving away from the crack).*

With the above approach, specific results are obtained using variational and numerical methods for the plates and cylindrical shells, and the results obtained are compared with the experimental results. The discussed results can be found in the published articles, which are indicated in the Introduction of Sect. 10.1 and partially in the monograph [2] (vol. 4, book 1).

As an example, let us give some results for the plates relative to the situation shown in Fig. 10.1. Using the *above-proposed approach* for the critical load value p_{cr} corresponding to the loss of stability, the following expression is obtained

$$p_{cr} = t_*(v) \frac{E}{6(1 - v^2)} \frac{1}{l^2}. \tag{10.1}$$

In (10.1) and below, the following designations are introduced:

E and v are Young modulus and Poisson ratio.
l is the dimensionless crack length referred to the plate thickness h.
$t_*(v)$ is the minimum eigenvalue of the formulated eigenvalue problem.

Within the framework of the Griffiths fracture mechanics [47], the value of the fracture load is determined by the following expression

Fig. 10.3

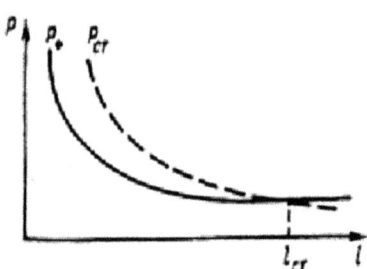

$$p_* = \sqrt{\frac{2E\gamma}{\pi h}} \frac{1}{\sqrt{l}}, \qquad (10.2)$$

where γ is the specific surface energy or, more strictly, is the density of the effective surface energy of brittle or quasi-brittle material.

In Fig. 10.3, the dependence is schematically shown for the quantity p_{cr} determined from expression (10.1) in the form of a dotted line and the quantity p_* determined from expression (10.2) in the form of a solid line on the dimensionless crack length l. The result of a qualitative nature follows from Fig. 10.3—*at the crack lengths $l > l_{cr}$, the loss of stability will precede failure.*

As follows from the expression (10.1), the accuracy of determining the value of the critical load p_{cr} corresponding to the loss of stability is determined by the accuracy of determining the minimum positive eigenvalue of the formulated eigenvalue problem following *the proposed approach* to solving the problem under discussion, which (approach) is summarized at the beginning of this section.

In connection with the above, some additional information is given about *the proposed approach* and the following specific results when applying it in the case of using the variational method. More detailed information is presented in the monographs [1] and [2] (vol. 4, book 1), which also analyze the various approximate schemes of the type shown in Fig. 10.2.

Following *the proposed approach*, formulated at the beginning of Sect. 10.3, the linearized Karman equations are used to study the local loss of stability of the thin-walled plates with cracks under tension. In this case, the increment of the total energy of the thin-walled plate Δ э is considered during the transition from the unperturbed state (at $w \equiv 0$) to the perturbed state ($w \neq 0$). The study is carried out for an "infinite" region (the exterior of the crack) and a variational method with minimization of the functional Δ э is applied. For the deflection w, a representation through coordinate functions $\left[\rho^{-n} \begin{matrix} \cos \\ \sin m\theta \end{matrix} \right]$ is used in the form

$$w = |\omega'(\zeta)|^4 \sum_{m=0}^{M} \sum_{n=1}^{N} \begin{matrix} A_{nm} \\ B_{nm} \end{matrix} \left(\frac{1}{\rho}\right)^n \begin{matrix} \cos \\ \sin m\theta \end{matrix}, \qquad (10.3)$$

Table 10.1

Ω	4	12	24	40
$t_*(\nu)$	25	5.945	5.516	5.508

Table 10.2

ν	0.10	0.20	0.25	0.28	0.30	0.33	0.35	0.40
$t_*(\nu)$	6.24	5.71	5.62	5.54	5.51	5.42	5.37	5.28

where A_{nm} and B_{nm} are the unknown constants and $\omega(\varsigma)$ is the function that implements the conformal mapping of the considered area (crack exterior, Fig. 10.1) to the exterior of a circle of the unit radius

$$z = x + iy; \; \varsigma = \rho e^{i\theta}; \; \omega(\varsigma) = \frac{l}{2}\left(\varsigma + \frac{1}{\varsigma}\right). \tag{10.4}$$

It should be noted that the used coordinate functions satisfy the attenuation conditions "at infinity" (at $\rho \to \infty$). The choice of the factor $\left|\omega'(\varsigma)\right|^4$ in expression (10.3) is discussed in detail and substantiated in the monographs [1] and [2] (vol. 4, book 1). The study uses the variational equation

$$\delta(\Delta\ \ni) = 0. \tag{10.5}$$

Let us present information about the particular results obtained using *the proposed approach* for the plate shown in Fig. 10.1. First of all, the information should be given about the accuracy of calculating the quantity $t_*(\nu)$—the minimum eigenvalue of the formulated eigenvalue problem when applying *the proposed approach*, which is included in the expression (10.1) to determine the critical load corresponding to the loss of stability.

For the case $\nu = 0.3$, the values of the quantity $t_*(\nu)$ are shown in Table 10.1, which are obtained for a different number of coordinate functions in the representation (10.3). At that, the number of coordinate functions in (10.3) in Table 10.1 is determined by a number Ω according to the designations [1] and [2] (vol. 4, book 1). It follows from the analysis of the results in Table 10.1 that the results for $\Omega = 24$ and $\Omega = 40$ differ in the third significant digit. Thus, in calculations, the number of coordinate functions can be limited to $\Omega = 24$. From the results presented in Table 10.1, the following general conclusion can be drawn.

The proposed approach is very effective since it allows to get fairly accurate results. Nevertheless, to obtain sufficiently accurate results, it is necessary to involve a sufficiently large number of coordinate functions (in the case under discussion, the required number of coordinate functions is 24).

In Table 10.2, the numerical values are given for the quantity $t_*(\nu)$ for the various values of Poisson's ratio (the quantity ν), which is included in the expression (10.1) to determine the value of the critical load corresponding to the loss of stability.

Presented in Table 10.2 results are obtained with an accuracy corresponding to the results shown in Table 10.1 (at $\Omega = 24$).

Thus, it can be assumed that for the plate in the situation presented in Fig. 10.1, the results (expression (10.1) and Table 10.2) solve quite consistently and strictly and relatively accurately the first question (according to the terminology of the middle part of Sect. 10.2) when constructing fracture mechanics for **Problem 8** (*Fracture of thin-walled bodies with cracks under tension in the case of preliminary loss of stability*). Additional information related to the study of *the first question* for the plates and shells can be obtained from the monographs [1] and [2] (vol. 4, book 1) as well as from the articles indicated in the Introduction of Sect. 10.1.

In the study of *the second question* (in the terminology of the middle part of Sect. 10.2) of the construction of fracture mechanics for **Problem 8** in the Department of Dynamics and Stability of Continua, the results of specially designed experiments of two types were analyzed. For the first and second types of experiments, the samples (plates and shells) of the same material and with the same geometric parameters were used.

In the first type of experiments, the loss of stability before the fracture was not excluded for those samples for which this phenomenon is possible.

In the second type of experiment, the preliminary loss of stability before the fracture was eliminated due to the deformation of the plate between two rigid guides, the distance between which did not change.

The difference between the values of the fracturing loads corresponding to these two types of experiments, in a certain sense, illustrates the influence of preliminary loss of stability (local loss of stability near cracks) on the fracture mechanism. The numerous specific results obtained for the various materials (metals and alloys) as applied to the various plates and shells are given in the monographs [1] and [2] (vol. 4,

Table 10.3

Materials	$2\,b$, mm	h, mm	l_0/b	ρ, MPa		δ (%)
				p_1	p_2	
Aluminum alloy AMg6M	250	0.90	0.32	119	126	8.8
	250	0.26	0.40	116	144	23.7
Aluminum alloy AMtsM	250	0.92	0.24	67	72	6.9
	250	0.92	0.40	52	57	8.8
Steel 65G	250	0.51	0.24	256	270	5.2
	250	0.54	0.24	210	220	4.5
Steel 20	250	0.54	0.32	189	200	5.5
	250	0.54	0.40	164	176	6.8
Steel H36	250	0.43	0.32	335	435	21.1

book 1). Table 10.3 shows, by way of example, the specific results for five materials in the case of a plate.

Table 10.3 introduced the following designations:

$2b$—the width of the plate (strip);

h—the plate thickness;

$2l_0$—the crack length;

p_1—the value of the breaking load, when the loss of stability was allowed;

p_2—the value of the breaking load for the same plate, when the loss of stability was excluded;

$\delta = (p_1 - p_2) \cdot p_2^{-1} \cdot 100\%$—the relative error.

It follows from the results presented in Table 10.3 that the difference between the above-mentioned breaking loads can reach 23%. The noted fact indicates **the essentiality of the phenomenon under consideration and the need for the development of this scientific direction.**

Note 10.2 In conclusion to this chapter, the following situation should be noted. This chapter devoted to a short review of the results on **Problem 8** (*Fracture of thin-walled bodies with cracks under tension in the case of preliminary buckling*) is formulated in the volume that is the smallest of all **Problems 1–8**. Besides, the material in this chapter is largely consistent with the material of the reviews [9] of 2006–2007 and the review [10] of 2009.

The above situation has formed, apparently, because, in the opinion of the author of this monograph, during the years that have passed since the publication of these reviews (from 2006–2009), no new, to a certain extent, significant results have been obtained on the **Problem 8**.

We restrict ourselves to the information set out above in this chapter when discussing **Problem 8** (*Fracture of thin-walled bodies with cracks under tension in the case of preliminary loss of stability*).

References

1. Guz, A.N., Dyshel, MSh., Kuliev, G.G., Milovanova, O.B.: Razrushenie i ustoichivost tonkikh tel s treshchinami (Fracture and stability of thin bodies with cracks). Naukova Dumka, Kyiv (1981)
2. Guz, A.N., (ed.): Neklassicheskie problemy mekhaniki razrusheniya, v 4 tomah, 5 knigah (Non-Classical Problems of Fracture Mechanics, in 4 volumes, 5 books. Naukova Dumka, Kyiv, (1990–1993), Kaminsky, A.A., (ed.), T.1. Razrushenie viazkouprugikh tel s treshchinami (Vol. 1. Fracture of Viscoelastic Bodies with Cracks) (1990), Guz, A.N. (ed.): T.2. Khrupkoe razrushenie materialov s nachalnymi napriazheniiami, (T.2. Brittle Fracture of Materials with Initial Stresses) (1991), Kaminsky, A.A., Gavrilov, D.N. (eds.): T.3. Dlitelnoe razrushenie polimernykh i kompozitnykh materialov s treshchinami (T.3. Long-Term Fracture of Polymer and Composite Materials with Cracks) (1992), Guz, A.N., Dyshel, M.Sh., Nazarenko, V.M., (eds.): T.4, kniga 1. Razrushenie i ustoichivost materialov s treshchinami (V.4, Book 1. Fracture and Stability of Materials with Cracks) (1992), Guz, A.N., Zozulya, V.V., (eds.): T.4, kniga 2.

Khrupkoe razrushenie materialov pri dinamicheskikh nagruzkakh (V.4, Book 2. Brittle Fracture of Materials under Dynamic Loads) (1993)

3. Guz, A.N., Dyshel, MSh.: Fracture and stability of notched thin-walled bodies in tension (Survey). Sov. Appl. Mech. **26**(11), 1023–1040 (1990)

4. Guz, A.N., Dyshel, MSh., Kuliev, G.G., Milovanova, O.B.: Fracture and local instability of thin-walled bodies with notches. Sov. Appl. Mech. **17**(8), 707–721 (1981)

5. Guz, A.N., Dyshel, MSh., Nazarenko, V.M.: Fracture and stability of materials and structural members with cracks: approaches and results. Int. Appl. Mech. **40**(12), 1323–1359 (2004)

6. Guz, A.N.: O neklassicheskikh problemakh mekhaniki razrusheniia (On non-classical problems of fracture mechanics). Fiziko-himicheskaya Mekhanika Materialov. **29**(3), 86–97 (1993)

7. Guz, A.N.: Study and Analysis of Non-classical Problems of Fracture and Failure Mechanics and Corresponding Mechanisms. Lecture presented at Institute of Mechanics, HANOI (1998)

8. Guz, A.N.: Description and study of some nonclassical problems of fracture mechanics and related mechanisms. Int. Appl. Mech. **36**(12), 1537–1564 (2000)

9. Guz, A.N.: On Study of Nonclassical Problems of Fracture and Failure Mechanics and Related Mechanisms, pp. 35–68. Annals of the European Academy of Sciences, Liège, Belgium (2006–2007)

10. Guz, A.N.: On study of nonclassical problems of fracture and failure mechanics and related mechanisms. Int. Appl. Mech. **45**(1), 1–31 (2009)

11. Guz, A.N., Dyshel, M.Sh., Kuliev, G.G., Milovanova, O.B.: Ustoichivost tonkikh plastin s treshchinami (Stability of thin plates with cracks). Dokl. Akad. Nauk USSR Ser. A. **5**, 421–426 (1977)

12. Guz, A.N., Kuliev, G.G.: K postanovke zadach ustoichivosti deformirovaniia tonkikh tel s treshchinami (On stating problems of deformation stability for thin bodies with cracks). Dokl. Akad. Nauk USSR Ser. A. **12**, 1085–1088 (1976)

13. Guz, A.N., Kuliev, G.G., Zeinalov, N.K.: Vypuchivanie rastianutoi plastiny s krivolineinym otverstiem (Buckling of a stretched plate with a curved hole). Izvestiya Akademii Nauk SSSR, Mekhanika tverdogo tela. **2**, 163–168 (1979)

14. Guz, A.N., Kuliev, G.G., Tsurpal, I.A.: Kontseptsii ustoichivosti v teorii khrupkogo razrusheniia (Stability concepts of brittle fracture theory). Annotatsii dokl. IV Vsesoiuz. sieezda po teor. i prikl. MekhanikeKyiv, (1976), p. 90

15. Kuliev, G.G.: O razrushenii deformiruemykh tel c centralnoy vertikalnoy treschinoy v odnorodnom silovom pole (Fracture of deformable bodies with a central vertical crack in a uniform force field). Dokl. Akad. Nauk USSR Ser. A. **8**, 714–717 (1978)

16. Kuliev, G.G.: O predshestvovanii processa poteri ustoychivosti vozle treschiny processu khrupkogo razrusheniya (Preceding of stability loss process near a crack to a process of brittle fracture). Dokl. Akad. Nauk USSR Ser. A. **5**, 355–358 (1979)

17. Dyshel, MSh.: Failure in thin plate with a slit. Sov. Appl. Mech. **14**(9), 1010–1012 (1978)

18. Dyshel, MSh.: Stability under tension of thin plates with cracks. Sov. Appl. Mech. **14**(11), 1169–1172 (1978)

19. Dyshel, MSh.: Fracture of plates with cracks under tension after loss of stability. Sov. Appl. Mech. **17**(4), 371–375 (1981)

20. Dyshel, MSh.: Stability of thin plates with cracks under biaxial tension. Sov. Appl. Mech. **18**(10), 924–928 (1982)

21. Dyshel, MSh.: Tension of a cylindrical shell with a slit. Sov. Appl. Mech. **20**(10), 941–944 (1984)

22. Dyshel, MSh.: Stability of a cracked cylindrical shell in tension. Sov. Appl. Mech. **25**(6), 542–547 (1989)

23. Dyshel, MSh.: Stress-intensity coefficient taking account of local buckling of plates with cracks. Sov. Appl. Mech. **26**(1), 87–90 (1990)

24. Dyshel, MSh.: Local stability loss and failure of cracked plates during the plastic deformation of materials. Int. Appl. Mech. **30**(1), 44–47 (1994)

25. Dyshel, MSh.: Tensile stability and failure of two-layer plates with cracks. Int. Appl. Mech. **34**(3), 282–286 (1998)

26. Dyshel, MSh.: Local buckling of extended plates containing cracks and cracklike defects, subject to the influence of geometrical parameters of the plates and defects. Int. Appl. Mech. **35**(12), 1272–1276 (1999)
27. Dyshel, MSh.: Influence of buckling of a tension plate with edge crack on fracture characteristics. Int. Appl. Mech. **42**(5), 589–592 (2006)
28. Dyshel, MSh.: Stability and fracture of plates with two edge cracks under tension. Int. Appl. Mech. **42**(11), 1303–1306 (2006)
29. Dyshel, MSh., Mekhtiev, M.A.: Deformation of tensioned plates with cracks with allowance for local buckling. Sov. Appl. Mech. **23**(6), 586–589 (1987)
30. Dyshel, MSh., Mekhtiev, M.A.: Failure of tensioned plates weakened by circular hole with radial cracks emanating from its contour. Sov. Appl. Mech. **25**(5), 490–493 (1989)
31. Dyshel, MSh., Milovanova, O.B.: Method of experimentally analyzing the instability of plates with slits. Sov. Appl. Mech. **13**(5), 491–494 (1977)
32. Dyshel, MSh., Milovanova, O.B.: Determination of the critical stresses in the case of tension of plates with a cut. Sov. Appl. Mech. **14**(12), 1330–1332 (1978)
33. Guz, A.N., Dyshel, MSh.: Fracture of cylindrical shells with cracks in tension. Theor. Appl. Fract. Mech. **4**, 123–126 (1985)
34. Guz, A.N., Kuliev, G.G., Tsurpal, I.A.: Theory of the rupture of thin bodies with cracks. Sov. Appl. Mech. **11**(5), 485–487 (1975)
35. Guz, A.N., Kuliev, G.G., Tsurpal, I.A.: On fracture of brittle materials from loss of stability near crack. Eng. Fract. Mech. **10**(2), 401–408 (1978)
36. Kuliev, G.G.: Theory of stability of bodies with a crack in the case of plane deformation. Sov. Appl. Mech. **13**(12), 1235–1239 (1977)
37. Kuliev, G.G.: Effect of the form of external loads on the loss of stability of the state of equilibrium of half-space near a central vertical crack. Sov. Appl. Mech. **15**(10), 1001–1002 (1979)
38. Kuliev, G.G.: Problems of stability loss of half-space with a crack of infinite depth. Sov. Appl. Mech. **14**(8), 815–819 (1978)
39. Milovanova, O.B., Dyshel, MSh.: Experimental investigation of the buckling form of tensioned plates with a slit. Sov. Appl. Mech. **14**(1), 101–103 (1978)
40. Milovanova, O.B., Dyshel, MSh.: Stability of thin sheets with an oblique slit in tension. Sov. Appl. Mech. **16**(4), 333–336 (1980)
41. Guz, A.N., Kuliev, G.G., Tsurpal, I.A.: On failure of brittle materials because of grippling near cracks. Abstracts of the 14-th IUTAM Congress, p. 90. Delft (1976)
42. Guz, A.N.: Osnovy teorii ustoichivosti gornykh vyrabotok (Fundamentals of the Theory of Stability of Mine Workings). Naukova Dumka, Kyiv (1977)
43. Guz, A.N.: Establishing the fundamentals of the theory of stability of mine working. Int. Appl. Mech. **39**(1), 20–48 (2003)
44. Pelle, D.A., Costello, R.G., Brok, J.E.: Vypuchivanie paneli s krugovym otverstiem pri rastya-jenii (Buckling of panel with circular hole under tension). Raketnaya Tekhnika Kosmonavtika **6**(10), 241–243 (1968)
45. Dal, Yu.M.: O mestnom izgibe rastianutoi plastiny s treshchinoi (Local bending of a stretched plate with a crack). Izvestiya Akademii Nauk SSSR, Mekhanika tverdogo tela. **4**, 135–141 (1978)
46. Cherepanov, G.P.: O vypuchivanii membran s otverstiyami pri rastyazhenii (On buckling of perforated membranes under tensile). Prikladnaya Matematika Mekhanika **27**(2), 275–286 (1963)
47. Griffith, A.A.: The phenomena of rupture and flow in solids. Phil. Trans. Roy. Soc., Ser. A. **211**(2), 163–198 (1920)

General Conclusion to the Monograph (Parts I, II, III)

The relevance of the investigations under discussion (**non-classical problems of fracture mechanics**) and a rather long period (**about half a century**) of their carrying out in the Department of Dynamics and Stability of Continua of the S.P. Timoshenko Institute of Mechanics of the NASU made for the formation (for the author of this monograph) of the conviction of the expediency, and to a certain extent, the need to prepare the monograph presented to the readers in the form of a review which consists of three parts, published by the journal "Applied Mechanics" in No. 2–4 of 2019 in Russian and the journal "International Applied Mechanics" in English.

The specialists in various scientific and technical areas can find in this monograph the information in a concise form on a wide aspect of non-classical problems of fracture mechanics. The first time, young researchers will find in the presented monograph the information on the relatively new problems of fracture mechanics with a fairly extensive list of references, numbering 597 titles.

Of course, the creation of this type of monograph, in addition to the availability of relevant scientific results, requires a long analysis of various aspects of research (long-term work over many months); in this regard, apparently, a large number of such publications will not appear in the near future.

It is worth noting that the author of the discussed publication was engaged in its preparation for more than two years before submission to the journal "Prikladnaya Mekhanika" of separate parts for the publication.

From the experience of preparing this monograph, it seems to the author, the conclusion follows that when creating this type of publication, a situation is preferable when the author is *only one scientist* since it is necessary to analyze *from a single general point of view* a significant number of results obtained over a long time.

Concerning this monograph, the consideration arose that in connection with the constantly expanding creation of new materials and, on their basis, new structural members, **the non-classical problems of fracture mechanics** may remain quite relevant.

A. N. Guz, *Eight Non-Classical Problems of Fracture Mechanics*,
Advanced Structured Materials 159,
https://doi.org/10.1007/978-3-030-77501-8

It seems to the author of this monograph that the discussed scientific results on *eight non-classical problems of fracture mechanics* obtained in the S.P. Timoshenko Institute of Mechanics of the NASU over the past 50 years, *by the generality and stringency of the formulation of problems, the accuracy, and generality of the specific results obtained, as well as the thoroughness and validity of the formulated conclusions related to the study of the analyzed mechanical phenomena,* **have no analogs in the modern world science in the problems under consideration.**

Bibliography

In Cyrillic

1. Guz, A.N.: Razrushenie odnonapravlennykh kompozitnykh materialov pri osevom szhatii (Fracture of unidirectional of a composite material with an elastic-plastic matrix). Mekhanika Kompozitnykh Materialov **3**, 417–425 (1982)
2. Dekret, V.A.: Ploska zadacha stiikosti kompozyta armovanoho dvoma paralelnymy korotkymy voloknamy (Plane problem of stability of a composite reinforced with two parallel short fibers). Dopovidi NAN Ukrainy **12**, 38–41 (2003)
3. Dyshel, M.Sh.: Uchet lokalnoy poteri ustoychivosti plastin s treschinami pri eksperimentalnom opredelenii koefficienta intensivnosti napryazheniy (Allowance of local buckling of plates with cracks upon experimental determination of the stress intensity rate). Dokl. Akad. Nauk USSR Ser. A. **11**, 40–44 (1988)
4. Panasiuk, V.V. (ed.): Mekhanika razrusheniya i prochnost materialov. Spravochnoe posobie v 4 tomah (Fracture Mechanics and Strength of Materials. Handbook in 4 volumes). Naukova Dumka, Kyiv (1988–1990), Panasyuk, V.V., Andreykiv, A.E., Parton, V.Z., (eds.): T. 1. Osnovy mekhaniki razrusheniia materialov (Fundamentals of Fracture Mechanics, vol. 1) (1988), Savruk, M.P., (ed.): T. 2. Koeffitsienty intensivnosti napriazhenii v telakh s treshchinami (Stress Intensity Factors in Bodies with Cracks, vol. 2) (1988), Kovchik, S.E., Morozov, Ye.M., (eds.) T. 3. Kharakteristiki kratkovremennoi treshchinostoikosti materialov i metody ikh opredeleniia (Characteristics of Short-Term Crack Resistance of Materials and Methods for their Determination, vol. 3) (1988), Romaniv, O.N., Yarema, S.Ya., Nikiforchin, G.N., Makhutov, N.A., Satdnik, M.M., (eds.) T. 4. Ustalost i tsiklicheskaia treshchinostoikost konstruktsionnykh materialov (Fatigue and Cyclic Crack Resistance of Structural Materials, vol. 4) (1990)
5. Sporykhin, A.N.: O neustoychivosti deformirovaniya sloistykh massivov v uprochnyauschikhsya plasticheskikh sredah (On instability of deformation of layered massifs in hardening plastic media). Izvestiya Akademii Nauk SSSR, Mekhanika Tverdogo Tela **1**, 63–65 (1975)
6. Uflyand, Ya.S.: Integralnye preobrazovaniya v zadachah teorii uprugosti (Integral Transformations in Problems of the Theory of Elasticity). Izdatelstvo Akademii Nauk SSSR, Moscow-Leningrad (1963)
7. Uflyand, Y.S.: Metod parnykh uravneniy v zadachah matematicheskoy fiziki (Method of Paired Equations in Problems of Mathematical Physics). Nauka, Leninrad (1977)

© The Editor(s) (if applicable) and The Author(s), under exclusive license
to Springer Nature Switzerland AG 2022
A. N. Guz, *Eight Non-Classical Problems of Fracture Mechanics*,
Advanced Structured Materials 159,
https://doi.org/10.1007/978-3-030-77501-8

In Latin

8. Akbarov, S.D., Rzayev, O.G.: Delamination of unidirectional viscoelastic composite materials. Mech. Compos. Mater. **39**(3), 368–374 (2002)
9. Bogdanov, V.L., Guz, A.N., Nazarenko, V.M.: Nonclassical problems in the fracture mechanics of composites with interacting cracks. Int. Appl. Mech. **51**(1), 64–84 (2015)
10. Dal, Yu.M., Litvinenkova, Z.N.: Hypercritical deformation of a plate with a crack. Sov. Appl. Mech. **11**(3), 278–284 (1975)
11. Cherepanov, G.P. (ed.): FRACTURE. A Topical Encyclopedia of Current Knowledge. Krieger Publishing Company, Malabar Florida (1998)
12. Guz, A.N.: Three-dimensional theory of stability of elastic-viscous-plastic bodies. Sov. Appl. Mech. **20**(6), 512–516 (1984)
13. Guz, A.N.: Theory of delayed fracture of composite in compression. Sov. Appl. Mech. **24**(5), 431–438 (1988)
14. Guz, A.N.: Construction of fracture mechanics for materials subjected to compression along cracks. Int. Appl. Mech. **28**(10), 633–639 (1992)
15. Guz, A.N., Guz, I.A., Menshikov, A.V., Menshikov, V.A.: Penny-shaped crack at the interface between elastic half-space under the action of a shear wave. Int. Appl. Mech. **45**(5), 534–539 (2009)
16. Guz, A.N., Knyukh, V.L., Nazarenko, V.M.: Three-dimensional axisymmetric problem of fracture in material with two discoidal cracks under compression along latter. Sov. Appl. Mech. **20**(11), 1003–1012 (1984)
17. Guz, A.N., Knyukh, V.L., Nazarenko, V.M.: Cleavage of composite materials in compression along internal and surface macrocracks. Sov. Appl. Mech. **22**(11), 1047–1051 (1986)
18. Guz, A.N., Knyukh, V.L., Nazarenko, V.M.: Fracture of ductile materials in compression along two parallel disk-shaped cracks. Sov. Appl. Mech. **24**(2), 112–117 (1988)
19. Guz, A.N., Knyukh, V.L., Nazarenko, V.M.: Compressive failure of material with two parallel cracks: small and large deformation. Theor. Appl. Fract. Mech. **11**(3), 213–223 (1989)
20. Guz, A.N., Korzh, V.P., Chekhov, V.N.: Instability of layered bodies during compression taking into account the action of disturbated surface loads. Sov. Appl. Mech. **25**(5), 435–442 (1989)
21. Guz, A.N., Korzh, V.P., Chekhov, V.N.: Surface instability of a laminar medium connected with a homogeneous half-space under multilateral compression. Sov. Appl. Mech. **26**(3), 215–222 (1990)
22. Guz, A.N., Korzh, V.P., Chekhov, V.N.: Stability of a laminar half-plane of regular structure under uniform compression. Sov. Appl. Mech. **27**(8), 744–749 (1991)
23. Guz, A.N., Zozulya, V.V.: Contact interaction between crack edges under dynamic load. Int. Appl. Mech. **28**(7), 407–417 (1992)
24. Guz, I.A.: Composites with interlaminar imperfections: Substantiation on the bounds for failure parameters in compression. Compos. B **29**(4), 343–350 (1998)
25. Guz, I.A., Chandler, H.W.: Bifurcation problem for ceramics compressed along interlaminar microcracks. In: Moore, R.R. (ed.) Abstracts of the 5th International Congress on Industrial and Applications Mathematics, ICIAM 2003, p. 311. University of Technology, Sydney, Australia, 7–11 July (2003)
26. Guz, I.A., Soutis, C.: Critical strains in layered composites with interfacial defects loaded in uniaxial or biaxial compression. Plast. Rubber Compos. **29**(9), 489–495 (2000)
27. Soutis, C., Flek, N.A., Smith, P.A.: Failure prediction technique for compression loaded carbon fibre-epoxy laminate with open hole. J. Compos. Mater. **25**, 1476–1498 (1991)
28. Soutis, C., Guz, I.A.: On analytical approaches to failure composites caused by internal instability under deformations. In: Proceedings of EUROMECH Colloquium 400, pp. 51–58. London, 21–29 Sept (1999)
29. Soutis, C., Guz, I.A.: Predicting fracture of layered composites caused by internal instability. Compos. A. Appl. Sci. Manuf. **39**(9), 1243–1253 (2001)

Lightning Source UK Ltd.
Milton Keynes UK
UKHW020626180822
407492UK00008B/719